GRAVITY

<u>IN</u>

RELATIVISTIC PARTICLE THEORY:

A PHYSICAL FOUNDATION
FOR THE LIFE SCIENCES

By

Harold J. Hamilton Ph.D.

First Edition

Printed in the United States of America.

Library of Congress Control Number: 2013908161
ISBN 978-0-9892572-4-4 (pbk)

SciPress Publishing Company
Olympia, Washington
SciPress@Outlook.com

Acknowledgements

Many people have helped with the decades of work represented here. I thank them all. In particular:

- Robert Noyce, Donald Webster, Ralph Burhoe, and Alexandr Zotin, all now deceased, encouraged me to seek the larger perspective, each in his own time and way.
- My son David helped me write this book, from commas to end notes, graphics and software glitches, and along the way, provided an understanding of, and contributions to its most important messages, and constantly pushed me to be precise and thorough.
- My wife of 53 years, LaVerne, kept me on course through one struggle after another on this long journey. I could never adequately express my gratitude for the unflagging support she has provided throughout those years. She sacrificed her "adobe home among olive trees" to my Ph.D. and set aside her retirement dreams with an unquestioning belief in my passion for this project.

CONTENTS

PART ONE:
GRAVITY <u>IN</u> RELATIVISTIC
THERMODYNAMICS

- All particles are **real**, characterized by internal relativistic structure and energy-density
- Emergence and limitations of Classical Thermodynamics Formal terms and definitions:

- The linear regime as defined by Onsager and Prigogine and the Calculus of Variations

- Weakly nonlinear systems—*evolving* beyond the confines of equilibrium

- Strongly nonlinear systems—*evolving* open subsystems far from equilibrium

General Principles of:

- Thermodynamic Selection

- Minimal Flow-Specific Entropy Production in Closed Thermodynamic Systems
- Maximal Energy-Density in Subsystems of Closed Thermodynamic Systems
- The Extended *2nd Law* and the Nonequilibrium Thermodynamic Imperative (NTI)

PART TWO:
GRAVITY <u>IN</u> RELATIVISTIC COSMOLOGY

- Problems with the Big Bang and Inflation-adjusted Standard Models

- "Constants" c and $\sqrt{G}$ coevolve as parameters with energy-density ρ of the Universe
- Leads to the Universe being causally connected at all time, nearly flat ($\Omega \geq 1$) in all expansion and Δc dilates in the Lorentz Transformation, instead of Δt, taken here as a Universal unit of time advance

- The Initial Quiescent Quantum State of the Universe in the quasi-infinite past.
- Primordial GSBP (Gravitational and Symmetry Bound) Quarks/Antiquarks, in an initial Quantum Euclidean space, thereby directly and self-consistently linking the demands of quantum, gravity and relativity theory

- Energy down-conversion (quanta number-doubling and energy halving) in each of 265 doubling epochs accounting for all the present *matter* particles in the Universe

- Scalar Expansion, Evolution of the Relativistic Quantum/Cosmic Universe
- Annihilation of all of the antimatter particles in the Plasma Event—a Spacetime-Distribution of Little Bangs lasting an epoch or more
- Expansion and cooling lead to Quarks Recombination, followed by passive Transmutation of some Neutrons into Protons and Positrinos, making no requirement for the Weak Force or the creation of W^-, W^+, Z^0 Particles
- NTI-Creation of atoms, as *enduring* open Thermodynamic Subsystems, conserving most of the mass-energy in and minimizing the rate of entropy increase in the Universe
- Nucleosynthesis of Hydrogen, Helium and Lithium
- The foregoing argues: There is no need for a Higgs Boson, the WEAK Force, DARK ENERGY or DARK MATTER, since the *Force* of Gravity $m\sqrt{G}$ is an absolute Universal constant, accounting for the stability of gravitationally flat structures, e.g., Spiral Galaxies
- Radiation Decoupling occurred six epochs ($\sim 10^{60}$ years) ago, followed by the creation of all the Cosmic Structure we now see, though our present look-back capability is limited to a single epoch (about 10^{10} years).
- Supernovae and Black Holes may have emerged only an epoch following Radiation Decoupling, as Open Subsystems of Galaxies which grew, maximizing their Energy–Density, while minimizing the Flow-Specific Entropy Production in the galaxies and the rate of Entropy Increase in the Universe.
- Planck's Constant h may play an important role in the cosmos comparable to that in the quantum world, i.e. in characterizing

the fine structure of the expanding Universe. A Gravitational Fine Structure Constant $A \equiv (m\sqrt{G})^2 / \hbar c$ (where $m\sqrt{G}$ is the Gravitational Coupling Constant) might define the relativistic velocity of stars and gas in stable Spiral Galaxies and other cosmic bodies.

- All such events and parameters in the expanding Universe are represented self-consistently in Fig. 9-1 and Table 9-1

PART THREE:
GRAVITY IN RELATIVISTIC QUANTUM MECHANICS

- Tracing the roots of quantum theory: of *wave-particle duality, point particles,*
- *Single particles* in interference phenomena, *superposition* and their philosophical roles in setting current quantum theory apart from classical mechanics.

- Since Primordial Quanta/AntiQuanta (GSBP physical particles) and their number expansion played an essential role in our scenario in the origin and evolution of the Universe, gravity played a key role in quantum theory from the earliest time, in contrast to current theory.
- In our scenario there is no need for three "Generations" or "Families" of quarks and leptons as in present theory. The higher-mass representations are simply higher *energy-density* states of such particles in an earlier time.
- The four fundamental forces in current theory are here reduced to two: gravity and the electromagnetic force. Gravity is mediated by the emission of real bosons from neutral or charged particles,

resulting in an attractive force via impulse-recoil action on the emitting particle or body. The same holds for the EM force acting on fermions to attract or repulse them.

- Spin-1 "massless" vector bosons like fermions also have GSBP internal dynamic structure though with the axis of photon gyration orthogonal to the boson's trajectory. Thus it is the internal symmetry that distinguishes bosons from fermions. The gyrating photon in the boson serves as a frictionless flywheel, conserving the energy, momentum, spin angular momentum and polarization of the boson over the life of the boson. The quantum and cosmic worlds thus are shown to have a common origin and theoretical foundation.

- General Relativity is inherent in the internal structure of the quarks and leptons as first-order GSBP SU(2) composite-particles.

- The hierarchical energy-order of the composite particles

- The above argues for a strictly particle representation of the organization and dynamics of the nucleus of atoms, in which the values of the parameters c, $\sqrt{G}$ vary as ρ^{-3} and m as ρ^{3} over four hierarchical energy-density ranges.

- Particles Diffraction and Interference From First Principles

CHAPTER 12: A GSBP-SCHRÖEDINGER RELATIVISTIC
QUANTUM (PARTICLE) MECHANICS   Pg 189

- The Challenge in developing a relativistic particle mechanics is to meld the well-established formalism and power of Schröedinger mechanics with the inherently relativistic structure and dynamics of GSBP particles, so as to directly satisfy the demands of general and special relativity. This may be accomplished straightforwardly and simply in the one dimensional Schröedinger Time-Independent equation by reducing the electric charge e in the same proportion as we reduce the electron's mass.

- The challenge also is to square GSBP-Schröedinger mechanics, self-consistently, with related fundamental theoretical issues including: The Lorentz Transformation, The Physical Basis for the Indeterminacy Relations, The Uniformity and Constancy of the Velocity of Bosons Emitted in the Decay of Atomic States, The Derivation of E=mc² and The Seamless Joining of Relativity and Quantum Theory with c and G evolving as parameters.

CHAPTER 13: GSBP-QUARK ELECTRODYNAMICS - THE ATOMIC NUCLEUS Pg 207

- The importance of *energy-density states* in bound GSBP-QED nucleons, as contrasted with *energy states* in QCD theories.

- The transmutation of quarks, key to such evolution in GSBP-QED, will be seen not as the work of "weak force" particles, (W⁺·, Z⁰), but simply as the underlying imperative of the extended *2nd Law* and the NTI, as the average energy-density of the Universe declined to a critical enabling energy-density level.

- As the average energy-density of the Universe declined to a critical level, the d-quarks in some neutrons passively reduced their mass and their electric charge from –1/3 to +1/6 and their mass proportionally, thereby transmuting the neutron into a proton (plus a positrino) and leaving the nucleon with a net positive unit charge which greatly increased its long term stability and opened the way to the creation of the light atoms. The accompanying violation of parity, in this transmutation, is readily explained in the GSBP model as a polarity inversion of the d-quark's angular momentum. In addition to accounting for small variations in the mass and charge of the system, the positrino may extend the fine structure corrections to the Bohr model of the hydrogen atom, pointing toward a precision physics of $(\alpha^2)^2$ order of magnitude, or about one part in 10^9.

- Further expansion and cooling would lead to more transmutations, enabling protons to combine with neutrons in the nucleus and the creation of the heavier atoms, all tending to decrease entropy increase in the Universe.

- Accounting for the *asymptotic freedom* and *containment* of quarks in the nucleus

- Present problems with the Standard Model, seen as non-problems in our scenario

- We build a GSBP-electron model of the hydrogen atom by first defining the parameters of the GSB Photon in an electron marginally-bound to a proton, in which the electron is nearly at rest (having near-zero KE and negative PE relative to the proton), with both its mass m_r and electric charge e_r reduced by the factor $m_p m_e/(m_p+m_e)$, where m_p is the mass of the proton and m_e the mass of the free electron).

- This leads us straightforwardly to an exact GSBP-electron model of the hydrogen atom self-consistent with general as well as special relativity.

- Accounting for the Lamb shift

- In radiation all boson emission occurs as the electron spirals down from one allowed discrete energy level to another, such that the time required to make an exact integral number of GSBP orbits in the electron coincides precisely with the closure of one orbit of the electron at relativistic speed about the nucleus and is equal to the inverse frequency ν^{-1} of the emitted boson.

- In hydrogenic, hydrogen-like atoms and ionized high Z atoms, the challenge for GSBP particle theory is to determine the manner in which the local energy-density of a system varies with the atom's immediate environment, enabling thereby the above corrections to the energy levels of bound electrons in a simple and straightforward way.

- We show that quantum mechanics, contrary to its reputation, can be inherently intuitive, rational and consistent with the demands of relativity and thermodynamics, once we take into account the internal dynamic structure of the fundamental particles and shed our long-held commitment to the notions of wave-particle duality, wave packets, wave function reduction, the single particle assumption and superposition.

- Hydrogenic Atom and Small Molecule Theory
- The problem that has faced theorists in modeling molecular systems has turned on the question: How and in what form can we include the corrections, needed to square theory with experiment in modeling increasingly complex molecules, so as to minimize computational costs, given "point particle" assumptions?
- Efforts toward this end have been focused on the development of Molecular Orbital Theory (MOT) in dealing with small molecules and on Density Functional Theory (DFT) to apply to more complex systems, including electron dynamics in solids. These are many-body problems in which the motion of one electron affects, and is affected by, that of all the others and so we should expect that the fundamental problem traces to the lack of physical dimension and internal dynamic structure of the electron, especially as they may involve many weak bonds. Consequently progress in these efforts have been slow and limited
- GSBP-Schrödinger theory should open the way advances in both theory and experiment.

PART FOUR:
GRAVITY IN
THE RELATIVISTIC MACROMOLECULES
AND SYSTEMS OF LIFE

CHAPTER 17: THE ORIGIN OF LIFE – A THERMODYNAMIC
                 PERSPECTIVE   Pg 278

- A century of theoretical speculation and laboratory research on the origin of life has shown how relatively easy it is to create the amino acids we find in all cells today from available precursor molecules and how all-but-impossible it is to make nearly all such amino acids left-handed stereospecific, as in present cells. As a consequence laboratory progress has been largely stalled for several decades, there being no way to create the high-mass proteins and enzymes from the racemic amino acids.

- A hydrothermal vent may be seen as a localized stable heat source and sink, in which the vortex-flow of hot saline water appears in part as a mesoscopic, localized, single Bénard cell. It is a kinetic, open thermodynamic subsystem of the larger energy-flow, geophysical system, which serves to minimize entropy production and maximize the efficiency of heat transfer in the system. It does this through the very long-range organization of molecular motion. But much more is going on in the "system" and the "subsystem;" the water cycles with it ions, other atoms, molecules and solutes, some small portions of which are newly added from within the earth in each cycle. The vortex is driven and constrained by mechanical (hydraulic) forces as well as thermal and gravitational gradients, and the existence and flow of ions results in electromagnetic forces and flows, giving rise to a very complex open subsystem involving the interplay of four forces and their respective flows.

- In such a system chlorine and sodium ions cycle in quasi-laminar flow and, since the mass of Cl^- is about 1.5 times that of Na^+, we should expect that chlorine ions would be accelerated (by thermophoretic forces) to a slower average velocity and on smaller vortex orbits than for the sodium ions, in the stationary state. The effect of this ion-flow asymmetry might be to create a moderate-to-strong magnetic field in the regions of relatively large net positive current flow and the asymmetry of reaction zones should play an important role in the simultaneous synthesis of reactants at one level and of reaction products at a higher level. Such relatively simple, ubiquitous and asymmetric physico-chemical systems in the prebiotic ocean could account for a significant enantiomer excess (ee) of amino acid monomers if the magnetic field strength, produced by the cycling net charge, were sufficient, relative to thermal noise, to suitably align the spins in the C—N bond in the vast majority of amino acid reactions, whether via a single-pass through the reaction zone or multi-cycle amplification of stereoisomers. The same applies to the carboxyl and side-chain groups and the peptide bond.

- Our coarse calculations support the above reasoning, as a way around the homochirality problem.

- No less important in this consideration is the realization that the greater the number and kind of L-amino acids available in the early phase of the polymerization process, the sooner would be the yield of stereospecific enzymes to catalyze other reactions leading to the creation of D-ribose sugars, bases, nucleosides, nucleotides and ATP-like energy carriers.

PREFACE

What began a half-century ago as a search for a rigorous physical underpinning of Darwinian theory, became over time, an understanding that existing physical theory cannot suffice as a theoretical foundation for the life sciences. The various theoretical frameworks, from Hamilton and Lagrangian mechanics to classical thermodynamics, quantum mechanics, relativistic quantum field theory, and now to cosmology, are each constrained mathematically to refer only to systems very near to equilibrium or a stationary state. In each field, theorists were challenged to invent or adapt extraordinary mathematical tools that were needed to make their theories conform to empirical evidence. In consequence, aspiring young physicists had to master those tools and in so doing may have been discouraged from questioning the accepted fundamental theory.

Though pragmatic insight into the workings of the world has been gained since the time of Newton, current theory is now seen by some as flawed at the most fundamental level. Only now is a small but increasing number of physicists beginning to confront the limitations and gross inconsistencies that call for a rethinking of long-held beliefs about the **fundamental constants** and their imposed constraints on the development of fundamental physical theory.

Here, we argue that: **all <u>organized</u> physical systems and subsystems in the Universe (cosmic, quantum and living) come into being and evolve, however slowly or rapidly, via <u>nonlinear</u> processes, from one relatively stable state to another, in response to imposed changes in their environment. At the most primitive level, this requires all physical particles to be inherently relativistic, an option not possible in current "point particle" theory.**

Classical thermodynamics and Darwinian theory emerged together in time in the early 19th century, but because they were so different in their conceptualization their common theoretical underpinnings were obscured. In my early effort to develop fundamental Nonequilibrium Principles, my greatest insight was found in the writings of Alexandr Zotin, a world-renowned developmental biologist in the USSR-Russian Academy of

Sciences. I sent him a copy of my first (unpublished) effort in formulating such principles. (This was an extension of my earlier monograph published by The University of Chicago Press[1], 1978: A Thermodynamic Theory of the Origin and Hierarchical Evolution of Living Systems, which in turn followed from my earlier interests in Neural Theory and Modeling, Stanford University Press[2], 1964). I was pleased to receive Professor Zotin's latest book and paper in response. In his letter he thanked me for helping him "greatly to understand problems of individual development of organisms" and he urged me to apply my reasoning, with caution, "not only to biological but also to physical or even cosmic problems." I likely would never have expanded my effort and had the courage to follow that path toward such a far-greater task, had it not been for Professor Zotin's counsel.

That path has enabled me to develop a self-consistent theory of:

- the creation, in the quiescent Quantum Universe, of all the existing matter particles (with no need for a massive Higgs boson) and the lack of antimatter particles;
- the time scale in our scenario being many orders of magnitude greater than in present theory;
- the accelerating adiabatic expansion of the Universe;
- the constancy of the electromagnetic and gravitational forces in that expansion;
- the non-problems of Dark Energy and Dark Matter and
- what I believe will be a major breakthrough in the stalled Origin-of-Life studies, potentially applicable to exoplanets in the Universe.

In my scenario, all of Creation, so evident to our eyes and instruments, is a natural evolutionary consequence of a universal driving force, i.e. a creative force, I call the Nonequilibrium Thermodynamic Imperative (NTI). The NTI is a direct result of my Extended Second Law and the Heisenberg Indeterminacy Principle.

The force of gravity defines the internal properties and dynamics of all the particles throughout this book. It is because gravity plays such a central role that I emphasize the word <u>IN</u> in its title and the headings of its parts.

To aspiring physicists and others seeking a deeper and wider understanding of contemporary science: the path to an understanding of our wonderful Universe and indeed to "precision physics" may not necessarily be found on blackboards and the pages of books filled with arcane mathematical equations. Some understanding of elementary calculus is needed to follow the formulation of nonequilibrium thermodynamic theory, however high school algebra suffices in the formulation of relativistic particle theory applicable to the quantum, cosmological and living worlds.

Those seeking an overview of the proposed sweeping revisions in fundamental theory and its origin may find it condensed in Conceptual Steps Toward a Self-Consistent Fundamental Particle Theory in the Appendix of this book.

Harold Hamilton

PART I

GRAVITY
IN
RELATIVISTIC THERMODYNAMICS

Classical thermodynamics emerged in the first half of the 19th century, along with the independent field studies of Darwin and Wallace. The emerging theory of evolution cast a new light on the creation of increasingly complex organisms in dynamic states far from equilibrium. Curiously, at the same time the Laws of thermodynamics quickly assumed a position of fundamental importance in all of science, notwithstanding the fact that these Laws referred only to isolated systems in, or very near to a state of equilibrium. It was puzzling also that such important Laws had nothing whatever to say about how all the complex dynamic systems in the universe, including Darwin's finches, etc., came into being in the first place. That state of inquiry remained at an impasse until the early half of the 20th century, when Lotka, Bertalanffy and others with interests in Darwinian theory began to search the physical sciences for insight into the nature of living organisms. The first advance in classical dynamics occurred with Onsager's formulation of the principle of minimum energy dissipation, which mandates the *evolution* of *linear* thermodynamic systems, very near equilibrium, toward a stationary state of minimum energy dissipation, under constant external constraints.[3]

But the principles of natural selection on the one hand and minimum energy dissipation on the other still had little constructive to say to one another. In the middle of the 20th century Prigogine[4] developed his theorem of minimum entropy production, which governs the evolution of *linear reversible* processes in systems close to equilibrium, toward stationary states of minimum entropy production. He and others also demonstrated that the *theorem of minimum entropy production* did not apply to systems in states driven far from equilibrium, but rather that the rate of entropy production in such systems must increase the farther they are from equilibrium. No less important, they showed that such systems could become unstable as constraints are varied, leading often to unpredictable bifurcations.

Thus, nonequilibrium thermodynamic theory currently consists of a rigorous part, applicable only to the domain of linear reversible processes very near equilibrium or a stationary state, and a second part referring to nonlinear, strangely behaving systems, far from equilibrium, but largely lacking in any general explanatory or predictive laws or tools. It is important to note here that the above limitation of the formal theory derives from the limitation of the Calculus of Variations used in the demonstration of both minimal energy dissipation and minimal entropy production principles, as it is not applicable to nonlinear systems, i.e., to systems for which the characteristic phenomenological relations between the generalized forces and flows cannot be represented by constants. Moreover, for systems involving multiple interacting forces and flows, there is no reason to believe that reciprocity of such phenomenological relations should prevail in nonlinear systems, as is the case for linear systems and as Onsager had demonstrated. In order to significantly extend the province of equilibrium dynamic theory, it would be necessary, not only to circumvent the restrictive formalism of the variational calculus, but more importantly, to understand that *the dynamics of nonlinear physical systems can only be comprehended in evolutionary terms*. That is to say, it is the *rate* at which such systems move (**evolve**) upwards toward stationary or nonequilibrium states (or in reverse) and the factors that govern such processes that we need to understand. Prigogine[5] argues along similar lines: "irreversibility is deeply rooted in dynamics. One may say that irreversibility starts where the basic concepts of classical or quantum mechanics (such as trajectories or wave functions) cease to be observables. Irreversibility corresponds not to some supplementary approximation introduced into the laws of dynamics but to an embedding of dynamics within a vaster formalism. Thus …there is a microscopic formulation that extends beyond the conventional formulations of classical and quantum mechanics and *explicitly* displays the role of irreversible processes." And these thoughts resonate also with those of Matsuno[6]: "Physics equipped with the adiabatic approximation conspires to make matter inanimate. If we free ourselves from adiabatic approximation, matter can be seen as an active agent simply by virtue of the fact that interaction changes propagate at a finite velocity. *If it is to be applied to material evolution, non-equilibrium thermodynamics must be extended so that it can rid itself of the fetters imposed by the adiabatic approximation.*(emphasis added) Matter aggregated in self-generated open systems, is not inanimate."

Thus our aim in Part I is to pursue the development of a rigorous general thermodynamic theory, applicable to stable systems far from equilibrium as well as those in or near a state of equilibrium, in the light of a *generalized theory of evolutionary processes applicable throughout this book*. And as the above Part I heading implies, *the substance of such systems will be taken as inherently relativistic particles*. In the first chapter, we commence by stating the fundamental assumptions underlying the arguments to follow.

CHAPTER 1

FUNDAMENTAL ASSUMPTIONS, ARGUMENTS AND DEFINITIONS

Energy dissipation in any thermodynamic system is identified with the flow of energy from regions of relatively high to relatively low energy-density in the system or to its environment, such that some portion of that energy flow is ultimately converted into relatively low energy photons – heat. If all the mass-energy in the system, including the heat, could be completely contained (conserved) within the boundaries of the system, it would be meaningless to speak of the system's environment and we would refer instead to an **isolated thermodynamic system**, governed by the 1st and 2nd Laws of thermodynamics. But such a system can only exist as a simplifying model in an expanding Universe in which mass-energy *cannot be completely bound* by any material barrier or by the force of gravity. In reality, all systems exist in some environment and so the 1st Law requires that some portion of the radiant energy leave the system and is therefore no longer available to do work in the system; the photons that leave have a relatively high probability of transporting energy to lower density regions *in* the Universe and a low probability of directly contributing to the expansion of the Universe. The microwave background radiation we find today is a remnant of such early dissipation contributing to the expansion of the Universe.

Entropy production in the system is inversely related to the absolute average temperature of the system and therefore inversely related to the probability that some very small portion of the photons leaving a system will ultimately contribute to the work of expanding the Universe. In every macroscopic region of the Universe, energetic processes result in some finite energy dissipation and entropy production, which contribute (directly or indirectly) to the marginal expansion of the Universe in spacetime. **Entropy** increases with time in quasi-isolated physical systems not because of some statistical dictum of which the system can scarcely be aware, but because energetic processes therein ultimately give rise to some few photons with momentum directed at the boundary of the slowly expanding universe making all such processes irreversible in time. As a consequence, there is no "equilibrium" state of maximum entropy toward which the Universe is evolving; entropy increases indefinitely. It follows that we should speak strictly only of *quasi-equilibrium, quasi-stationary* and *quasi-isolated* states. In most discussion we will ignore this nuance. All energetic processes are constructive; what we regard as dissipation is to be seen in the large not as the waste of energy but as the employment of some small portion of energetic activity to the ongoing expansion (construction) of the Universe. All of this implies that there is some minimal value of entropy production per unit energy flow governing the origin and evolution of the Universe that is fundamentally associated with Planck's Constant h, about which we will have much more to say in chapters 7 and 8.

A **closed thermodynamic system** refers to any system that can be considered closed to the exchange of matter with its environment, but open to the exchange of energy. This effectively limits the class to systems that transport thermal or kinetic energy in a stationary state. In the transport of electrical energy through a wire, for example, matter (electrons with mass) in a source, characterized by a relatively high density of electrons in the conduction band transfer kinetic energy to a sink having a lower density of electrons in the conduction band. Since it is only the kinetic energy of such matter transport particles (atoms, electrons, molecules, etc.), not identifiable with massive particles themselves that enter and leave, the system is closed to the exchange of such particles.

An **open thermodynamic system** refers to any system in which matter as well as energy may be exchanged with its environment or to a subsystem of a larger thermodynamic system that is open to the exchange of both matter and energy with the larger system.

Matter-energy exchange is conservative in all irreversible processes and the flow of matter-energy in thermodynamic systems occurs at finite velocity.

The flow or exchange of matter-energy, in any thermodynamic system or macroscopic region thereof, occurs in response to some thermodynamic gradient or chemical affinity. In general, a local gradient or affinity may vary from one macroscopic region to another and with time, and so a system may be characterized by a *thermodynamic gradient structure* (TGS) and/or a *thermodynamic affinity structure* (TAS) in spacetime, which reflects the spacetime distribution of energy and matter. Moreover, where more than one kind of gradient exists in a macroscopic region, the system must be characterized by a corresponding number or set of TGS and/or TAS. However, in this case, it is necessary to refer to *effective gradients* and *effective affinities*, which take into account interference phenomena associated with multiple forces and flows or multi-component chemical reactions, as defined below. Thus, the notion of a TGS or TAS is associated with any macroscopic arrangement or distribution of matter and energy distinct, on average, from that which would characterize an equilibrium state. Generally, in order to simplify discussion, we will take forces or gradients to include the notion of affinities and TGS to encompass the concept of TAS. It has been the practice to refer to such varied gradients or affinities and associated flows or reaction velocities as **generalized forces and flows**.

Local phenomenological functions (local p-functions), define the relative ease or constraint with which matter-energy is exchanged with and enabled to (a) flow through a macroscopic region in response to a local thermodynamic effective gradient or (b) participate in chemical reactions in response to a local affinity, i.e., they define the relations between local effective gradients (affinities) and conjugate flows (reaction velocities). (In general, the local p-functions may vary in space, time and direction and may therefore be represented by three-vector parameters as in field theory.) Where m kinds of effective gradients and flows are involved, matter-energy flow through the region is characterized by m local p-functions. Similarly, m local p-functions would relate to m local effective affinities and conjugate reaction velocities.

The effective thermodynamic gradient, which drives a conjugate flow in any macroscopic region of a thermodynamic system, is equal to the algebraic sum of all imposed gradients driving the flow in the same or opposite direction, whether relating directly to an ordinary externally imposed force or resulting from interference phenomena associated with other imposed forces and resulting flows, i.e.,

$$X_i' = X_i + \sum_{k-1}^{m} X_{ik} \qquad (k \neq i) \qquad (1.1)$$

where X_i' is the effective thermodynamic gradient of the ith kind which drives the thermodynamic flow J_i , X_i is the ordinary imposed thermodynamic gradient and X_{ik} (k≠i) are the transverse gradients of the ith kind generated by interference phenomena associated with effective gradients X'_k and flows J_k for all k≠i. Here we assume, to simplify discussion, that all gradients and flows are one dimensional. Also the J_k are assumed to be well-behaved (typically weak) functions of X_{ik} for all k≠i. Similarly, the effective affinity which drives a conjugate chemical reaction is equal to the algebraic sum of all imposed affinities, including those resulting from interference phenomena, and expressed as above in terms of generalized forces X_{ik}. Such transverse gradients are created at the same time as, and by the action of, the externally imposed gradient, i.e., the flow of some form of matter-energy J_i, generated by an imposed gradient X_i, gives rise to an ancillary gradient X_{ik}, adding to or opposing X_i and altering J_i accordingly. As an example, let a wire be in a state of equilibrium with a thermal/electrical source and sink. A sudden increase in the temperature of the source would create at the same time a kinetic energy gradient and a charge density gradient of electrons in the conduction band in the region proximate to the source, causing electrons to drift toward the sink, transporting both kinetic and electrical energy. In the stationary state, an imposed emf, equal to but opposite in sign, would deny electron flow while enabling the flow of kinetic energy. When the imposed thermal gradient is small, the charge density gradient is small, as is the flow of kinetic energy and electrical charge, such that the p-functions relating the two gradients and their conjugate flows may be treated as constant coefficients in a linear representation. However, a larger imposed thermal gradient may result in greater than linear increases in the flow of both kinetic and electrical energy because the force acting between electrons in the conduction band is a nonlinear function of the average local density of electrons. All electrokinetic phenomena, including the Seebeck, Peltier and Thompson effects and the relations between electroosmotic pressure and streaming currents through porous walls, are subject to similar analysis. Differential diffusion in a multi-component system and thermal diffusion (Soret and Dufour effects) may be examined in a similar manner. Multi-component chemical reactions close to equilibrium may generate either no transverse affinities or those for which the respective p-functions may be taken as constant coefficients. Far from equilibrium, formal analysis would become much more complex, but the essential point is this: **All such reactions to a change in imposed constraints can only evolve in time**.

The **rate of entropy production** $(dS_i/dt)h,t$ in any macroscopic region of a thermodynamic system h, at any time t, is equal to the sum of the products of all *effective gradients* and conjugate flows and the sum of products of *effective affinities* and conjugate reaction velocities, i.e.

$$(dS_i/dt)_{h,t} = T_{h,t}^{-1} \sum_{i=1}^{m} (J_i X_i')_{h,t} =$$

$$T_{h,t}^{-1} \sum_{i=1}^{m} (L_i X_i'^2)_{h,t} \qquad \text{for all } k \neq i \qquad (1.2)$$

where $T_{h,t}$ is the absolute temperature in region h at time t (T can be assumed uniform over any small region at time t) and L_i is the value of the p-function relating J_i and X_i' in region h at time t.

The **rate of entropy production** in a thermodynamic system at any time t is equal to the sum of entropy production over all n macroscopic regions of the system at time t, i.e.,

$$(dS_i / dt)_t = \sum_{h=1}^{n} (dS_i / dt)_{h,t} =$$

$$\sum_{h=1}^{n} T_{h,t}^{-1} (J_i X_i')_{h,t} = \sum_{h=1}^{n} T_{h,t}^{-1} (L_i X_i'^2)_{h,t} \geq 0 \qquad (1.3)$$

This may be expressed as

$$(dS_i / dt)_t = \sum_{h=1}^{n} T_{h,t}^{-1} J_{it} (X_{it} + \sum_{k}^{m} X_{ikt}) =$$

$$\sum_{h=1}^{n} \sum_{i=1}^{m} (J_i X_i')_{h,t} = \sum_{h=1}^{n} \sum_{i=1}^{m} (L_i X_i'^2)_{h,t} \geq 0 \quad (1.4)$$

for all $k \neq i$, applicable to all thermodynamic systems.

If m local p-functions in any macroscopic region of a thermodynamic system are identical to those in all other regions, the system will be referred to as a **homogeneous thermodynamic system**. If all such local p-functions may be characterized by m constants relating the flows and their conjugate effective gradients, the system will be referred to as a **linear, homogeneous thermodynamic system**. A thermodynamic system is linear over some range of imposed constraints if there exists no correlation, on average, of the behavior of matter-energy flow or exchange in any macroscopic region relative to that of its neighbors or elsewhere in the system. That is, there exists no short- or long-range coordination or organization of matter-energy exchange or flow.

A system in which there exists, at most, short-range coordination of the behavior of matter-energy flow or exchange in any region, relative to that of its neighbors, such that all local p-functions are everywhere well-behaved, i.e., they are either constants or continuous and single-valued functions over a range of imposed constraints, will be referred to as a **weakly-nonlinear system**. In all such systems the relations between effective gradients (affinities) and conjugate flows (reaction velocities) are characterized by p-functions having positive (negative), first-derivatives, i.e., the flow (reaction velocity) increases (decreases) at a greater than (less than) linear rate with increasing effective gradient. This reflects some constrained positive (negative) "autocatalytic" or "crosscatalytic" activity among neighboring elements of the region, whether only kinetic in nature, in the case of transport mechanisms, or involving chemical reactions. Thus, for example, the mutual reinforcement of directed molecular flow in a viscous fluid subject to boundary constraints may be taken as a form of positive "autocatalytic" activity in energy transport.

Linear and weakly nonlinear systems will be referred to, collectively, as belonging to the class of well-behaved systems. Note that there is nothing in such nonlinear behavior that prevents us from accounting for energy dissipation and entropy production, as above, in terms of the sum of products of *effective gradients (affinities)* and conjugate flows (*reaction velocities*), allowing a release from previous theoretical constraints.

A system in which a p-function in at least one region may be multi-valued, over some range of imposed constraints, will be referred to as a **strongly-nonlinear thermodynamic system or subsystem**. This marks a transformation to a state characterized by relatively *unconstrained autocatalytic* or *crosscatalytic* activity. Later, it will be shown that the use of effective gradients in accounting for transverse effects in energy dissipation and entropy production applies generally to all thermodynamic systems.

For simplicity, all of the following representations of the evolution of thermodynamic systems, involve only transport processes.

CHAPTER 2

EVOLUTION OF THE STATIONARY STATE IN LINEAR THERMODYNAMIC SYSTEMS

Consider a simple, thermal conductor of uniform cross section and conductivity μ, in thermal equilibrium with reservoir T at absolute temperature T, and imagine the conductor to be divided into many cells, as shown in Fig. 2-1 (panel a), such that the average temperature in each cell is T. If, at time t_0, we attach a reservoir, at temperature T'>T, the immediate result will be a relatively large temperature gradient and thermal energy flow into cell 1 (panel b). (In different terms we would say that the temperature in the cell increases rapidly, while other cells remain momentarily in the equilibrium state T, due to the finite velocity of energy transport.) Subsequently, the temperature in the first cell declines from a maximum state, that gave rise to a relatively large gradient and energy flow in cell 2. (A small portion of that energy decline is dissipated in the form of low energy radiation and another portion is conserved in the transport of energy to and through cell 2, leaving the remainder in cell 1 in the form of a molecular Thermodynamic Minimal Gradient Structure.) Repetition of this process from cell to cell leads in time to the evolution of a stationary state in the system.

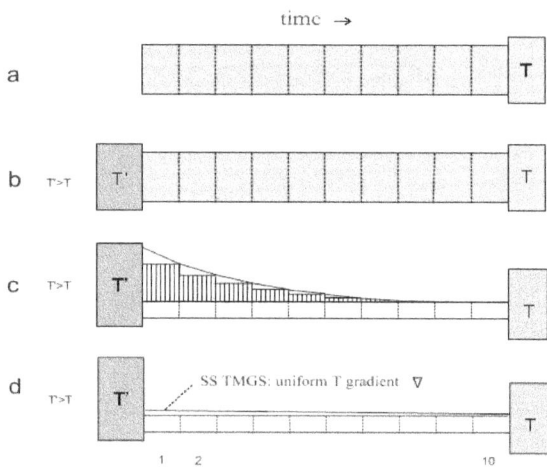

Fig. 2-1 Creation and maintenance of the stationary state (SS) in a simple nonequilibrium thermodynamic system, initially in a state of equilibrium with reservoir T (panel a). At time t_0 (panels b and c) a reservoir at temperature T'>T forms a heat transport system. The initial high temperature gradient in cell 1 results in a maximal rate of energy flow from reservoir T'. A small portion of the total energy flow in cell 1(white triangle) is dissipated (lost from the system). A much larger portion (hashed rectangles) is temporarily stored in an initial molecular Thermodynamic Minimal Gradient Structure (TMGS), creating a gradient in cell 2. The remaining portion of energy flow (white rectangles) represents the constant energy transport through the system in the SS. The curvature generated by the discrete model of dissipation arises from the fact that dissipation varies as the square of the energy flow. Ultimately the system behaves over time as a closed linear thermodynamic system, assuming a constant uniform conductivity coefficient (panel d). The dissipation in the SS is relatively large or small, depending on the conductivity μ. Doubling the T'/T gradient would cause the dissipation Ψ to quadruple, while only doubling the flow-specific dissipation Ψ_f and the rate of energy flow through the system. The stationary state would be characterized by a uniform radiation γ of low energy bosons.

Thus, both the gradient and flow at the interface with reservoir T' decline from initial maximal values, as the gradients and flows elsewhere in the system rise. Ultimately the system relaxes into a stationary state in which the thermal energy flow into and out of each cell is exactly equal, on average, and therefore the flow into the system from reservoir T' is exactly equal, on average, to the flow out of the system into reservoir T. *Such behavior is in conformance with the requirement on conservation of energy, as evidenced by the constant energy-density in every cell (zero energy divergence in every cell).* The process is characterized by the *evolution* of a **Thermodynamic Minimal-Gradient Structure (TMGS)**, i.e., the selection of those paths and processes of energy exchange with matter, which tend on average to minimize the gradient in each cell, consistent with imposed constraints, in this case culminating in a uniform minimum temperature gradient throughout the system. In consequence, the stationary state is characterized by a maximal rate of thermal energy transport and minimal rates of energy dissipation and entropy production $\propto J^2/\mu$ (nearly the same in all cells), consistent with the imposed constraint and the uniform constant coefficient of thermal conductivity μ. It is characterized in general also by a unique physical structure or dynamic (in this case kinetic) organization brought about by processes of **thermodynamic selection** (might we say **natural?**) and maintained by constant energy flow J. The net effect of this is to **minimize the rate of entropy production per unit of energy flow $\propto J/\mu$ and thus, ultimately, the rate of entropy increase in the system's environment.** And finally we see that **a maximal quantity of energy is stored in the TMG structure thus created (here, thermal energy) for as long as the energy flow is maintained, the greater is the value of μ and the smaller the dissipation energy γ.**

The above characterizes the system as a closed, linear thermodynamic system. It behaves in accordance with the assumption that in every cell (more generally in every macroscopic region of the system) the greater the deviation $\Delta\psi$ of dissipation from that which characterizes the stationary state ψss, the greater should be the rate of change of that deviation, i.e.,

$$-d\Delta\psi/dt = \lambda\Delta\psi \qquad (2.1)$$

where λ is a positive constant. Integration of (2-1) yields

$$\psi(t) = \psi_{ss} + C_e{-}\lambda t, \qquad\qquad (2.2)$$

where $\psi(t)$ is the rate of dissipation at time t and ψ_{ss} is the constant minimal rate of dissipation in each cell in the stationary state and C_e is a positive constant.

Thus the system relaxes exponentially from a state characterized by maximal rates of energy-flow and energy dissipation to a stationary-state characterized by maximal energy transport and minimal dissipation and entropy production in the fastest possible way, consistent with the laws governing the exchange of energy with matter in the system. The stationary state is the dynamic equivalent in nonequilibrium thermodynamics of the equilibrium state in classical thermodynamics, with vanishing average variance in energy flows everywhere in the system equivalent to vanishing flows and gradients in the equilibrium state. In both cases the internal energy of the system is constant, consistent with the requirement of the 1st Law. If relation (2-1) is expressed in the form

$$-d\Delta\psi/dt = \lambda\Delta\psi(\psi) = \lambda(\psi{-}\psi_{ss}), \qquad\qquad (2.3)$$

then $[\,\lambda\,(\psi{-}\psi_{ss})]^2$ may be taken as a Lyapunov function, valid for all $\psi > 0$, establishing the stability of the system at ψ_{ss}, according to Lyapunov's first theorem for stability.

What has been said above applies not only to simple ideal systems but to all real, *linear,* single-component systems to the extent that our assumption of constant p-functions is valid. These arguments apply as well to all *linear,* multi-component systems, when the system parameters are expressed in terms of thermodynamic *effective gradients* and conjugate flows or *effective affinities* and rates of corresponding chemical reactions. The constant local p-functions for the various components will have appropriate forms defining "conductance" of one kind or another, diffusion or effusion rates of matter, mechanical energy flow, or chemical reaction rates. The stationary state of the system is approached by the evolution of TMGS (TMAS), characteristic for each component, and in general at different rates, as in (2.3). Again, it is only necessary that the local p-functions be defined everywhere by suitable constants and that the dissipation be defined in terms of *effective gradients* and conjugate flows or reaction velocities.

In our simple thermal system, the evolution of entropy production dS/dt proceeds in a similar manner to that of energy dissipation, except that it is necessary to take into account the temperature decline in all macroscopic regions, accompanying the decline of energy dissipation in the approach to the stationary state. This may be accomplished by including the contribution to the evolution of entropy production in each cell at time t, made by a decline in absolute temperature -ΔTt from their maximal values. In many cases (especially if the mean temperature of the system is well above absolute zero), $\Delta T_t << T_{mean}$ in all cells, and so we may represent the evolution of entropy production in the system by $d(\Delta S/dt) = \lambda \Delta \psi (1-\Delta T)/T_{mean}$ and the corollary of equation (2.1) would be

$$S(t) = S_{ss} + C_e^{-\lambda t(1-\Delta T)/T} \qquad (2.4)$$

This assumes negligible *retained* heat of dissipation in the system. The effect of retained heat of dissipation is to further diminish the *rate-of-decline* of entropy production in the approach to the stationary state, resulting in a somewhat higher level of stationary-state entropy production than would otherwise be the case. This holds for all imposed constraints, even when an abrupt change of constraints on the system would leave the level of stationary-state dissipation unchanged. Since the rate of decline of dissipation may vary significantly from the rate at which thermal transport or radiation occurs between the system and its environment, the rates at which dissipation and entropy production approach stationary states may vary significantly.

Thus, any evolutionary loss of heat in the system following a change of constraint, whether owing to the evolution of dissipation or outflow of heat to the environment, would result in a relatively higher stationary-state value of dS/dt. The higher the nominal absolute temperature of a system, the lower would be the effect of ΔT_t on dS/dt in the stationary state. Conversely, for T→0 in the system, a small decline in the retained heat of dissipation could result in relatively high entropy production in the stationary state.

As was the case for energy dissipation, the above arguments apply equally well to the evolution of entropy production in all linear, multi-component thermodynamic systems, under conditions for which $\Delta T_t << T_{mean}$ holds. And it is important to note that our concern in the foregoing is with entropy production in the closed system and *not* with the flow of entropy to or from the system.

One may see in the above a fundamental distinction between the concerns of theorists in the development of classical or quantum mechanics, on the one hand, and those with the advancement of nonequilibrium thermodynamics, on the other. The former sought to simplify analysis by avoiding nuances arising from friction and other dissipative processes, as exemplified in Hamilton's principle in classical mechanics and its application in quantum mechanics. In so doing, analysis was largely restricted to the study of systems close to equilibrium. The latter effort now seeks to cope with the challenges of the real world in the analytical and descriptive language of nonequilibrium thermodynamics. In so doing, it must be content with much less precise or detailed predictive capability in return for a much greater insight into the workings of systems far removed from equilibrium. Such is revealed in the structural creativity and evolution even in the above simple linear system, that results from the constant flow of energy, some small portion of which is thereby diverted from further dissipation, so long as the energy-flow is sustained. *Such process underlies the creation and evolution of everything in the Universe, as we shall see in all that follows in this book.*

It is interesting that Hamilton formulated his principle in 1827, about the same time that others (Carnot, Joule; etc.) were working toward the development of classical thermodynamic theory, each essentially confined to the analysis of systems in or near equilibrium. It was a century later before serious consideration was given to the possibility of extending thermodynamic theory to deal with systems in the nonequilibrium regime and only now to contend with systems far from equilibrium. Toward this goal, quantitative analysis of the dissipation losses in the *creation* of a TMGS in a stationary state could greatly extend the Hamiltonian understanding of the emergence and evolution in space and time of various systems in the real world. We see this possibility even in the creation of a simple linear nonequilibrium thermodynamic system (Fig. 2-1) and the more so in the creation and evolution of systems far removed from equilibrium, as discussed in the following three chapters and indeed in all of the remaining chapters of this book. Lastly we note that the arguments presented above for the evolution of a stationary state in our simple linear closed thermodynamic system, grounded on the 1st Law, define the essence of the calculus of variations that underlies Lagrangian and Hamiltonian mechanics and are therefore more general, as will be shown in the next chapter.

CHAPTER 3

EVOLUTION OF THE STATIONARY STATE
IN WEAKLY NONLINEAR SYSTEMS

In a more general energy-transport system, some or all of the local p-functions, relating the flows to their *effective gradients*, may vary moderately from one region to another and in time, reflecting relatively short-range correlation or organization of molecular interaction (dependent on external constraints and imposed gradients) in a closed thermodynamic system. Such systems may be referred to as *weakly nonlinear* or *weakly organized systems* and the *p-functions* as *single valued and well behaved*. Nevertheless, the matter-energy flow and exchange processes in the system will act in a similar, albeit more dynamically-complex, manner to adjust thermodynamic effective gradients throughout the system so as to minimize the rates of thermodynamic flows and the rate of energy dissipation in the stationary state, consistent with constraints. Also, the stability of the stationary state is assured, over all regions of the state space in which all p-functions are well-behaved, in accordance with Lyapunov's first theorem, as will be shown below.

In every macroscopic region, in the stationary state, each component of thermodynamic flow must satisfy two requirements: First, the rate of flow into and out of the region must be exactly equal on average, as required by the 1st Law. That is, the divergence of all thermodynamic flows must vanish, on average, in all regions in the stationary state. Second, for each component in the region, the rate of flow on average must be exactly that determined by the local p-function and the local *effective gradient*. Here, the assumption is made that the region is sufficiently small so that the gradient does not vary significantly over the region. So long as the p-function is everywhere single-valued and well-behaved for all flows, all effective gradients in every macroscopic region must be minimal and stable (over small perturbation limits) in the stationary state, and therefore the rates of flow and energy dissipation for each kind of flow must be minimal and stable. Any departure from those minimal values could only occur as a damped fluctuation. The effect of weak nonlinearities and interference phenomena in such systems is thus only to modify the exponential decline of effective gradients, flows and energy dissipation and their minimal stationary-state values.

Since the effect of correlation or organization of molecular behavior is to more closely couple the relations between gradients and flows, i.e., to increase the conductance of mass-energy flow, we should expect the rate of flow and energy dissipation to increase, for a given gradient, in proportion to the extent of short-range molecular organization. However, it will be shown in chapter 4 that, *while the energy dissipation increases, the dissipation per unit energy flow (the flow-specific energy dissipation) decreases, as a result and to the extent of the organization of molecular behavior.* From a different perspective, we see that *mass-energy flow in the system is employed with increasing efficiency in the creation and maintenance of increasingly complex subsystems, with increasing rate of energy flow.* In the case of relatively short-range organization, such subsystems may be numerous, evanescent and highly sensitive to fluctuations, though in the whole establishing a relatively stable stationary state.

Clearly, demonstration of minimal energy dissipation in the stationary state in weakly nonlinear systems is not contingent on assumptions about: linear relationships between gradients and conjugate flows, reciprocal relationships between coupled phenomenological interference coefficients, the proximity of transient thermodynamic states to an equilibrium or stationary state, time invariance of the motion of microscopic particles, or "local equilibrium." as was required by the variational calculus required in the linear analyses of Onsager and Prigogine. Contrary to that contingency, the existence of a local gradient is here explicitly assumed. The only requirement is that the local p-functions be single-valued and well-behaved in every region and for all thermodynamic flows. In all cases, the stability of the stationary state in well-behaved systems is guaranteed by the 1st Law requirement on the evolution of TMGS, consistent with given constraints and physical makeup of the system. This may be demonstrated also in terms of Lyapunov's first theorem for stability. Here, it is reasonable to assume, as a first approximation for the evolution of the system:

$$-d\Delta\psi/dt = \lambda\Delta\psi(1+\alpha\Delta\psi), \qquad (3.1)$$

where α is a small positive constant. This reflects an increasing conductivity in the system, resulting in increasing flow and dissipation.

(The term *conductivity* is used here generally to refer not only to various energy transport mechanisms but also to the processes of matter diffusion.) Consequently, the stationary state is approached through a quasi-exponential relaxation process *at a maximum possible rate* that is somewhat more rapid than is the case for the linear system. As above, equation (3.1) may be rewritten relative to the stationary-state dissipation ψ_{ss} to define a Lyapunov function, valid in some immediate vicinity of the stationary-state value of dissipation, thus assuring stability for all admissible perturbations.

However, the foregoing ignores important considerations relating to the effects of magnetic fields and Coriolis forces. These forces, where operative, alter the evolutionary path and approach to the stationary state, since they give rise to a component of thermodynamic gradient that is perpendicular both to the flow and the field or axis of coordinate rotation. As such they further complicate the spatiotemporal evolution of TMGS. Nevertheless, the exchange of energy with matter in these skewed flows must result, in the stationary state, in TMGS and minimal rates of flow and energy dissipation, albeit different from those that would otherwise prevail. This assumes only that the p-functions relating all flows and forces, including those associated with magnetic fields or Coriolis forces, are everywhere and at all times single-valued and well-behaved.

What has been said in the previous chapter, regarding the evolution of entropy production in linear systems, holds for weakly nonlinear systems, assuming again that the p-functions in all regions of the system remain well-behaved in the course of any change of heat content and temperature in the system.

In summary, in the approach to the stationary state in any linear or weakly nonlinear closed thermodynamic system, i.e., in any well-behaved thermodynamic energy-transport system, following a step-function change of constraints or perturbation, the rates, at which energy is dissipated and entropy is produced, decrease monotonically, on average, at maximal, monotonically decreasing rates, toward minimal stationary-state values. That is $\psi>0$, $d\psi/dt<0$ and $d^2\psi/dt^2<0$ and $S>0$, $dS/dt<0$ and $d^2S/dt^2<0$. In the stationary state, energy dissipation and entropy production are positive, minimal and on average unchanging., i.e., $\psi>0$, $d\psi/dt=0$, and $S>0$, $dS/dt=0$. The flow specific parameters ψ_f and dS_f/dt evolve in a similar manner except toward somewhat lower stationary state values, relative to a linear system.

Similar analysis would apply to the behavior of a linear or weakly nonlinear mass-transport (open) system, driven by one or more externally imposed concentration gradients, in which chemical reactions among the constituents may be considered negligible under given constraints.

Whether in open or closed systems, i.e., in energy-transport or mass-transport non-chemical systems, it remains for us to account physically for the short-range ordering or organization of weakly nonlinear systems. In such systems, we could speak of the probability of any matter particle in the system to be free to transport thermal energy in the direction of the imposed gradient, dependent on its interaction with its immediate neighbors, on average a distance d away. The larger d, the smaller the density of such particles and their interaction and the lower the probability of mutually-directed energy transport, as in the case of an ideal gas system. The reverse is true in systems where d is relatively small and their potential for interaction greater, enabling a spontaneous, short-range "cooperative" organization of energy transport, as exemplified in weakly nonlinear systems, whose entropy production per unit energy flow decreases as the rate of energy flow is enabled to increase. Thus the density of matter particles (free to transport energy), the mutual interaction between such particles and the rate of energy flow, are seen as defining parameters in distinguishing nonlinear from linear behavior in such systems in which chemical interactions are negligible. A particularly important form of particle interaction in energy transport systems is associated with the viscosity of the medium (especially of matter in a liquid state).And we see in this "cooperative" energy transport a similarity with autocatalysis in accelerating, if only marginally, the transition toward a *diminished* stationary state.

With the above in mind, consider again the system in Fig. 2-1, this time in the language and analysis of Hamilton's principle, which in this case would be to characterize the system in its stationary state. In that analysis, there is no reason or recourse (formal or otherwise) to conjure panel (c) in Fig. 2-1, at an early period in the path to the stationary state. It is sufficient to describe the stationary state in experimental terms, for example Fick's law of diffusion or Ohm's law of electric current flow.

Nevertheless our understanding of the system is incomplete, especially since its origin and evolution is microscopically unique and associated with a relatively high state of stress, all of which becomes increasingly important the farther such systems are from equilibrium. Moreover the same arguments should apply in regard to our understanding of the decline of the system, once its sustaining energy-flow is suddenly removed. In effect we are arguing that a more complete understanding can only be obtained via the bundling of nonequilibrium thermodynamics, in theory and experiment, with that of Hamiltonian or Lagrangian mechanics or better, as we shall argue later in the case of conservative systems, with GSBP-Schröedinger mechanics. Prigogine argued along similar lines when he wrote: "irreversible processes are as *real* as reversible ones; they do not correspond to supplementary approximations that we of necessity superpose upon time reversible laws. …irreversibility is deeply rooted in dynamics. One may say that irreversibility starts where the basic concepts of classical or quantum mechanics (such as trajectories or wave functions) cease to be observables. Irreversibility corresponds not to some supplementary approximation introduced into the laws of dynamics but to an embedding of dynamics within a vaster formalism. Thus…there is a microscopic formulation that extends beyond the conventional formulations of classical and quantum mechanics and *explicitly* displays the role of irreversible processes."[7]

The farther the system is from equilibrium, the greater the need is to understand its origin and evolution, as revealed in the next chapter.

CHAPTER 4

EVOLUTION OF THE STATIONARY STATE IN STRONGLY NONLINEAR THERMODYNAMIC SYSTEMS

We consider next an incipient strongly nonlinear thermodynamic energy-transport system in a stable state far from equilibrium but below that rate of energy-flow which would give rise spontaneously to a critical instability. (This would characterize, for example, a thermal Bénard system in a state just below the emergence of vortex convection.) Let the system be subject to a constrained thermodynamic flow J (analogous to a current-limited electrical source), and let the level of constraint be increased incrementally such that eventually a fluctuation engenders the transformation of a local p-function, characterized by an *apparent* discontinuous increase in *conductance* $\mu(J)$ in some macroscopic region (Fig. 4-1). This will result in a small increase of energy flow in the region and a local state transformation, via the nucleation and limited growth in space and time of a subsystem, an organized dynamic structure characterized by increasingly long-range coordination (organization) in the matter-energy exchange process. Such organization is brought about by the propagation of viscous-driven mutually-directed molecular motion from one microscopic region to another, leading to long-range organization of kinetic activity. (Where such long-range organized activity cannot evolve, for whatever reason, the system can only pursue more chaotic or destructive evolutionary paths.)

The emergent subsystem, which may be viewed as being bounded by a real or virtual membrane distinguishing the subsystem from its environment, grows (the volume and/or mass of the subsystem increases), as this transient increase of local gradients and flows induces a transformation in neighboring regions, with consequent increase of energy dissipation in the subsystem. However, such limited subsystem growth can occur only if and to the extent that energy dissipation in the system as a whole decreases, since thermodynamic flow through the system is held constant and there occurs an increase $\Delta\mu$ in the conductivity of energy-flow through the system, i.e., $\Delta\psi = J^2[1/(\mu+\Delta\mu)-1/\mu] < 0$. This implies that the energy dissipation associated with each unit of matter decreases, as it leaves the environment (the system) and is incorporated in the subsystem. That is, the *dissipation* per unit mass and/or volume (dissipation per unit density Ψ_d) in the subsystem decreases with each added unit of matter. In effect, a unit of matter is selectively transported through the "membrane" in response to an imposed *density-specific dissipation-gradient* $\nabla\Psi_d$ at the boundary; the transition occurs more or less readily to the extent that the subsystem's energy-density continues to increase. Following the transition, the freedom of the unit of matter to behave relatively independently becomes constrained to follow the more organized behavior of matter within the subsystem. The increased energy dissipation in the subsystem occurs at the expense of a somewhat greater decrease of dissipation in the subsystem's environment (the system). In the stationary state, the subsystem is characterized by minimum dissipation per unit density; this is equivalent to saying that the mass-energy-density is maximal in the stationary state of minimal dissipation, under given external constraints.

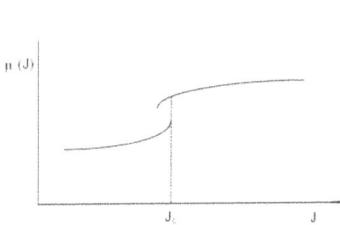

Fig. 4-1 Discrete increase of system conductivity $\mu(J)$, at critical instability J_c with incremental increase in constrained energy flow J in the system.

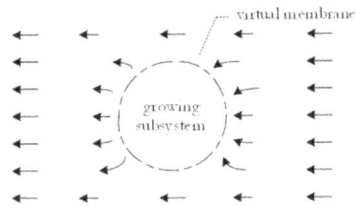

Fig. 4-2 Energy matter flow (←) through a virtual membrane, in the origin and evolution of a growing open subsystem.

Thus, the nucleation and growth of the subsystem resulted in an increase in the energy-density of the subsystem, while at the same time effecting a decrease in the energy dissipation in the system as a whole. Given the constraint on J, the latter implies a decrease in the energy dissipation per unit flow of energy in the system, which we may henceforth refer to as the *flow-specific dissipation* Ψ_f. From this we see that the resultant thermodynamic gradient on the system as a whole must necessarily have decreased, a result wholly consistent with the assumed imposition of constrained flow and increase of conductivity and with the

42

requirement for a decrease in the TGS of the system toward a minimal state.

But the above model and dynamics, with the implied assumption of a one-way street through the "membrane", are too simplistic. More realistically, especially if two or more molecular species of different mass or properties are involved, the "membrane" would likely be selectively permeable to the outflow as well as inflow of matter, thereby characterizing subsystems in general by the exchange of both matter and energy with their environment. In effect, there would evolve an increasing, negative *density-specific dissipation gradient* $-\nabla_{lu}(\Psi_d)$, which would drive a small but increasing outflow of low-utility matter-energy, whose behavior poorly serves the growth and development of the subsystem. At the same time there would evolve a positive *density-specific dissipation gradient* $\nabla_{hu}(\Psi_d)$ driving an initially large but decreasing inflow of high-utility matter-energy, which would better serve the subsystem's growth (see Fig. 4-2). The subsystem would behave as if there were an initially large but declining *net specific dissipation gradient* $\nabla_n(\Psi_d) = \nabla_{hu}(\Psi_d) - \nabla_{lu}(\Psi_d)$ at the boundary, resulting in rapid initial but declining growth of the subsystem. Ultimately, growth of the subsystem (in volume and/or density) would cease when the exchange of matter with the environment would no longer result in any decrease in net density-specific energy dissipation, i.e., when $\nabla_n(\Psi_d)$ vanishes everywhere at the boundary. This does not imply that the circulation of matter through the subsystem would cease, only that the net mass-energy of the subsystem remains constant or quasi-constant. (Where suitable material resources exist in the environment, the size, conformation or other constraints on the subsystem may lead not to a cessation of growth but to some structural metamorphosis or the spawning of new, similar subsystems via fission, budding or other processes. Such structural considerations are closely related to the buildup of stress in the subsystem and are matters of fundamental importance in the study of cosmological and quantum systems, as well as living organisms and the origin of life. Indeed we might be prompted, on occasion, to refer to certain such subsystems as primitive *inanimate* organisms.) The processes sketched above reflect a marginal decline in the rate of increase of thermodynamic conductivity of the system, with growth of the subsystem, which in turn reflects a marginal decrease in the effectiveness or efficiency with which matter-energy exchange may be spatiotemporally coordinated, as the subsystem's boundary expands. In the case of a Bénard system subject to constrained thermal flow, the growth of the subsystem might be limited to one or a few localized vortex cells.

Along with these processes, characterizing the nucleation and growth of the subsystem, the internal energy of the subsystem U_{sub} and of the system U_{sys} evolve to maximum possible values in the stationary state, while the decline of Ψ_d in the subsystem and Ψ_f in the system reflect the evolution of $TMGS_{sub}$ and $TMGS_{sys,}$ respectively.

Thus, the fluctuation-induced nucleation and constrained growth of an organized subsystem has resulted not in an increase in energy dissipation in the system, which would appear as a violation of the principle of minimum energy dissipation, but the opposite, a decrease in energy dissipation, in consonance with that principle. To be sure, this result holds only for the case of *constrained* energy flow (4.1). Within the constraints imposed, the subsystem acts to maximize its density and internal energy, as well as energy flow through and energy dissipation in the subsystem, while minimizing its TGS_{sub} and density-specific energy dissipation.

This concept may be expressed in the form of a *principle of maximal capture and minimal density-specific dissipation* of available free energy and materials from the environment of an organized subsystem: *Once nucleated, organized subsystems will act to maximally increase their energy-density, internal energy, energy flow and energy dissipation so long as the net incorporation of matter in the subsystem results in a marginal decrease of density-specific dissipation in the subsystem and a concomitant marginal decrease of flow-specific dissipation in the system, subject to constant imposed constraints. The system will thereby proceed toward and remain in a stationary or quasi-stationary state of minimum energy dissipation per unit energy flow, under given constraints. This principle may be seen as a generalization and extension of Lotka's concept that living matter tends to maximize the utilization of available energy flux in the environment[8] (4.2). As such it underscores an important focus on the efficiency of the energy acquisition and growth process. All such subsystems evolve so as to maximize the utilization of available free energy, in the growth, development and maintenance of the subsystem, and to accomplish those ends in the most efficient way.*

The behavior of the system in the immediate vicinity of the critical instability may be better understood by reference to Fig. 4-3, showing the rate of dissipation in the system $\psi_{sys}(J)$, the dissipation associated with the subsystem ψ_{sub} and the density-specific dissipation ψ_d of the subsystem, with incremental increase of constrained flow J. As J is allowed to rise incrementally, $\psi_{sys}(J)$ increases steadily, but at a decreasing rate. At the point of critical instability J_c, $\psi_{sys}(J)$ undergoes an *apparent* discontinuous transition from a higher to a *lower*, stationary state of dissipation. This reflects the nucleation and limited growth of the organized subsystem. Following the transition, the system is stable about J_c for small perturbations over a region in which all transformed p-functions remain well behaved. Further growth of the subsystem requires some increase in ψ_d. The requirement on conservation of energy provides assurance that a Lyapunov function can be defined, valid in the immediate vicinity of the stationary-state value of ψ_{ss}, and demonstrating the stability of the system for small perturbations. A first approximation for perturbation behavior in the vicinity of the stationary state would take the form:

$$-d\Delta\psi/dt = \lambda'\Delta\psi(1 + \alpha'\Delta\psi + \beta\Delta\psi^2), \qquad (4.1)$$

where λ', α' and β are positive constants characterizing the matter-energy exchange processes in the *lower* dissipation regime. The stationary-state ψ_{ss} is an "attractor state" which is defined, subject to limitations on the range of perturbation, by the minimum of a *flow-specific dissipation potential*. Any small fluctuation over this region will result in countervailing gradients and flows at the boundary of the subsystem and elsewhere, thereby returning the system to the stationary state at J_c. (We neglect here the possibility of aperiodic or oscillatory behavior about an average minimal value of thermodynamic flow.) ψ_{sub} of the subsystem rises rapidly from creation of the organized subsystem, followed by slower growth with incremental advance of J, while the course of $\Psi_d(J)$ is similar but inverted.

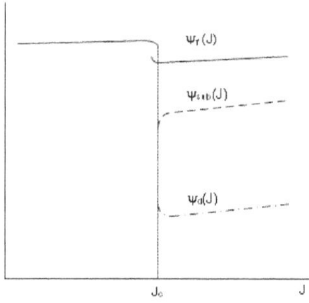

Fig. 4-3 Flow-specific energy dissipation of the system $\Psi_r(J)$, dissipation of the subsystem $\Psi_{sub}(J)$ and density-specific dissipation of the subsystem $\Psi_d(J)$, with incremental increase in constrained-energy flow J.

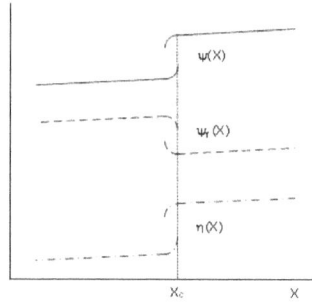

Fig. 4-4 Energy dissipation $\Psi(X)$, flow-specific dissipation $\Psi_r(X)$ and energy efficiency $\eta(X)$, for the system, with incremental increase in constrained fixed-force. X_c: point of critical instability.

Thus, while energy dissipation in the system must increase, as the subsystem is enabled to grow beyond J_c, it will do so, neglecting fluctuations and damped oscillations, along a path of *minimum dissipation increase*. The emphasis in this statement is meant to underscore the fact that, among the various possible thermodynamic paths in the new regime, only that which results in the *least dissipation increase* is consistent with the requirements of the first Law. A fluctuation may result in a critical instability and admit new thermodynamic paths and a transformation of state, characterized by increased energy flow and dissipation, but the system must necessarily pursue those particular options which tend to minimize the resulting gradients and flows, in compliance with the mandate of the 1st Law.

Consider now the behavior of the system, subject to incremental increase of fixed-force constraint, as illustrated in Fig. 4-4. At X_c the dissipation in the system $\psi(X)$ undergoes an *apparent* discontinuous transition to a *higher* stationary state. In a similar manner to that discussed above, stability of the stationary state at X_c is guaranteed over a range of perturbation in which all transformed p-functions, characterizing the new regime, remain well-behaved, and an admissible Lyapunov function may be defined. Again, that stationary-state value of ψ_{sys} is an attractor state, which is defined by the local minimum value of $\psi_f(X)$ at X_c. Any small fluctuating increase (decrease) of thermodynamic flow and dissipation in the subsystem would result in a countervailing decrease (increase) of flow and dissipation in the system such that $\nabla_n(\Psi_d) \to 0$ everywhere at the boundary of the subsystem, in compliance with the requirements of the 1st Law. All of this is similar to the evolutionary nature of well-behaved systems, except that the principal gradient and flow adjustments occur at the boundary of the subsystem. A first approximation of this behavior, valid in the region, might take the form:

$$-d\Delta\psi_{nl}/dt = \lambda'\Delta\psi_{nl}(1 + \alpha''\Delta\psi_{nl} - \beta'\Delta\psi_{nl}^2), \qquad (4.2)$$

where λ', α'' and β' are positive constants, characterizing matter-energy exchange processes in the higher-dissipation regime. Also shown in Fig.4-4 is the energy efficiency parameter $\eta(X)$, which measures the efficiency with which energy-flow in the system is employed in the creation and maintenance of *dynamically organized* states of matter-energy in subsystems, given by:

$$\eta(X)=[U_{sys}(X)-U_{lin}(X)]/U_{lin}(X)=[\psi_{f(lin)}(X)-\psi_f(X)]/\psi_f(X)], \qquad (4.3)$$

where $U_{sys}(X)$ is the internal energy of the system and $U_{lin}(X)$ and $\psi_{f(lin)}(X)$ are the internal energy and flow-specific dissipation, respectively, of a comparable linear system, all as functions of incremental increase in X. At each incremental step $\eta(X)$ increases marginally and, at X_c, $\eta(X)$ undergoes an apparent discontinuous increase from a lower to a higher locally stable, stationary state value. Thus, the efficiency with which similar systems promote the evolution and maintenance of dynamic subsystems could be compared and analyzed.

The behavior of thermodynamic systems subject to decreasing incremental constraint of energy flows or forces would follow along the lines previously discussed and illustrated, except with the time course and arrows of incremental advance of J and X reversed and associated with a different critical instability.

From the foregoing, we see that there is nothing inherent in the nucleation and growth of an organized subsystem, following a critical instability, that mandates an increase in the energy dissipation in the system, though an increase should be expected if growth of the subsystem is *enabled* to continue under fixed-force constraint, since additional matter-energy flow and exchange in the system is necessary in order to expand and maintain long-range organized behavior in the subsystem. The efficiency of this process is always maximal or quasi-maximal in the stationary state, under given constraints, whether or not the system undergoes any state transformation. While the efficiency may be relatively large, in the case of strongly nonlinear systems, the stability tolerance for external change of constraints or internal fluctuations may be relatively small. In this regard interference phenomena associated with multiple forces and flows, interacting regions and forms of nonlinear behavior, as well as the existence of magnetic fields or Coriolis forces, should be expected to play important roles in determining the stability of the system. The transformation initiated by a critical instability, resulting in a new stationary state, may be thought of as the nonequilibrium counterpart of the emergence of a phase transition in equilibrium thermodynamics.

Thus, we may conclude that *all* thermodynamic systems, subject to constant constraints, evolve on average along a path of minimum *flow-specific energy dissipation*, in the fastest possible way, toward more- or less-strongly stable stationary or quasi-stationary-states. The stability of the stationary state is assured in accordance with the requirements of the 1st Law, though the tolerance to perturbations may vary greatly. In this regard Zotin[9] has drawn attention to the fact that the range of stability may be significantly extended, where control loops and regulatory processes, as exemplified in living organisms, enable strongly-nonlinear systems to behave in a quasi-linear manner in the vicinity of stable states (4.3).

The evolution of entropy production in the approach to a stationary state in a strongly nonlinear system may be examined in the same manner as before. Any decrease in the heat content and temperature of the system accompanying the evolution of energy dissipation in the approach to a stationary state, following a critical instability and a state transformation, would serve only to moderate the decline of entropy production from what would otherwise prevail in the approach to the stationary state. This assumes of course that all p-functions remain single-valued and well behaved over the range of temperature change. And, as before, the effect would be the more pronounced the lower the absolute temperature of the system, other things being equal.

Thus, the evolution of entropy production in the region of a critical instability would have the same general character, though moderated in form and dynamic behavior, as that for energy dissipation, as depicted in

Figs. 4-1 and 4-2, under similar constraints. In the same manner, it may be demonstrated that an organized subsystem evolves toward a state of minimum *density-specific entropy production* dS_d/dt, when the contribution of subsystem heat content is taken into account.

The above reasoning applies also to open thermodynamic systems driven by molecular concentration gradients in which chemical reactions are allowed to occur. All such reactions are nonlinear in character, with the reaction rates in simple systems increasing with imposed concentration gradients. At low levels of concentrations and molecular flow, the system behaves as a near-linear system, with low probability of chemical reactions. For a system enabling a state transformation at some critical imposed concentration level, it should be possible, as above, to explore the dynamic evolution of the transformation under conditions of molecular flow constraint. Analogous to the abrupt increase in thermal conductivity in the above system at the onset of subsystem creation, we should expect an abrupt increase in reaction rate, signaling the nucleation and limited growth of a subsystem in the open system and marking an abrupt marginal *decrease* in an imposed critical molecular gradient. Again this holds only for constrained molecular flow. Thus, consistent with the above analysis, we see that the dissipation per unit molecular flow must decrease, with increasing imposed constraint and displacement from equilibrium, some portion of the molecular mass flow being locked up in the creation of relatively stable high-energy molecules. The increase in reaction rate is analogous to the increase in conductivity in the Bénard convection system.

The evolution of entropy production in the subsystem follows that of energy dissipation over a non-critical temperature range. As above, we could equate increasing energy-density in the evolution of the subsystem with decreasing energy dissipation and the same may be said with respect to entropy production. That is to say, *the subsystem evolves toward a state of maximal energy-density* and *minimal entropy production* under given system constraints.

We are now in a position to better appreciate the distinction between linear, weakly nonlinear and strongly nonlinear thermodynamic systems in regard to their displacement from equilibrium, their complexity and their potential for the creation and evolution of increasingly large and more complex systems, as suggested in Fig. 4-5. Panel (a) is included to draw attention to the fact that the gravitational force (G grad), though very weak, plays a fundamental role in the origin and evolution of TMG structures in the stationary state, as revealed by a vertical molecular density gradient in a reaction chamber at temperature T_0/T_0. A correspondingly weak thermal force at temperature T_1/T_0, opposing G-grad, results in uniform molecular density (panel b) and a linear thermal TMGS in the stationary state. A larger applied thermal gradient (panel c, T_2/T_o) gives rise to a weakly nonlinear system, characterized by numerous evanescent short-range molecular interactions, to marginally increase the thermal conductivity of the system, the more so the greater the viscosity of molecular flow. Further increase in the applied thermal gradient may lead eventually to a critical point resulting in the nucleation, growth and evolution of an open subsystem, characterized by very long range ordering (vortex motion) of molecular flow, in which gravity and the viscous force play a special role (panel d, T_3/T_0). Looking forward to our concerns in chapter 17, panel (e, T ~350^0c/3^0c, 300bar) draws attention to the vortex activity in submarine vents, which likely paid a critical role in the origin of life on earth.

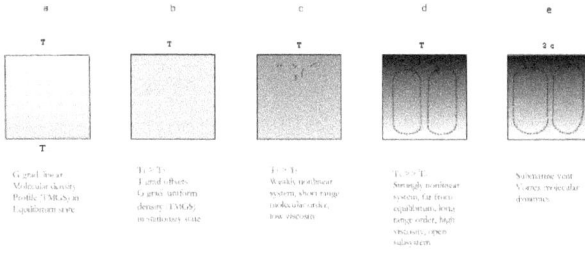

Fig. 4-5 Comparative stationary-state features of linear, weakly-nonlinear and strongly-nonlinear molecular energy-transport systems. Panel (a) is included here to underscore the fact that the gravitational force (G-grad), though very weak, plays a fundamental role in the origin and evolution of TMG structures in the stationary state, as revealed by a molecular density gradient. A correspondingly weak thermal force, opposing G-grad, results in uniform molecular density (panel b) and a linear thermal TMGS in the stationary state. A larger applied thermal gradient (panel c) gives rise to a weakly nonlinear system, characterized by numerous evanescent short-range molecular interaction, to marginally increase the thermal conductivity of the system, the more so as the viscosity of the medium increases. Further increase in the applied thermal gradient may lead eventually to a critical point resulting in the nucleation, growth and evolution of an open subsystem, characterized by the very long range organization (vortex motion) of energized molecules, in which gravity and viscous force play a special role (panel d). Looking forward to our concerns in chapter 17, we add panel (e), in which vortex activity in submarine vents is of extraordinary importance.

Nucleation and Evolution of Similar or Distinct Species of Open Subsystems in a Closed Thermodynamic System Under Constrained Energy Flow

Given various forms and amounts of elemental matter and fundamental forces acting upon them, characterizing a closed thermodynamic system, there is reason to believe that multiple similar open subsystems could nucleate and evolve toward quasi-stationary states, under constrained energy flow through the system. However the dynamic behavior of the system, as well as that of the open subsystems, would likely be much more complex due to competition between energy-flow paths in the system and between open subsystems for available matter-energy in the system. Also we should expect the critical onset of nucleation to be sensitive to small variations and fluctuations in the energy landscape of the system – some subsystems would emerge and quickly vanish at the expense of rapid growth in others nearby. These and numerous other considerations could be explored via computer models of the emergence and evolution of closely similar (single species) types of open subsystems in a closed thermodynamic system, e.g., Bénard convections, where the density, gravitational force, etc., vary marginally from one region to another.

The same applies to the modeling of numerous species of open subsystems (an ecosystem) in a single closed system, though the complexities would mount rapidly as different kinds of subsystems compete for the available matter and energy resources.

Nucleation and Evolution of an Open Subsystem Within an Open Subsystem of a Closed System, Subject to Constrained Energy Flow.

The study of the nucleation and evolution of an open subsystem within a subsystem of a closed thermodynamic system, subject to constrained energy flow, presents a still greater challenge, one involving the interaction of subsystems at hierarchically different levels, time scales and stability demands. Just as the constrained flow of energy through the system is necessary to the nucleation and evolution of a subsystem toward a stationary state of minimal density-specific entropy production in the subsystem and minimal flow-specific entropy production in the system, so the constrained flow of matter and energy through the subsystem is necessary to facilitate the nucleation and evolution of daughter subsystems. The nucleation and evolution of galaxies in galaxy clusters and of stars in galaxies are examples of this in the cosmos as is the symmetry-breaking transformation of neutrons into protons, electrons and neutrinos leading to the emergence and evolution of hydrogen atoms in the Universe. Gravity constrains the directed flow of energy in the former examples with fluctuations giving rise to large-scale gravity-bound structures and structures within structures having relatively large energy-density. In the latter example, it is the decline of local energy-density, in which neutrons reside in the expanding Universe, that eventually drives the electromagnetic transformation and creation of wholly new quanta, whose stability and combinatorial range gave rise to the diversity (in kind and scale) of the quanta-inhabitants of the Universe (see chapter 9).

CHAPTER 5

EVOLUTION OF A QUASI-ISOLATED OR QUASI-ADIABATICALLY EXPANDING UNIVERSE

The arguments that have been presented provide the basis for the following principles:

General Principle of Thermodynamic Selection, applicable to all closed thermodynamic systems.

> The approach to the stationary state in any closed thermodynamic system, under constant constraints, is characterized by the evolution (thermodynamic selection over space and time of that path of mass-energy flow that tends to minimize, on average, the effective-gradient structure for each component of mass-energy transport), as required by the 1st Law. The stationary state structure is uniquely defined by the external constraints and the stationary state p-functions over the space of all microscopic regions of the system.

General Principle of Minimum Flow-Specific Energy Dissipation, applicable to all closed thermodynamic systems.

> In the approach to the stationary state of any closed thermodynamic system (subject to constant constraints), in which the local p-functions are everywhere single-valued and well-behaved, the system evolves in the fastest possible way along a path of *minimum flow-specific energy dissipation*, i.e., the flow-specific dissipation Ψ_f is positive and decreases monotonically from some initial imposed state and, on average, at a maximal, monotonically-decreasing rate: $\Psi_f > 0$, $d\Psi_f/dt < 0$ and $d^2\Psi_f/dt^2 < 0$. In the stationary state, $\Psi_f > 0$ and $d\Psi_f/dt = 0$.

General Principle of Minimum Flow-Specific Entropy Production, applicable to all closed thermodynamic systems.

In the approach to the stationary state of any closed thermodynamic system subject to constant constraints, the system evolves in the fastest possible way along a path of *minimum flow-specific entropy production*, i.e., the flow-specific entropy production Ψ_f is positive and decreases monotonically and, on average, at a maximal monotonically decreasing rate: $\Psi_f > 0$, $d\Psi_f/dt < 0$ and $d^2\Psi_f/dt^2 < 0$. In the stationary state, $\Psi_f > 0$ and $d\Psi_f/dt = 0$.

General Principle of Minimum Density-Specific Entropy Production, applicable to all open thermodynamic subsystems in closed thermodynamic system.

In the approach to the stationary state of any open thermodynamic subsystem, under closed system constant constraints, the open subsystem evolves in the fastest possible way along a path of minimum density-specific entropy production, i.e., the density-specific entropy production Ψ_d is positive and decreases monotonically and, on average, at a maximal monotonically decreasing rate: $\Psi_d > 0$, $d\Psi_d/dt < 0$ and $d^2\Psi_d/dt^2 < 0$. In the stationary state, $\Psi_d > 0$ and $d\Psi_d/dt = 0$. The foregoing is equivalent to saying: An open thermodynamic system evolves in the fastest possible way along a path toward maximal mass-energy-density, as its density-specific entropy production declines toward a minimal value in the stationary state: $\Psi_d > 0$, $d\Psi_d/dt > 0$ and $d^2\Psi_d/dt^2 < 0$. In the stationary state, $\Psi_d > 0$, $d\Psi_d/dt = 0$. Coincident with the emergence and evolution of an open subsystem in a closed thermodynamic system under constant constraint, the system evolves along a path of minimum flow-specific entropy production toward a quasi-stationary state of minimal entropy production.

Since any long-term decline of constraints on a closed thermodynamic system would cause the system to evolve as rapidly as possible toward a lower stationary state of minimum flow-specific entropy production, we may conclude that a quasi-isolated thermodynamic system would evolve, on average, along a path of minimum flow-specific entropy production, and therefore entropy would increase monotonically, on average, along a path of minimum possible rate of entropy increase, i.e., in the *slowest possible* way. Similarly, the Universe, considered as an adiabatic-expanding thermodynamic system, would evolve, on average, along a path of minimum flow-specific entropy production, and therefore entropy in the universe would increase monotonically, on average, in the *slowest possible* way. As such expansion necessarily implies, the flow of mass-energy from a smaller earlier volume of space to the subsequent expanded volume, as discussed in chapters 6 and 7. Consequently, the general principle of minimum flow-specific entropy production serves as the basis for the following extended formulation of the *2nd Law* of thermodynamics:

Postulate:
> In the evolution of a quasi-isolated thermodynamic system or of the Universe, taken to be an ongoing, adiabatically-expanding thermodynamic system in spacetime, entropy is positive and increases monotonically, on average, and at a minimal, monotonically-decreasing rate, i.e., $S>0$, $dS/dt>0$ and $d^2S/dt^2<0$. Thus, entropy increases without limit in all such systems.

The foregoing extends the stature of the 1st Law of Thermodynamics: in addition to its requirement on conservation of energy, *it imposes a requirement on the 2nd Law that entropy in the Universe increase at a minimal pace. The above postulate and General Principles, together with Heisenberg's Indeterminacy* relation *(see chapter 7), may be referred to as the* **Nonequilibrium Thermodynamic Imperative (NTI)**.

The evolution of TMGS, in well-behaved and strongly nonlinear thermodynamic systems, is the consequence of *thermodynamic selection* processes set in motion by changes in external constraints or by internal fluctuations. Real systems may never attain a projected stationary state, inasmuch as the external constraints are not in fact strictly constant but only evolve more slowly than the pace of evolution of the systems they energize. Thus, in all such cases systems evolve *toward* some quasi-stationary state. This caveat has special significance for developmental and evolutionary biology and for the study of geophysical and cosmological evolutionary processes, as will be discussed in later chapters. In particular, ***thermodynamic selection and the nonequilibrium imperative* will be seen as a unique fundamental physical process (force) driving *Darwinian natural selection and the evolution of living systems*.**

While auto catalytic or coupled reactions routinely characterize the behavior of chemical or bio-molecular systems, similar fundamental phenomena account for nonlinear behavior in thermal, electromagnetic or mechanical systems. And similar "auto-catalytic" or "cross-catalytic" phenomena are operative in social, political, economic or ecological relationships, though the "reactions" or interactions are often mediated via symbolic exchange of information and mental images. Indeed, it is reasonable to imagine that some form of neural "auto-catalytic" or "cross-catalytic" activity is involved in the ontogeny of the central nervous systems of all organisms.[10] In the more complex chemical, bio-molecular, ecological and social systems, a critical instability may present not one but several possible paths of state transformation, depending not only on the system parameters but also on the evolutionary history of the system.

To the extent that the rates of energy dissipation and energy flow and temperature change may be accurately measured in various experimental studies of weakly- and strongly-nonlinear thermodynamic systems, it should be possible to test the validity of the proposed principles of minimum flow-specific energy dissipation and entropy production. Where practicable, the use of constrained thermodynamic flows, as discussed previously, should facilitate such investigations. Thus, incremental increase of constrained heat-flow in the study of Bénard convection should enable step-by-step observation of the nucleation and spatial growth of an organized convection subsystem, while measurements of the temperature gradient across the system, as the *constrained heat flow* is increased incrementally, should reveal a pattern of dissipation *decline* similar to that of ψ_{sys} shown in Fig. 4-3. Such experiments could be undertaken with a controlled, low thermal-mass, radiant-energy source. (The kinetic equivalent of unconstrained autocatalysis, i.e., strong mutual reinforcement of directed molecular motion in convection patterns, is the mechanism underlying this transformation. The organized subsystem nucleates and grows marginally as a means of minimizing energy dissipation in the system under constrained heat-flow. Bénard convection not only illustrates the important role played by the near-infinitesimal gravitational gradient, but the difficulty of freeing any experiments from such force.)

If indeed the nucleation of a convection cell and the "budding" or "fissioning" of "daughter" cells could thus be demonstrated, would it be constructive to think of such subsystems and their chemical counterparts as precursors of the simplest sort of self-reproducing organisms in which the instructions for self-replication are embodied in the characteristic internal and external constraints on auto-catalytic or cross-catalytic activity in the subsystem? Would it be possible to demonstrate, experimentally, the mutation and evolution of variant Bénard subsystems in response to changes in the physical environment, lending further credence to their inclusion as simple representations of a broad class of "lifelike subsystems?" In a similar way, we might be able to explore the emergence, growth and evolution of organized subsystems in other non-equilibrium systems under conditions of constrained thermodynamic flow. As we suggest in chapter 17, there is a need to define "lifelike subsystems," because of their very large range of complex structure and behavior, in terms of some appropriately large parameter space.

Since the initial period, following any rapid, significant increase in the boundary conditions of a system, may be characterized by relatively high dissipation, the appearance of an organized subsystem in some local region might be ephemeral, i.e., might soon collapse as energy flow in the region rapidly declines to unsustainable levels. And, since the initial period in any sustainable state transformation offers the greatest options for selection among alternative paths of structural evolution, we might expect that the most significant and rapid structural or dynamic change would occur at such times, both in the sense of growth and decay. This appears to be the case in the evolution of living matter, as the paleographic evidence suggests. The process of speciation may be better characterized as punctuated evolution toward a new quasi-stationary state defining a new species, in response to relatively rapid changes in external or internal constraints imposed on an existing one.[11]

Just as the entropy formulation of the *2nd Law* appears in need of extension, so Kelvin's statement of the *2nd Law* may be extended as follows: No process is possible in which the *sole result* is the absorption of heat from a reservoir and its *complete conversion* into work [or the *complete lack thereof*]. A corollary statement might read: No process is possible in which the *sole result* is the *complete dissipation* of mechanical energy. These statements simply assert that the conversion of energy from one form into another cannot be accomplished in any finite process with either 100% or 0% efficiency. The argument that mechanical energy may be completely dissipated in frictional processes, is untenable, since in any such process some work has been done in the creation of ordered, if only ephemeral, TMGS and the brief storage of free energy in associated physical structures, all of which serve, however minutely, to slow the increase of entropy in the Universe. More significantly, the burnished rubbing surfaces or work-hardened elements of mechanical systems bear relatively long-term witness to the fact that work has been done in the creation of such stable features.

Lastly, there is no further need for discourse about Maxwell's Demon, since there can be no state of equilibrium in the Universe in which such a Demon might spontaneously conjure order from disorder.

Thus, irreversible processes always give rise simultaneously to destruction and construction of ordered structures and processes in the Universe, though the entropy decrease associated with the latter is always exceeded by the entropy increase associated with the former. The decay of order in the one region provides the driving force, energy flow and materials necessary for the creation of order in the other region. This is in fact the way in which all structures and systems come into being in the Universe, providing evidence of their role in slowing the rate of entropy increase from what would otherwise occur in their absence. Nature abhors *excessive* energy dissipation and entropy production and is thus attracted, post-haste, to a quasi-stationary state and reluctantly toward the much maligned "heat death."

All ordered structures or organized systems, created and maintained by the flow of energy through matter, not simply those which evolve in strongly-nonlinear thermodynamic systems far from equilibrium, are to one degree or another, "dissipative structures," if only ephemeral. However, this expression in the literature poorly serves our needs, since it provides little insight into the important properties that distinguish the one class of systems from the other. As Lynn Margulis and Dorion Sagan have observed: " 'dissipative structures' [is] a rather awkward term because it focuses on what the structures--actually, systems, not structures--throw away rather than what they retain and build..."[12] The wonder is not so much that dynamic structures evolve as a consequence of dissipation processes but that they do so in the most efficient way, making maximal use of thermodynamic gradients and flows in the creative process. Indeed, the evidence suggests that the greater the extent and complexity of hierarchical organization of matter-energy and corresponding demands on energy flow, the greater is the value of the efficiency parameter $\eta(X)$. In this hierarchical range, multiple feedback controls may result in stable quasi-linear behavior, as Zotin[13] has argued.

All physical (massive) particles and organized systems thereof come into being and evolve as open subsystems (from the quantum to the atomic, molecular, the living state and cosmic levels), which serve thereby to minimize the entropy production per unit density, as will become apparent in subsequent chapters. All such systems and subsystems are created and maintained for a time by this **Nonequilibrium Thermodynamic Imperative** (**NTI**), which we associate here with Heisenberg's uncertainty principle, taken as a *determinacy* principle in driving the expansion of the Universe (chapter 7).

PART II

GRAVITY IN
RELATIVISTIC COSMOLOGY

Part II explores the implications of nonequilibrium thermodynamics, as developed in Part I, for cosmology and relativity theory. Astronomical evidence suggests that the Universe may be considered a quasi-adiabatically expanding, nonequilibrium thermodynamic system. Yet such view of the origin and evolution of the Universe would be in fundamental conflict with the current widely accepted Big Bang/Inflation scenario. Chapter 6 probes the underlying assumptions, on which present theory was founded, in part to clarify the requirements on any possible alternative scenario. The presumed absolute, universal constancy of the speed of light c and Newton's gravitational constant G emerges at the heart of this conflict.

Chapter 7 sketches a radically different cosmology, in which c and $\sqrt{G}$ are constant over multiple epochs in an initial quiescent quantum state (raising questions about Planck's Constant h), but vary inversely with the average energy-density of the Universe ρ in the subsequent scalar expansion era.

In chapter 8, gravity (relativity), symmetry, quantum indeterminacy and the *extended 2nd Law* (*2nd Law*) are key to characterizing the physical nature of the quantum Universe. An evolutionary process, consistent with the requirements of the *2nd Law*, accounts for: the early particle-number expansion of the quiescent quantum Universe (the creation of all of the matter particles in the Universe), the transition to an extremely hot state, and the present observed ratio of free photons to matter particles.

And chapter 9 examines the physical expansion of the Universe (following an annihilation-plasma event) with particular focus on the role of the evolving parameters c, ε *(the permittivity of space)* and $\sqrt{G}$. (As explained in chapter 8, the parameter $h(\rho)$, in the quiescent quantum era, emerges as an absolute universal constant with the transition from the quantum to the scalar expansion universe, commencing with the plasma epoch.) In all considerations we are required to square our reasoning with the principles and extended *2nd Law* of thermodynamics as presented in chapter 5.

Given the importance we assign to the extended *2nd Law* (its creative as well as destructive role), we shall often simply refer to the *2nd Law* in the remainder of this book, with that implication.

CHAPTER 6

CONFLICTS AMONG FUNDAMENTAL ASSUMPTIONS UNDERLYING BIG BANG/INFLATION COSMOLOGY

Our purpose in chapter 6 is to: (1) examine conflicts among the fundamental assumptions, theory and empirical evidence leading to the demise of the standard hot big bang model of the Universe; (2) examine possible conflicts of fundamental assumptions underlying inflation scenarios crafted to circumvent problems with the standard big bang model and (3) determine whether the inflation concept is consistent with the thermodynamic principles and *2nd Law*, as presented in chapter 5.

We proceed in this effort by sketching the evolution of twentieth-century relativistic cosmology, focusing on the principal assumptions and turning points.

EVOLUTION OF RELATIVISTIC COSMOLOGY

Einstein's special and general theories of relativity led rapidly, in the early twentieth century, to the joining of observational astronomy with physics and the development of what has come to be labeled astrophysics. Based on Einstein's field equations, in which c and G were (and still are) taken as absolute universal constants, various quantitative models of the Universe soon appeared, including an attempt by Einstein to create a static model of the Universe, as he believed it to be unchanging in size and form. To his dismay, a static solution to his equations was shown to be unstable, leading him to add a term to his equation incorporating a constant Λ (now referred to as a "cosmological constant"). In a letter to Besso in 1918, Einstein explained: "Either the universe has a centre, has a vanishing density everywhere, empty at infinity where all the thermal energy is gradually lost as radiation; or, all the points are equivalent on the average, and the mean density is everywhere the same. In either case, one needs a hypothetical constant Λ, which specifies the particular mean density of matter consistent with equilibrium. One perceives at once that the second possibility is more satisfactory, especially since it implies a finite size for the universe. Since the universe is unique, there is no essential difference between considering Λ as a constant which is peculiar to a law of nature or as a constant of integration."[14] As it turned out, such a term, depending on its nature, enables the development of dynamic models, and in particular models of an expanding Universe, and hence is incorporated in various current models in one interpretation or another. (A remark by Coles and Ellis may be seen as a **first potential inconsistency**: **"A non-zero Λ affects all matter in the universe but is affected by nothing – violating the fundamental principle that anything that acts on matter should also be acted on by matter."[15]** But Hubbles's plot of redshift data in 1931 showed the Universe to be expanding, ending any further belief in a static universe on the part of Einstein. (One might speculate that the above events should have led Einstein to a first small doubt about the constancy of the speed of light. But that was not likely in view of all the empirical evidence to the contrary at that time.)

Further, as Kragh notes "The Friedmann-Lemaitre theory was the first step toward not only an expanding universe but also the idea of a big-bang origin. … It was only in 1931 that the first proposal of what can reasonably be called a big-bang theory was made, namely, in Lemaitre's hypothesis of what he called the primeval atom."[16] In a letter that appeared in *Nature* in 1931, Lemaitre said:[17] "the present state of quantum theory suggests a beginning of the world very different from the present order of Nature. … Thermodynamic principles from the point of view of quantum theory may be stated as follows: (1) Energy of constant total amount is distributed in discrete quanta. (2) The number of distinct quanta is ever increasing. *If we go back in the course of time we must find fewer and fewer quanta, until we find all the energy in the universe packed in a few or even in a unique quantum."* (emphasis added) In the letter Lemaitre succinctly characterized a quantum origin of the universe[18]: "If the world had begun with a simple quantum, the notions of space and time would altogether fail to have any meaning at the beginning; *they would only begin to have a sensible meaning when the original quantum had been divided into a sufficient number of quanta.* (emphasis added) If this suggestion is correct, the beginning of the world happened a little before the beginning of space and time. I think that such a beginning of the world is far enough from the present order of nature to be not at all repugnant…we could conceive the beginning of the universe in the form of a unique atom, the atomic weight of which is the total mass of the universe. This highly unstable atom would divide in smaller atoms by a kind of super-radioactive process." The idea of a primeval atom, as Kragh put it, "was grafted upon one particular solution of the relativistic equations for which R=0 for t=0, but it could equally well have been applied to other solutions with the same property, such as the Einstein-de Sitter model."[19] This was the first big bang model of a physical nature in cosmology.

The notion that the Universe had a beginning in time and therefore a limiting age led to a **potential inconsistency in *time scales*, suggesting a possible conflict between the age of the Universe and that of its constituents.** De Sitter was deeply worried. "I am afraid," he wrote, "all we can do is to accept the paradox and try to accommodate ourselves to it, as we have done to so many paradoxes lately in modern physical theories. We shall have to get accustomed to the idea that the change of the quantity R, commonly called 'the radius of the universe,' and the evolutionary changes of stars and stellar systems are two different processes, going on side by side without any apparent connection between them. After all the 'universe' is an hypothesis, like the atom, and must be allowed the freedom to have properties and to do things which would be contradictory and impossible for a finite material structure."[20] *This reflects the extraordinary willingness of the times to accept the inadequate explanations for the strange behavior of the cosmic and quantum worlds.* Kragh notes Tolman's warning against identifying the beginning of the expansion with the beginning of the Universe: "The difference between the time scales for stellar evolution and nebular expansion suggests that no definiteness could now be attached to any idea as to *the* beginning of the physical universe. Indeed, *it is difficult to escape the feeling that the time span for the phenomena of the universe might be most appropriately taken as extending from minus infinity in the past to plus infinity in the future.*"[21] (emphasis added)

The connection between cosmology and quantum physics commenced in the 30's, with the advances in nuclear physics. "In late 1939, Gamow entertained the idea of a big-bang universe, at least in some vague sense. ... It was the realization that the stars were unable to produce a wide range of elements in the right ratios that convinced some astrophysicists of the necessity to focus on a pre-stellar formation and then opened up for them the possibility of a big-bang universe."[22] [The essential idea of a big bang model, building on a combination of nuclear physics and the Friedmann-Lemaitre equations, emerged with Gamow in the fall of 1945. Gamow viewed equilibrium theories as hopelessly inadequate. *"The only way of explaining the observed abundance-curve lies in the assumption of some kind of unequilibrium process taking place during a limited interval of time."*[23] (emphasis added) The development of the standard hot big bang theory reached a climax in a paper in 1953, with the work of Alpher and Herman in collaboration with James Follin (p 127), in which the assumed homogeneity of the early Universe would consist of causally unconnected parts because the horizon distance ct would be much smaller than the radius of the Universe. This appears to be the first time that the basis for the *"horizon problem"* was explicitly stated.

These and other problems led to the demise of the standard big bang model, and with it attention turned to a steady-state alternative model on which rested a most radical assumption "that matter is continually created throughout the universe, emerging spontaneously out of nowhere." As Kragh notes: "Apart from being an important cosmological theory in its own right, the steady-state theory provoked a major controversy in cosmology by questioning standard assumptions of the evolutionary, relativistic theory. This had two consequences, which together contributed significantly in changing cosmology from a mathematical game to a scientific theory. By offering a radically new world picture, steady-state theory forced astronomers and physicists to think more deeply and critically about the foundations of cosmology, and because the alternative had direct observational consequences, different from those of the evolution theory, it was an important factor in the emergence of new observational methods and practices."[24]

While the competition between the steady-state and evolutionary theories of the Universe did generate a great deal of provocative discussion, the debate hinged ultimately on the relative capabilities of the two theories to account for the *helium problem*, the abundance of helium and deuterium in the Universe, and the cosmic microwave background radiation. The latter, as Kragh writes, was "the final blow in an already dying theory. As early as 1948 Alpher and Herman had predicted the existence of a blackbody-distributed radiation of a present temperature of about 5K, and interpreted it as a remnant of the decoupling between matter and radiation in the early universe. However, neither their original prediction nor their and Gamow's several repetitions of it attracted interest among cosmologists. In the early 1960s it was effectively forgotten, a most remarkable fact."[25]

The hot big bang model, despite serious deficiencies, survived the long controversy with various versions of the steady-state theory, in the late 60's. It was a decade later before a variation of that model emerged to deal with the problems of the standard big bang model.

CONFLICTS IN THE STANDARD MODEL

The fundamental assumptions underlying the standard Model had their origin in the time of the Michelson-Morley experiments in 1887, demonstrating the constancy of the speed of light, independent of the velocity of the sending or receiving apparatus, and a half-century later in the empirical and theoretical advances in high energy physics. Maxwell's calculation of the speed of light (from measured values of the permittivity and permeability of free space), subsequently supported by direct measurement, led to the acceptance of the speed of light c as a fundamental absolute constant of Nature and a bedrock of the special theory of relativity. A decade later in 1915, Einstein's field equations, incorporating c, together with Newton's gravitational constant G (also taken as a universal absolute constant), gave rise to early models of the universe, as sketched above. Hubble's discovery in 1931 that the Universe was expanding at the speed of light (based on redshift measurements of light from distant galaxies) supported the notion of some form of positive cosmological constant and implied that all of the matter-energy of the universe had, at its origin, to be contained in a radius of quantum dimension. This in turn implied extreme energy densities, temperatures, and gravitational force, in keeping with quantum and relativity theory. The standard model of hot big bang cosmology quickly evolved.

We confront now the first conflicts in fundamental assumptions. The Standard Model of particle physics, like quantum electrodynamics, derives from relativistic quantum field theory, which emerged in the 1930's to square the notion of point particles (e.g., electrons, having no physical size or internal structure) with relativity theory. The standard model aimed at the unification of the four fundamental forces (electromagnetic, electroweak, strong force and gravity) in the early history of the Universe when the temperature was 10^{28}K or greater. **Most importantly, however, the force of gravity has remained logically and empirically apart from such unification, which might have been seen, early on, as a fundamental self-inconsistency.** From empirical estimates of the present average energy-density and corresponding background radiation temperature (3K) of the Universe, the radius of the Universe at a time when its temperature was 10^{28}K would have had to be about 3mm, some 10^{17} seconds ago, extrapolated from expansion at the speed of light. (This assumes, in keeping with thermodynamic requirements, that the temperature halves as the radius of the Universe doubles.) The problem thus presented is that the maximum distance over which any region could be causally connected to any other region would have been only $\sim 10^{-23}$m. As noted above, the *horizon problem* denied any possible accounting for the isotropy and homogeneity that characterizes the large-scale Universe, as we should expect from classical thermodynamic arguments. **Here we see Nature telling us that our model of the expansion of the Universe lacks empirical support (self-consistency) by some twenty orders of magnitude!** Could such disclosure, two to three years before Einstein's death have changed his mind regarding the absolute constancy of the velocity of light and perhaps that of Newton's gravitational constant G?

These and other problems led to the demise of the hot big bang model: The *flatness problem* (failure to account for the density parameter Ω being so close to unity, as it now appears to be, in which spacetime is quasi-Euclidean.) Similarly, the model failed to contend with the *initial-conditions* problem (Why did the Universe begin in a state of extreme temperature? What started the expansion and why apparently at the speed of light?) It failed also to account for the absence of magnetic monopoles in the Universe, as predicted by grand unified theories. And the model implied a very short *age-of-the-Universe*, relative to the estimated age of galaxies and other bodies in the Universe. Why? Likewise, the time required to seed and evolve large-scale structures we see in the Universe today leaves much unexplained. However, in the minds of many, that model garnered a certain support in accounting for the nucleosynthesis of the light elements.

CONFLICTS IN THE INFLATION SCENARIO

Alan Guth proposed a way around the most serious of these problems in the early 1980's, that depended ultimately on the existence of energy in the vacuum, a quantum field theory concept that played a critical role in the development of quantum electrodynamics (QED). Just as the "physical vacuum" was tapped for energy to create and annihilate virtual photons, electrons and positrons, etc., in QED a half-century earlier, that energy source (quantum field fluctuations in the vacuum) was called on to empower a very brief period of superluminary expansion of the Universe. Most important, energy in the vacuum played a fundamental critical role in dealing with the horizon and flatness problems, just as it did in QED in accounting for the Lamb shift. Guth reasoned that the physical vacuum might energize a brief period of accelerated expansion of *space* from a quantum scale beginning, sufficient to leave it expanding at the speed of light c and with a radius of about one centimeter. This process was viewed as a phase transition in which the Universe supercooled some 28 orders of magnitude below a critical temperature. In the course of the transition, the latent heat would be released, resulting in an *extreme rate of entropy production* and leaving the Universe once again close to the critical temperature, though now at a suitably larger radius. As Guth said: "Such a scenario is completely natural in the context of grand unified models of elementary-particle interactions."[26] Guth saw the requirement of the standard model, that the universe expand adiabatically, as the underlying problem and said **"one can see that both problems could disappear if the assumption of adiabaticity were grossly incorrect."** And in broader perspective he argued **"In order for this [inflationary] scenario to work, it is necessary for the universe to be essentially devoid of any strictly conserved quantities."** (emphasis added) These are clear and direct acknowledgments of very deep inconsistencies in fundamental physical reasoning. Guth was quick to point out that: *the horizon wouldn't actually disappear, rather that it would simply be moved out of the observational range of the visible universe.) But this argument covers-over yet another inconsistency? If we can have knowledge of only a minute portion of the early universe, how can we be certain that that portion is representative of the whole Universe?* Stated differently: *the Universe we see today appears isotropic and homogeneous because it represents only a very small patch of relatively uniformly distributed mass-energy.*

Guth saw the inflationary scenario as a "natural and simple way" to deal with the horizon and flatness problems, though he acknowledged two areas of concern: "the difficulty in finding a smooth ending to the period of exponential expansion" and the randomness of the bubble formation process (in the termination of the phase transition) that leads to gross inhomogeneities.

Thus, the inflation scenario may be characterized by the following theoretical assumptions, conflicts (inconsistencies) and questions deserving rational answers:

- Like the standard model, the inflationary scenario assumes a hot start to the Universe. **Why and how did it start that way? What happened before the bang? Did the Universe expand before, as well as immediately after inflation, at the speed of light?** These questions deserve serious discussion if only to draw our attention explicitly to the *singularity* problem and our dismissal of it as being beyond the province of science.

- **How did gravity conspire with radiation pressure to ensure such a precise rate of expansion?** The inflation scenario, like all big bang models is fundamentally dependent on the Hubble constant H_0 and the assumption that the expansion rate is exactly c and that c is absolutely constant. **What is there about the value of this constant that warrants such absolute authority, especially since we are now seeing evidence that the speed of light has been marginally increasing as we look into the distant past of the cosmos? Ought we not expect $c \rightarrow 0m/s$, under such conditions of extreme energy-density, as near the horizon of a black hole? But if the speed of light is not absolutely constant, how do we account for the linear red shift data and how do we square all of this with relativity?**

- **Did the gravitational constant inflate, along with the velocity of light, so as to preserve the constancy of J in the field equations? The constancy of c and G appear to be at the heart of the inconsistencies in the standard model of cosmology and far from disposed of in inflation scenarios.** It has been assumed that the energy of the physical vacuum, behaves as an antigravity force, a negative pressure resulting in the expansion of *space* at an accelerating rate, greatly in excess of the constant velocity c, while not violating special relativity. **However, it is inconsistent to say that it is only space that is expanding faster than light, when space itself is defined by the distribution of matter-energy in it; some work is required to expand the Universe, accompanied by some dissipation of real physical energy and some minimal increase in entropy. How does inflation circumvent the 1st Law?**

• **How do we calculate the high frequency cutoff of the quantum fields filling all space, so that the energy-density of such fields is just sufficient and no more for the task of expanding the Universe? What direct empirical support is there for such energy, other than the Casimir effect? Is there an alternative explanation for that? (See chapter 11 for a different, energy-density account of the Casimir effect.)** As we noted above, inflation would allow the flatness, initial conditions, monopole and horizon problems to be sequestered to other possible universes, about which we would have been and would remain causally disconnected for all time.

• **What are we to say about a science in which we discard all but one part in 10^{83} of the energy, matter, etc., that may or may not have helped to define the Universe in which we live? Since we could not have influenced the evolution of all these other worlds, how were they allowed to shape ours? The *multiverse* problem, thus envisioned, would appear to be a very costly way to resolve the above problems.** These questions and apparent inconsistencies have received relatively little attention in the rush to accept the inflation solution to several long-debated difficult problems. With evidence accumulating that the Universe has been and is expanding at an *increasing* rate, the inflation scenarios are further confronted with what may be called the **dark energy** and **dark matter** inconsistencies:

• **How can one fine tune the marginally-increasing expansion rate and how can we include the unobservable immeasurable dark matter in the fine tuning process? (We will show (chapter 9) that these concerns derive straightforwardly from the assumption** that c and G evolve covariantly with the inverse average energy density of the Universe and local variation thereof.) Finally one sees that inflation scenarios grossly violate the Principles and extended *2nd Law* of nonequilibrium thermodynamics set forth in Part I and deny even the authority of classical thermodynamics. How could the creation of the Universe and all subsystems in it emerge in such an inconsistent model?

We should require that all of the above-noted conflicts, inconsistencies, and unresolved questions be dealt with rationally in any alternative cosmology.

CHAPTER 7

A RADICALLY DIFFERENT COSMOLOGY: EVOLUTION OF c AND G

Chapter 7 explores three kinds of questions. First, can a rational argument be made for the co-evolution of the fundamental "constants" c and G with ρ, the average energy-density, characterizing the scalar expansion of the Universe in spacetime? What implications would that have for relativity and quantum theory? Second, could the Universe have begun in the quasi-infinite past in some nonsingular, quasi-stable quantum state (Lemaitre's primeval atom?), nearly perfectly bound by the forces of gravity (relativity) and symmetry? How could the singularity problem be resolved and what governs the orderly expansion of the Universe? Third, would such a scenario be broadly consistent with the principles and extended *2nd Law* of nonequilibrium thermodynamics, as set forth in chapter 5?

There is precedent for concerns about the fundamental constants, as Barrow[27] reminds us: "Each really major advance in physical science goes hand in hand with a revision or extension of our understanding of some constant of Nature." He emphasizes this point by noting that despite the vast complexity of everything made out of atoms and molecules, together with the vast range of properties that straddle the states of matter from gases to liquids to solids, the gross features of this entire world of materials is determined by the values of just two numbers: the mass of the proton divided by the mass of the electron 1836 - and the 'fine structure constant' – 1/137.036, obtained by combining the measured values of the three constants that comprise it. "We do not know why these two numbers take the precise values that they do. Were they different, our Universe would be different, perhaps unimaginably different. … The sizes of planets and stars are not random accidents or the pre-programmed result of particular initial conditions at the Big Bang. Rather, they arise as equilibrium states between opposing forces of Nature. These forces come into balance only when the aggregate of particles involved attains a certain size." And Barrow goes on to say that the 'Planck length' (the basic unit of length in the elementary-particle and gravitational worlds) "is the only quantity with the dimensions of a length that can be built from the three most fundamental constants of Nature: the velocity of light c, Planck's Constant h, and Newton's gravitational constant G. It is given by $l_p = (Gh/c^3)^{1/2} = 4 \times 10^{-33}$cm. This tiny dimension encapsulates the attributes of a world that is at once relativistic (c), quantum mechanical (h), and gravitational (G)."[28] We will find answers to these puzzles later in Part II.

Evolution of The Fundamental Constants

We offer, as our most fundamental assumption, that the dielectric "constant" ε and permeability μ of free space each vary as the cube-root of the average energy-density of the Universe $\rho(t)$. (We are assuming here that the measured values for ε_0 and μ_0 hold in our local interstellar region and time of the Universe. The fact is we have no way of knowing or accurately measuring the rate at which those parameters are varying with cosmic time. ε_0 is a constant in Coulomb's Law, relating the electrostatic force F_C in free space between two charges q_1 and q_2 and their separation r, as $F_C = q_1 q_2 / 4\pi\varepsilon_0 r^2$. Similarly, μ_0 is a proportionality constant in Ampere's Law relating the force F_A, between two very long parallel wires, carrying currents i_1 and i_2 a distance l unit apart, as $dF_A/dl = 2\mu_0 i_1 i_2 / l$.) Accordingly, the speed of light $c \propto (1/\varepsilon\mu)^{-1/2} \propto (\rho^3 \rho^3)^{-1/2}$ would therefore vary as ρ^{-3}. The above assumptions are equivalent to arguing that the distributed uncharged mass-energy in the Universe (the average energy-density) acts as a dielectric to reduce the Coulomb coupling strength of charges q_1 and q_2, and also as a permeable medium to decrease the coupling strength between parallel paths of charge flow, in the early, high-density Universe. That is to say, the charge coupling between two charged particles or bodies may induce electric dipole moments in the matter particles, whose effect would be to decrease the coupling strength of the charged bodies, the more so the greater the density of uncharged matter particles, as sketched in Fig. 7-1. (As will be shown in the next chapter, we assume the distributed particles defining the energy-density are neutral composites of charged particles and therefore subject to polarization in an external electric field.) But we should expect the internal energy-density (charge-density) in the charged bodies to rise proportionately with the average energy-density, since otherwise the force acting on them would apply uniquely to the vacuum state, as indicated in Fig. 7-1. Similar reasoning would apply in regard to the coupling strength between parallel paths of charge flow. (See chapter 11 for a new understanding of the electromagnetic and gravitational forces and the Casimir effect. Thus the foregoing provides a fundamental accounting of the constancy of the electromagnetic force in terms of the average (or local) energy-density of the Universe. *The curvature of space thus implies a warping of the dielectric and permeability parameters of space, which thereby restricts the maximum velocity of any motion or signal in spacetime in any epoch, the more so in the earlier epochs of the physically expanding Universe.*

On the basis of these arguments, we take the speed of light c to vary inversely with the cube of the average energy-density of the Universe. That is, $c(\rho) = c\rho_0/\rho^3$, where c is the 'constant' measured velocity of light in the vacuum of our immediate spacetime neighborhood of the Universe, in which the average energy-density is the present value ρ_0. (Maxwell interpreted his equations in terms of two fundamental assumptions which we now challenge: that radiant energy propagates as waves and that ε_0 and μ_0 are absolute universal constants. And on that basis he estimated the speed of light. But we now suggest from the above that his equations apply equally well in a strictly particle theory of light and be inherently relativistic in general as well as special relativity.) **It is this variation in the speed of light that ensures that all regions of the Universe are causally connected, at all times in the expansion of the Universe and indeed that the Universe is nearly flat ($\Omega \rightarrow 1$) in all epochs of *physical* expansion.** (Recent redshift data, involving light from type 1a supernova several billion lightyears distant, reveals a small increase in the velocity of light that has accompanied the expansion of the Universe as that light traveled to us and as our assumptions would lead us to expect, consistent with a corresponding decrease in the average energy-density.)

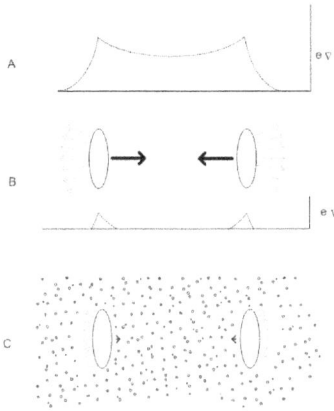

Fig. 7-1   Variation of Coulomb gradient (e∇) at oppositely charged bodies in vacuum (A), resulting in a strong mutual attraction via recoil momentum (black arrows) from photon emission (grey arrows) and in a dielectric gas (B), resulting in diminished gradients at the surface of the charged bodies (C).

Moreover we should expect G to vary with the average energy-density and indeed it would have to vary as ρ^6, if the gravitational charge (coupling strength $m\sqrt{G}$) were taken as a constant in any epoch (as would be the electric charge $q\sqrt{\varepsilon_0}$ and the magnetic charge $I\sqrt{\mu_0}$) and Einstein's field equations were to characterize a nearly static, Euclidean Universe *for all time*, as the radius of the Universe R varies as ρ^3 *adiabatically. In that way the* **force** *of gravity between particles (or aggregate bodies), separated by a fixed distance, would be invariant*, i.e., independent of variations in the average energy-density. This follows since m varies as ρ^3, while $\sqrt{G}$ varies as ρ^3. In consequence, we postulate that $\sqrt{G}$ *varies as* ρ^3. Are there fundamental physical arguments to support this postulate, as for the variation of the dielectric 'constant' and permeability with energy-density? In Fig. 7-1, we are assuming that every particle making up the energy-density gradient is capable of contributing a gravitational dipole moment (under an imposed gravitational gradient) as well as an electric dipole moment (under an imposed electric gradient), though the actual coupling-strength values of the two forces may differ greatly.

Indeed this common action at the micro level enables both **forces** at the micro, macro and cosmic levels to remain constant as the average energy-density of the Universe declines. In each case – the electromagnetic or gravitational force – a global or local decrease in mass-energy-density - would act effectively to conserve the interaction of massive bodies charged or neutral) via recoil impulse of emitted bosons, as shown in Fig. 7-1 and discussed in chapter 11. The variation in $\sqrt{G}$, at the cosmic scale, is thus explained also at the quantum scale in terms of the induced gravitational dipole moments of the neutral massive particles in the Universe. And *gravity would be seen as a gauge force or field at the quantum level, which would thereby necessarily make quantum gravity subject to the uncertainty principle*. This directly addresses Hawking's central concern: "The main difficulty in finding a theory that unifies gravity with the other forces is that general relativity is a 'classical' theory; that is, it does not incorporate the uncertainty principle of quantum mechanics."[29] But fluctuations of the order of Planck's Constant would have little if any effect where the local energy-density (masking the interaction) may be much larger, and so *we might expect that the value of Planck's Constant h should increase with the energy-density of particles in the quantum Universe*, as discussed later and in chapter 8.

The above necessarily requires that the linear dimension of a particle (its wavelength λ) or of the distance between two particles or massive bodies, must vary as ρ^{-3}, neglecting local variations in energy-density, and that time be taken as the ticking of a constant fundamental Universal clock, measuring the rate of radial expansion of the Universe. All of this, in turn, argues that it is the velocity of light that is dilated in relativity rather than the ticking of a clock. Strictly speaking then there can be no system or system-behavior immune from time's demand for physical change. Much remains for further discussion on this very important matter in chapters 8 and 9 and in Part III.

But it would be reasonable to expect the local energy-density ρ_l to be somewhat larger near massive stars, neutron stars, black holes and the central regions of galaxies, while the average energy-density in any cosmic epoch remains essentially constant. This addresses the "dark energy" and "dark matter" problems, about which there will be more in chapter 9. We should also expect that c, $\sqrt{G}$ and $\sqrt{\varepsilon}$ would also vary appropriately in the *local* high energy-density environment of the nucleus of atoms, as will be discussed in Part III.

In this scenario, the Universe would have evolved smoothly, from some initial quantum state characterized by the values of $\rho \to \infty$, $\sqrt{G} \to 0$, and the $\sqrt{\varepsilon}$, $\sqrt{\mu}$ and $c \to 0$. It would have had a marginally negative spacetime curvature ($\Omega \cong 1$) from the beginning (as we shall argue later and in chapter 8). This would be consistent with Hawking's "no boundary" proposal, which asserts that the quantum Universe "could not have been completely uniform, because that would violate the uncertainty principle of quantum theory. There had to be small fluctuations in the density and velocities of particles."[30]

The proposed evolutionary model, also addresses Peebles' concern about the causal expansion of the Universe: "If the expansion [of the universe] does trace back to a dense state, the expansion rate at high redshift and high density is quite unaffected by the cosmological constant, and we need some other resolution to the mystery of what caused the universe to expand from whatever it was like before it was expanding, whatever that means."[31] The forgoing arguments also resonate with Tolman's early cautionary views: "'The difference between the time scales for stellar evolution and nebular expansion suggests that no definiteness could now be attached to any idea as to *the* beginning of the physical universe. *Indeed, it is difficult to escape the feeling that the time span for the phenomena of the universe might be most appropriately taken as extending from minus infinity in the past to plus infinity in the future.*"[32] (emphasis added)

In our epoch the distance to galaxies and galaxy clusters could only be measured in a spacetime metric in which both the velocity of light and the spatial separation of bodies vary with cosmic time. Spatial separation appears to be dominant now, but it should make a marginally declining contribution, as we look back farther and farther in time. Thus, in the low density Universe we now live in, nearby galaxy-clusters should be receding from one-another in spacetime at speeds somewhat less than that defined by the Hubble Constant. The above scenario is consistent with the current linear expansion data, since we are only now able to discern a significant departure from linearity at the farthest range of visibility. (The departure from linearity likely commenced closer to us, but we lack observational precision to reveal it.) However, while the inflation scenario places the GUTs era 13.7 billion years in the past, that sketched above may place it orders of magnitude farther back and perhaps lasting many millions of years in cosmic time. We will return to this in later chapters. But how does relativity shed light on the above arguments and how in turn might those arguments speak to relativity? And the same questions would need to be directed to quantum theory. The latter would appear to be by far the most problematical; in this chapter we offer a beginning in pursuing these questions.

Einstein's field equations have the general form $\boldsymbol{G_{ij}} = J\boldsymbol{T_{ij}}$, which means that each component of the metric tensor $\boldsymbol{G}$ is equal to the corresponding component of the energy momentum tensor $\boldsymbol{T}$ multiplied by the absolute universal constant J, where $J=4\pi G/c^2$. However, if G and c^2 each vary as ρ^{-6}, we see that the adjustment of both 'constants', now seen as parameters, preserves the absolute constancy of J and the validity of the field equations from epoch to epoch. Consequently, the Einstein equations would remain as a valid model of a quasi-static Universe of quantum dimensions but expanding very slowly and at a marginally increasing pace, assuming some means of marginally opposing the force of gravity. Special relativity would be preserved, with the value of $c(\rho)$ taken as a constant, in any particular epoch, corresponding to the average energy-density in that epoch.

A Quiescent Quantum Beginning

The foregoing implies that the radius R, whatever its early value, could only evolve, however slowly, to a somewhat larger value. But while this may accord with relativity, it leaves open, questions relating to quantum fluctuations and the uncertainty principle. A valid model for the origin and evolution of the Universe must articulate some means to avoid a singularity while ensuring the orderly expansion of the Universe. Here is where we seek to make a start toward conjoining nonequilibrium thermodynamics, cosmology, relativity and quantum theory.

The indeterminacy principle in quantum theory denies the possibility that a finite quantity of energy can be squeezed to a singular point in space and time, which is the same as saying that fluctuations must inevitably result in some physical expansion over time. This principle can be reformulated in terms of the indeterminacies in energy-density and time. Accordingly, given the requirement on the constancy of energy in the Universe, the Heisenberg uncertainty relation may take the form $\Delta\rho\Delta t$ $\geq \hbar$, in which total energy of the Universe E is taken as an absolute constant, $\Delta\rho$ is the variation (reduction) in energy-density of the Universe over time Δt and $\hbar = h/2\pi$, represents some *minimal action* that ensures the ongoing expansion of the Universe, and h is Planck's Constant. This would be equivalent to arguing that the Universe could never be in a state of critical density Ω and further that Planck's Constant would drive ongoing expansion of the Universe. But here, there is need for caution, as the magnitudes of ρ and therefore of $\Delta\rho$ appear to be all out of proportion to the magnitude of Planck's Constant, as will be dealt with later. (We note in passing that the above inequality could be represented equally well in terms of $\Delta p \Delta R \geq \hbar$, where Δp is a change in boson momentum or pressure acting to extend the radius R of the Universe an amount ΔR.) If the Universe commenced in a state of extreme energy-density, $\rho \to \infty$ ($\iota \to 0$ and $\sqrt{G} \to 0$), a fluctuation could lead either to a singularity or expansion, as a static solution to the Einstein equations, in which J is taken as an absolute universal constant, is unstable. But the *2nd Law*, which demands an increase in entropy with the passage of time, disallows negative values of time and time progression. Thus the uncertainty principle would constitute an external constraint on the field equations, disallowing any fluctuation leading to a contraction of the Universe and a state of infinite energy-density (a singularity). *Indeed the above inequality might be considered in a larger perspective as a causal determinacy relation, guaranteeing the orderly expansion of the Universe.*

In the early Universe such fluctuations might be manifested as a decline in the local energy-density ρ_{local} of some region, followed by an extended period of equilibration and static stability of the Universe, during which time the Universe would be spacetime flat, in accord with Einstein's original equations (See Fig. 7-2). This process would then repeat, as discussed in greater detail in the next chapter, but our purpose here is to draw attention to the physical interdependence of $\Delta\rho$, Δt, Δc and $\Delta\sqrt{G}$. Since we must now look back in spacetime close to ten billion light years to detect any decrease in the velocity of light, the foregoing arguments would imply a "beginning" of both time and the Universe when $\Delta t \to \infty$s in the past, and a time in the future when the Universe would lack any physical meaning, as $\Delta t \to 0$s, at which time $\Delta\rho$, ΔR *and* Δc $\to \infty$, in agreement with Tolman's view. (We need to keep in mind here that all physical processes progressed extremely slowly in the earliest times, as $\iota \to 0$m/s, though very fast in the distant future, as $\iota \to \infty$m/s.)

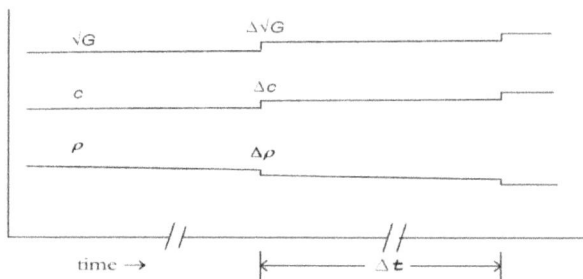

Fig. 7-2 Depicting the relation $\Delta\rho\Delta t \geq h_{bar}$ in which a fluctuation (decrease) in local energy-density ρ_{local} is followed by an extended period (Δt) of large scale equilibration of ρ, c and $\sqrt{G}$ and static stability of the Universe, during which time the Universe would be spacetime flat, in accord with Einstein's original equations. Δt defines an epoch of particle number expansion in the quantum Universe. This process repeats many times in the quantum era, as explained in chapter 8.

The inequality relation implies a discrete emission of radiant energy from one extreme energy-density state of the Universe to a lower state, just as the energy of an atom in an excited state is decreased by the emission of a boson as a discrete quantum of energy. Such outflow of energy may be viewed as a dissipation of energy, since it no longer exists as available free energy *in* the static energy-density-state of the Universe, or better it might be seen as energy associated with the ongoing work of building (expanding and structuring) the Universe. The result is a progressive decrease in the rate of entropy production per unit of radial energy flow (through unit area of surface expansion); i.e., the flow-specific rate of entropy production and the rate of entropy increase in the Universe decrease, in keeping with our postulates and the extended *2nd Law*. Given the nonlinear envelope of expansion and the uncertainties in our determination of astronomical distance (relative to the precision of measurements of c in our region), we have no way of accurately determining the present period in which the Universe may be assumed to be in a quasi-stationary state. The radiation era would have commenced hundreds of billions of years earlier than that estimated from linear extrapolation, and the period of nucleosynthesis would have occurred correspondingly farther in the past, perhaps extending over millions of years. In this scenario, the era of Grand Unified Theories would have occurred still further in the past and lasted perhaps for millions of years.

All of this shows how the time scale in the inflation and quintessence scenarios is turned on its head in the present view. (See the timeline chart Fig. 9-1 in chapter 9.)

The above inequality, together with the parametric adjustment of J for the values of $\sqrt{G}$ and c, thus function somewhat like a cosmological constant or quintessence ("dark energy"), causing the Universe to expand at a marginally accelerating rate from an extremely slow and small beginning, while making only a relatively simple formal adjustment in Einstein's equations. However, unlike the inflation or quintessence scenarios, there is no need for fine-tuning to square the present state of the Universe with previous epochs. There is no place or way of inserting a Planck-like term in the field equations; the Universe is ratcheted from one quasi-static state of average energy-density to another of marginally lower density, as relativity, quantum theory and thermodynamics demand. Relativity and quantum mechanics are different, the one structurally nonlinear and the other strictly linear (at least within any cosmic epoch), but they are partners in avoiding a singularity and ensuring an ever-expanding Universe. Planck's Constant may be seen both to govern and derive from a minimal negative spacetime curvature-imposed dispersion (dissipation) of energy in the early Universe. Such curvature-imposed dispersion would be today mirrored in a corresponding electromagnetic curvature-imposed minimal dispersion (dissipation) of energy in the ground state of the hydrogen atom that would be evidenced in the epoch-to-epoch expansion of the ground-state radius of the atom and the decrease in its energy-density, were we able to measure it.

In the foregoing, only the Coulomb charge $q\sqrt{\varepsilon_0}$, magnetic charge $I\sqrt{\mu}$, gravitational charge $M\sqrt{G}$, the ticking of the cosmic clock Δt and the numerical value of J are considered to be absolute, universal fundamental constants, while c, $\sqrt{\varepsilon_0}$ and $\sqrt{G}$ may be taken as constants in any given epoch, thereby satisfying the requirements of special and general relativity in any stable epoch. (The constancy of $q\sqrt{\varepsilon_0}$ and $M\sqrt{G}$ guarantees the constancy of the electromagnetic *force* between charged particles and the gravitational *force* between particles or bodies of fixed separation d, i.e. guarantees their independence of variations in ρ.) In the next chapter we will see that the fine-structure constant, which governs the relativistic spin and orbital dynamics of leptons, in the scalar-expansion era, and the

Rydberg constant must be added to the above list of absolute universal constants. (As we cautioned earlier, regarding the relative magnitude of $\hbar$, we will see in chapters 8 and 9 that $\hbar$ must vary directly with the average particle energy-density, in the quiescent quantum state of the Universe.) However, c, ε_0, $\sqrt{\mu}$ and $\sqrt{G}$ should vary marginally in any epoch wherever the local energy-density is significantly greater or less than the average density. We will return to this in chapter 9 and again in Part III.

Since the speed of light must decline, where the local energy-density exceeds the average energy-density, we see that "time dilation" is no longer a valid argument or have any meaning in relativity theory. Such would be the case, whether the cause of the lower maximum value of c is due to greater local energy-density, i.e., density of energetic particles, or to acceleration of a particle or body, as the Principle of Equivalence demands; indeed such acceleration results in an increase in the local energy-density of a particle or body. This leaves us with cosmic time closely related to the 1st Law, where the energy of the Universe is constant and the energy and frequencies of bosons, created epochs ago in the macro expanding Universe, remain unvarying throughout their journey to the present and serve as a universal clock; only the wavelength stretches from one epoch to another, as its local energy-density declines.

In such a Universe a nearly-ageless sentient being would sense only the dissipation and entropy production that characterizes the infinitesimal expansion of the Universe. The energy-momentum and angular momentum of massive particles are conserved, if unperturbed in any arbitrary macro-region of spacetime in an expanding Universe; only their energy-momentum density declines. The Lorentz transformation is only approximate in such regions and the light cone only marginally demarcates the boundary between spacelike and timelike regions in spacetime.

In summary:

Requiring c, $\sqrt{\varepsilon_0}$, $\sqrt{\mu}$ and $\sqrt{G}$ to vary with the average energy-density of the Universe led to a relativistic model of the Universe, smoothly consistent with the demands of quantum theory, the field equations and a self-consistent cosmology.

Adiabatic expansion of the quantum Universe is driven by a quantum indeterminacy relation (better a determinacy relation?), as an external adjunct to the field equations, in keeping with the requirements of the extended *2nd Law* of thermodynamics.

As such the Universe has been quasi-flat, homogeneous and isotropic from the beginning and has remained so throughout its physical expansion, in keeping also with the requirements of nonequilibrium thermodynamics.

CHAPTER 8

PHYSICAL NATURE AND EXPANSION
OF THE QUANTUM UNIVERSE

Thus far nothing has been said regarding the physical nature and process of expansion of the quantum Universe; this is our principal concern in Chapter 8. Einstein's equations and empirical evidence tell us that the trajectory of bosons may be bent by the force of gravity, even to the point of confinement, as in a black hole. The conventional view is that massless bosons as well as massive particles and bodies are subject to the gravitational force, mediated by the exchange of "gravitons," though the existence of such particles has yet to be confirmed experimentally. If relativity fully accounts for the force of gravity at quantum as well as cosmological dimensions, there would be no need for gravitons, within any static period. But in the larger spacetime picture, variations due to differences in local energy-density can only propagate at the speed of light, characteristic of a particular epoch and its average energy-density. (In chapter 11 we propose a very different dynamics for the forces of gravity and electromagnetism.) Here, we seek to: (1) specify the initial quantum state of the Universe, (2) characterize its evolution, (3) account for the creation and mass of all of the matter particles now present in the Universe and the ratio of free bosons to matter particles and (4) test these quantum propositions for their consistency vis-à-vis the proposed principles and Laws of nonequilibrium thermodynamics and the requirements of relativity and quantum theory.

Initial State of The Universe

In a simplistic model of the quantum Universe, the total energy E of the Universe might be partitioned among two triplets of three each photons of wavelength λ_i, traveling in circular orbits of radius $r_I = \lambda_i/2\pi$, at the initial speed of light $c_i \to 0\text{m/s}$, around Euclidean axes X, Y and Z. The photons of one triplet orbit in a left-handed direction (left helicity), while those in the other triplet orbit in the opposite direction with right helicity, such that the planes of opposing photon orbit lie in the negative and positive domains of each axis, separated by a small distance d, see Fig. 8-1. Gravity would act to bend the photon trajectories into very small circular orbits and to minimize d, while the angular momentum (charge) of the opposing orbiting photons would set a lower limit on d (if the Coulomb force were comparable in coupling strength to that of gravity) and fluctuations introduce no bias in one direction or the other. But this simplistic model would not leave the quantum Universe in exact angular momentum null, unless some additional requirement prevailed, as theory would require. Such would result if the Gravity and Symmetry-Bound Photons (GSBP) were taken as internal fundamental constituents of massive composite particles (Primordial Quarks), which instead would orbit the Euclidean axes, as sketched in Fig. 8-2. If the P-quarks orbiting the Z axis had +2/3 positive electric charge and those orbiting the X and Y axes had –1/3 negative charge, angular momentum of the system could be maintained null dynamically by fluctuating adjustments of the X, Y and Z ground-state radii a of the P-quarks. That is to say, it is the angular momentum of the P-quark in orbit about the Z axis that must be cancelled by the sum of the angular momentum of the P-quarks in orbit about the X and Y axes, suggesting an analogy with the electron's null angular momentum in the ground state of the hydrogen atom. Just as the ground state of the hydrogen atom requires corrections to achieve dynamic stability (eliminate wobble of the masses) and to make spin-orbit coupling inherently relativistic (chapter 14), the same must be achieved in the much more complex dynamics of the GSBP quark (antiquark) triplets. If that were possible, we could speak of the ground state of the initial matter and antimatter primordial quarks in which the total angular momentum of the Universe and that of each triplet would be null (see chapter 13). (We call attention, here and elsewhere in Parts II and III, to the equivalence of the angular momentum of quarks orbiting about their local Euclidean axes, *subject to SU(3) symmetry constraints, and their required discrete electric charges. Electric charge is a quantized measure of the orbiting angular momentum of massive particles and antiparticles and in the case of leptons SU(2) symmetry. The same applies to the spin angular momentum of the spinning photons in the leptons and their antiparticles in reference to their discrete charges.*) Thus characterized, the initial quantum state of the Universe may be seen as a primordial matter super-neutron gravito-symmetry bound to an antimatter super-neutron in a nearly-exact quantum Euclidean space, with zero total angular momentum, subject only to fluctuations.

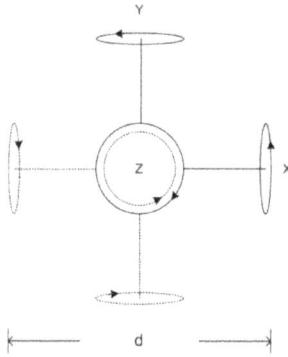

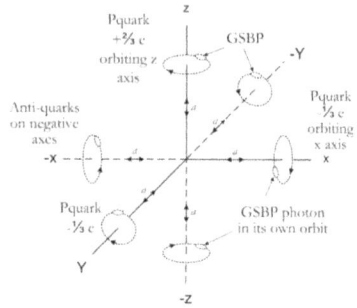

Fig. 8-1 Orbits of gravitationally-bound photons (solid lines) and antiphotons (dotted lines) about Euclidean axes in the early quantum Universe and the partial cancellation of total angular momentum. Separation of bound photons and bound antiphotons d, but this theory does not work as explained in the text.

Fig. 8-2 Orbits of GSB primordial photons about Euclidean axes creating zero angular momentum in the early quantum Universe. *a* is referenced in chapter 14.

Given the inequality relation in chapter 7, the energy-density of the Universe may be seen to be nearly constant throughout an extended energy down-conversion process, as the number of particles doubled and h halved with each doubling, keeping the total energy constant, as explained next. All of this could occur in a nearly-lossless adiabatic expansion of the *number* of particles which now make up the Universe.

Quantum-State Evolution via Energy Down-Conversion

Here we address three questions: How should the parameters c, $\sqrt{\varepsilon}$, $\sqrt{G}$, r_q and R vary with respect to the decline of ρ *on the quantum scale*, how should we define an epoch on that scale and in the cosmic Universe, and how could all of the massive particles in the Universe come into existence in number relation to the free bosons?

Classically, on macroscopic scales, we may argue that c, $\sqrt{\varepsilon}$, $\sqrt{G}$, r_q and R should coevolve *exactly as* ρ^{-3}, since the energy of a spinning GSB photon and that of the Universe must be conserved from epoch to epoch; but doubling the edge of a cube in quantum Euclidean space does not increase its volume quite as much or as rapidly as doubling the radius of a sphere that just encloses the cube in spherical space. Consequently, an incremental decrease in the energy-density $\Delta\rho$ of a quantum system generates a differential increase in the values of these parameters and a buildup of shear stress. That is, the values of r_q and R would have to slightly more than double in order that the densities of the P-quarks and that of the quantum Universe decline exactly by $10^{-2\times3}$, or six orders of magnitude. The stress increases, as fluctuations lead to the decline of energy-density until it reaches a critical value, giving rise to a phase transition which reduces the values of r_q and R and relieves the stress. However, as implied above, the transition might take the form of a down-conversion of the energy of the gyrating GSB photons in each triplet into two nearly-identical gyrating photons, whose total energy almost exactly equals that of the parent photon with λ doubling. But this would result in a rapid (geometric) increase in the number of GSB photons and P-quarks and P-antiquarks, and it would leave $c, r_q, R, \sqrt{\varepsilon}, \sqrt{G}$ and S increasing very fast in cosmic time, contrary to our assumption of adiabatic expansion of a quantum-state Universe. *This inconsistency derives from the assumption that the Planck value is an absolute universal constant.* Recall our caution, in chapter 7, about the magnitudes of E and ΔE relative to the value of $\hbar$. *If the value of $\hbar$ decreased from a very large value by one-half with each down-conversion, all of the above parameters would remain essentially constant, as would the wavelength of each newly created particle; only the energy-density of each newly-created particle would decline, leaving the Universe in a quantum state throughout the repeating down-conversion process (the quantum era).* But that could have happened only if we require $\hbar$ to double, looking back with each down-conversion event from that at the plasma time. Thus, in our scenario, we see that $\hbar$ should also be taken as a parameter in the quantum era, halving directly with the energy-density per unit particle ρ_p, accompanied by a marginal increase in dissipation Ψ and a decrease in dissipation per unit particle Ψ_p, as illustrated in Fig. 8-3. Fig. 8-5 depicts a typical down-conversion event, revealing the buildup of stress, culminating in a transition which relieved the stress (γ dissipation) and resulting in a decrease in entropy production per unit energy flow in the Universe. Clearly this would have major implications throughout quantum theory, which we leave for later discussion in Part III. Here we need to deal with three questions: (1) What is the physical significance of $\hbar$? (2) Is its variation consistent with the inequality postulated in chapter 7 and the avoidance of a singularity and of the "beginnings problem?" (3) Does $\hbar$ continue to decline from epoch to epoch as the Universe *physically* expands, following the plasma event?

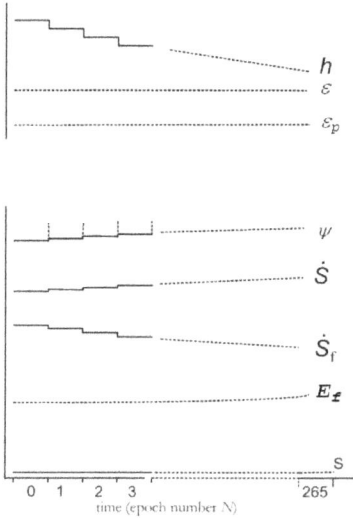

Fig. 8-3 Evolution of the principal quantum parameters with each down-conversion, showing the halving of Planck's Constant h and particle energy E_h, *the* slow increase in dissipation ψ, entropy production $\dot{S}$, entropy production per unit energy flow $\dot{S}_f$, the rate of energy flow E_f and the constant near-zero entropy S of the quantum Universe, characterizing the down-conversion process.

In regard to the first question, each newly created quark pair (antiquark pair) must fit in the same space of its progenitor. But that is precisely what characterizes bosons (in this case two GSB photons spinning in the space previously occupied by one gyrating photon having twice the energy), sanctioned by the law of spin and statistics, which requires identical bosons to occupy a common, lowest energy quantum state. This law, *defining the down-conversion era*, would require some 10^{80} gravitationally bound photons, having nearly-identical energy-momentum, to occupy the same space of the initial spinning photon of extreme-energy, such that the values of R, r_q, λ, c, $\sqrt{\varepsilon}$, $\sqrt{G}$, S, T and ρ remain quasi-constant, as the particle energy-density and the values of $\hbar$ and m halve, throughout the down-conversion, quantum era (see Fig. 8-3). As such it sets an upper limit on R_i the initial radius of the quantum Universe:

$\sim 10^{-5}$m $\approx (10^{25}/(10^{88})^{-3}$, where 10^{88} derives from 294 doublings of the number of particles. (see chapter 9). If we assume the GSB photons in today's quarks have a thickness or axial length of h ($\sim 10^{-33}$m), then 2^{265} epochs in the past (a factor of 10^{80} in volume and $10^{80/3} \cong 10^{26}$ in radius) the axial dimension of the initial P-quarks would have been about 10^{-7}m, well within the initial radius of the Universe. Thus the above assumption argues that the doubling of two GSB bosons would occupy not the exact *same* space but twice that of their progenitor in the particle doubling process, i.e., a doubling of the axial dimension (2h) of the combined GSB photons (see Fig. 8-4). In consequence, the halving of h in the down-conversion process, as the particle number doubles, conserves the total energy E and the average energy-density ρ of the Universe in the quantum era.

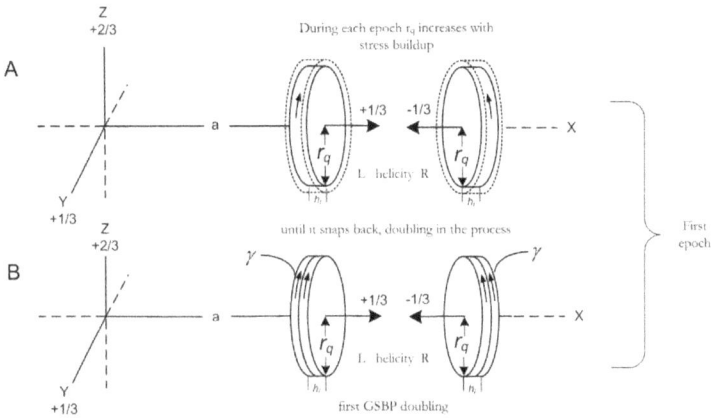

Fig. 8-4   Showing the $-1/3$ quark spin angular momentum cancellation of the $+1/3$ quark angular momentum (from the $+2/3$ Z charge) relative to the X Euclidean axis, in the first GSBP doubling epoch. Stress buildup (panel A) initiates the down-conversion doubling process, when the radius of GSB photon gyration reaches a critical level, resulting in h_i halving and the doubling of GSB photons (panel B). Similar process in the Y axis results in exact null angular momentum of the first epoch of the quantum Universe and in the many epochs to follow, as h_i and quanta fluctuations decline.

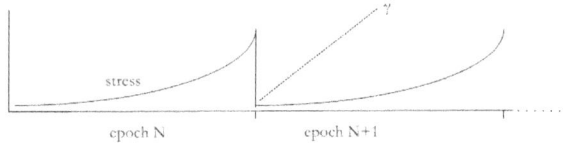

Fig. 8-5 Evolution of stress in each epoch culminating in the relief of
that stress with the emission of low energy photons γ of dissipation,
making the down-conversion process quasi-adiabatic.

It is important to note here that the 10^{80} massive particles in our
Universe came into being as stable open subsystems via the nearly-
dissipation-free down-conversion process as a first of many physical
creations of the Nonequilibrium Thermodynamic Imperative (NTI); as
such they locked up most of the energy of the Universe, minimizing its
rate of entropy increase.

In our scenario $\hbar$ is closely associated with the notions of particle
internal stress and quantum energy fluctuations as a necessary requirement
for the relief of that stress, as Heisenberg formulated the relationship. But
Planck's Constant emerged initially in an effort to resolve the black body
problem, lacking any apparent connection with matters of stress or
energy-density fluctuations; only in the larger picture can we now make
those associations. This brings us to the second question: What would be
the implications of $h_i >> h$, where h_i is our assumed "initial state" value of
h? The impossibility of such an earlier epoch, associated with the
requirement for a near-infinite fluctuation or an epoch of near-infinite
duration, would appear to be consistent with the notion of an "initial
state." Such requirement for extreme but gradually decreasing
fluctuations, as the number of particles doubled, would also be consistent
with our assumption of a very slow but marginally accelerating evolution
of the early Universe.

In answer to the third question, recall that the decline of h in the
down-conversion process was strictly associated with the increase in
particle *number* density, not average energy-density. *The post-plasma era is the
exact opposite, in which the number of massive particles remains constant as the particle*

energy-density declines, and so we should expect the value of h to be an absolute Universal constant in scalar expansion time.

How do the variations in ε and c with ρ determine v/c (the ratio α of electron velocity in the ground state of the hydrogen atom to that of light) with the decline of ρ? In our epoch $\alpha \equiv e^2/(4\pi\varepsilon_0 \hbar c) \cong 1/137$. Since e is an absolute constant, we see that ε decreases and c increases in proportion as ρ declines, making α a fundamental constant in the expanding Universe. Thus we conclude that the fine structure constant is, like $\hbar$, an absolute Universal constant in scalar expansion time. Similarly, we see that the Rydberg constant R = [$m/4\pi c$ h³]·[e²/4πε]² must be taken as an absolute Universal constant, since the first bracket would require R to vary as $\rho^{2/3}$, while the second would require R to vary as $\rho^{2/3}$, thus leaving R constant from epoch to epoch.

We are thus left with Planck's Constant and the fine structure and Rydberg constants as absolute Universal constants in the scalar-expansion era, while c, $\sqrt{\varepsilon}$, m and $\sqrt{G}$ must be seen, in the larger perspective, as parameters.

Our characterization of the evolution of the quantum-state of the Universe would result in a slight ongoing *decrease* in the period of static compliance with the field equations and a correspondingly marginal *increase* in the rate of energy flow accompanying the particle number expansion of the Universe. While the entropy production must necessarily increase, because the down-conversion process is not entirely free of dissipation, the flow-specific entropy production and the rate of entropy increase in the Universe would marginally *decrease*.

The import of all of this is that the number of GSBP triplets (of decreasing energy) would increase geometrically, with multiple epochs of marginally decreasing duration, all contributing to a progressive decrease in the rate of flow-specific entropy production and the rate of entropy increase in the Universe, in keeping with our postulates and the *2nd Law*. This process resulted in the quasi-adiabatic expansion of the Universe, culminating in an epoch-long annihilation of 10^{79} matter and antimatter particles and leaving about 10^{80} matter particles and 10^{90} free bosons in a plasma state, as shown in Fig. 9-1. For this to hold it is only necessary that the number of matter triplets exceed those of antimatter triplets by a factor of about 2×10^{80} at the termination of the down-conversion process.

The down-conversion process, which preserves symmetry, repeated over many epochs, as the number of particles increased geometrically until the process terminated, as the flow of energy became insufficient to drive a succeeding down-conversion transformation. This occurred first with the lower energy-density antimatter particles failing to down-convert one more time after epoch 265, leaving their parameters to physically expand while the matter particles' number doubled one more time to become free, enabling the quarks in each pair to annihilate one another. This left 10^{80} half-energy matter particles in a very hot plasma environment in an emerging macro-Universe, as depicted in Fig. 8-5.

An epoch in the quantum Universe would then be characterized by a marginally shorter period of static compliance (constant parameters) in the field equations. But in a deeper sense it may be thought of as a period Δt in which shear stress, associated with energy dispersion, builds to a critical level, culminating in a quantum transition, relief of the stress and the number doubling of massive particles, having increasingly long-term stability. These succeeding periods define a time when matter and antimatter particles proliferated, free of interaction, in the early quantum Universe. The emergence of each new particle-pair in this down-conversion process may be viewed as the creation of two new open thermodynamic subsystems, which serve to minimize the density-specific entropy production (maximize the energy-density) of the subsystems and minimize the flow-specific entropy production in the Universe, with each incremental increase of particle number expansion. The buildup of stress and its subsequent minimization, associated with some minimal entropy production, is inherent in all nonlinear processes and a driving force in the creation of ordered (organized) subsystems in the Universe. Indeed such stress might be seen as an early phase in the process of subsystem creation and evolution.

Note: there is nothing in this scenario that suggests a "quantum-foam" or "fireball" state of the early quantum Universe, as presumed in the big bang models. As yet temperature has little or no meaning and the Universe might be thought of as a nearly-quiescent "primeval atom," in the words of Lemaitre, who was the first to introduce quantum theory in modeling the evolution of the Universe: "Thermodynamical principles from the point of view of quantum theory may be stated as follows: (1) Energy of constant total amount is distributed in discrete quanta. (2) The number of distinct quanta is ever increasing. If we go back in course of time we must find fewer and fewer quanta, until we find all the energy in the universe packed in a few or even in a unique quantum"[33]

While classical thermodynamics was inadequate to the task, the proposed scenario, like that of Lemaitre, characterizes the early Universe as being in an initial state of minimum possible configurations with the entropy S→0 and the number of distinct quanta necessarily increasing, though not "ever increasing." Lemaitre's prescient insight was remarkable, nearly three-quarters of a Century ago. Each triplet of GSB photons, comprising the 10^{80} left-handed set might bear close resemblance to a contemporary neutron if the gyrating photons were taken to be quarks and the force binding the quarks together in triplets of SU(3) symmetry related simply to the gravi-electrosymmetry energy-density warping of quantum Euclidean space, conserving angular momentum (electric charge) and symmetry of the triplet (see chapter 13). The same applies to the triplets making up the large right-handed set, with reference to antiquarks and antineutrons, though we should expect, as per our earlier assumption, the total energy of the antineutrons to be somewhat less than that of the neutrons. The quarks and antiquarks in present neutrons and antineutrons are characterized by different "flavors," just as sketched above for the primeval quarks (P-quarks). This suggests that the present up, down and strange quarks, are very low *energy* states that condensed out of matter particles in higher *energy* states, under conditions in which ρ was much greater in the distant past. We leave these thoughts for further discussion in chapter 9.

To illustrate this epochal process, we note that it would require only 265 epochs of geometric increase in the number of antineutrons and one additional doubling of neutrons to result in $\sim 10^{80}$ surviving matter particles, following the annihilation epoch. Thus the ratio of matter (baryons) to that in the form of free bosons (about one part in 10^9), that is required by the Standard Model in order to account for the known density of hydrogen, helium and lithium in the Universe today, after further expansion and cooling, could be accounted for in a straightforward way, requiring no fine-tuning or resort to the Anthropic Cosmological Principle.[34] [35] [36] This follows from the nature of geometrical progression; a variation of one or a few doublings, of a total of 265 would still leave the ratio in the right order of magnitude. As discussed in the next chapter, further expansion and the *2nd Law* served to convert many of the neutrons into protons (plus electrons and antineutrinos), resulting in an estimated 10^{80} baryons in the Universe we now observe in which there are about 10^9 bosons for every baryon.

The above processes transformed a very cold quantum Universe into an extremely hot physically expanding Universe of quantum dimensions, in the very distant past. However this should not be confused with a "big bang" event as commonly thought, since the increase in energy dissipation and entropy production, associated with the creation of free particles, was offset by the increase in absolute temperature and the increase in energy flow per unit R, in keeping with the principle of flow-specific entropy production. Whereas the big bang inflation scenario assumes that the Universe commenced in a hot state of pure energy and that matter particles emerged only with subsequent cooling and phase transitions, in the present offering nearly all of the initial energy was bound in matter and antimatter quanta in quiescent states. Only *during* the annihilation, lasting one or more epochs and consisting of many little-bangs, did free bosons and a hot Universe emerge along with a residue of matter particles (Fig. 8-6). These particles did not come into being initially via a condensation process; there was no requirement for super-massive X or Higgs particles or fields to generate the matter particles in existence today. Nor was there any need to create massive magnetic monopoles, as required in the standard model. (These massive particles, like "virtual" photons, electrons, etc., were the necessary inventions of relativistic field theory, on which the Standard Model was founded.) Instead, the plasma epoch initiated an unending period of expansion and cooling in which matter particles were dispersed, yet constrained by the warped spacetime of relativity. It was therefore also a very early period of initial seeding of the large-scale structure of the Universe, in a competition between the thermodynamic forces, tending to make the Universe isotropic and homogeneous, and the gravitational force, tending to create local agglomerates of matter particles. It needs reminding that the time scale is radically different from the conventional view; the time for both forces to work was incomparably greater than that postulated for the Big Bang model. It is necessary also to remember that the 'constants' c and $\sqrt{G}$ continued evolving (increasing) incrementally, along with the expansion and decline of the average energy-density of the Universe. The increasing value of $\sqrt{G}$ ($\propto \rho^3$) served to preserve a constant gravitational coupling strength $m\sqrt{G}$ as ρ declined, while the increasing value of c was both a measure of the rate of expansion, with the tick of the cosmic clock, and a relativity requirement that the Universe remain nearly flat in spacetime.

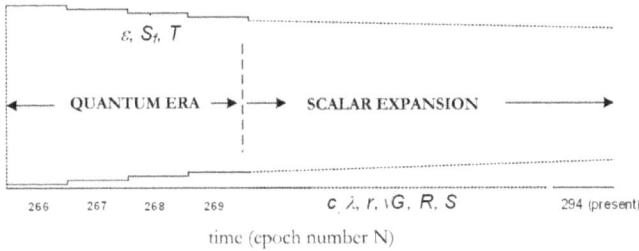

Fig. 8-6 Transition from quantum-number-expansion era to a scalar expanding Universe in which the quantum state in epoch 265, with the annihilation of 10^{80} matter and antimatter particles and leaving 10^{81} matter particles in a plasma state with free high-energy photons in a physically expanding Universe. This transition was marked by an extreme temperature increase followed by a marginal decline in ρ, T and entropy of the Universe S_f (per unit energy flow.)

While the period Δt of static compliance with the field equations has been declining, we find nevertheless that the velocity of light is increasing at a rate far too small to measure over any distance or time less than about 10^{10} light-years, and the accuracy of any such measurement is made uncertain by the fact that we cannot know what part of the measured redshift is due to the lower value of c, when the boson was emitted in the past, vis-à-vis the contribution owing to the expansion of the Universe. For the same reason we cannot know the precise age of the Universe.

Chapter 9 will consider the implications of the evolution of c and $\sqrt{G}$ for the Standard Model and subsequent eras of expansion and cooling of the hot Universe leading to the present time.

In summary:

The initial state of our cosmological model and the requirements of relativity (the field equations), quantum theory, nonequilibrium thermodynamics and symmetry (null angular momentum) are seen to be mutually self-consistent.

The evolution of quanta in the adiabatic expansion of the Universe is consistent with the above requirements. An epoch in this expansion is defined as a period of static compliance with the field equations, accompanied by the buildup of shear stress. Lacking any more precise measure of cosmic time, we take the look-back time to distant supernovae, at which we now detect a small decrease in the velocity of light, as the measure of an epoch in the present macro-state of the Universe. On this basis our timeline would place the annihilation period about 28 epochs, or about 10^{38} years in the past, as compared with $\sim 13 \times 10^9$ years in contemporary (constant c and G) cosmology.

All the matter quanta in the early Universe evolved down in energy-density and up in quanta number in a sequence of energy down-conversion epochs, the termination of which led to the complete annihilation of all antimatter quanta and half of all matter quanta, such that the number of such residual matter quanta (after further expansion and cooling), relative to the number of bosons present, was automatically "fine tuned" in that epochal sequence, consistent with the requirements of the Standard Model for the creation of the light elements, though on a very different time scale.

Self-consistency, in the proposed scenario, requires $\hbar$ to vary directly with the particle energy-density *in the quantum era* from an initial value $\propto 10^{80}$ times Planck's Constant ($\sim 10^{80/3} \times 10^{-33} \cong 10^{-7}$Js) to that in the plasma time and thereafter. This requires the dielectric parameter to vary as ρ^3, in the scalar expansion of the Universe, in order that the *force* between two charged particles remains constant. It also preserves the constancy of the fine-structure constant, as the ground state radius of the hydrogen atom must vary as ρ^3. Likewise the Rydberg constant is seen as an absolute constant in scalar expansion time.

Matter (antimatter) particles are seen in the quantum beginning as composite particles having internal dynamic structure in the form of gravito-symmetry-bound photons.

The concept of adiabatic energy down-conversion is abstract as is the notion of a quantum Universe, but it is also rational and simplifying in the extent to which it *self-consistently* resolves so many fundamental issues in thermodynamics, cosmology, relativity and quantum theory.

CHAPTER 9

EXPANSION AND EVOLUTION OF THE QUANTUM TO COSMIC UNIVERSE EVOLUTIONARY TIMELINE

Chapters 7 and 8 have sketched a radically different scenario relating to the very early evolution of the Universe, which circumvents many of the problems associated with the Big Bang-Inflation Standard Model, as discussed in chapter 6. Here, we sketch the principle features of the evolution and adiabatic physical expansion of the Universe to the present, referenced to the quantum-state evolution of all the massive particles, in a timeline chart (Fig. 9-1). Time is depicted in the quantum era in terms of the number N of geometric doublings of particles of proportionately reduced energy, commencing with that of the initial state (first doubling, N=0) and ending with the annihilation/plasma event (N = 266) temporal epochs of about 10^{10} years each. (The latter marked a transition from quanta *number* expansion *in* the quantum Universe to scalar physical expansion of the Universe. Likewise the plasma event marked a transition from the extreme values of T, M, ρ, k and k', created in the plasma event, to a factor of 10 epochal decline in their values.) The Quantum Era parameters h_N, m_N, k_N *(particle Coulomb charge) and k'_N (particle magnetic charge) halve* with each integer increase in N, as the number N of particles of half-energy m/2, doubles. The initial energy-density of the Universe ρ_i (~10^{61}kg/m³), radius of the Universe R_i, radius of gyration of the quarks r_{qi}, velocity of light c_i, $k_i\sqrt{\varepsilon_i}$ (the Coulomb coupling strength), $k_i'\sqrt{\mu_i}$ (the magnetic coupling strength), $m_i\sqrt{G_i}$ (the gravitational coupling strength, the temperature T_i (→0K) and entropy S are all essentially constant in the quantum era.

The scalar era parameters T, m, ρ and k, k' decline by a factor of ten with each integer increase in N, as R, r_q, c, λ (wavelength of the bound photons) $\sqrt{G}$, $\nu\varepsilon$ and $\nu\mu$ *increase to their present values*. Note: the values of the above parameters may vary in any scalar epoch in which the *local* energy-density ρ_l varies significantly from ρ, while the fine structure constant α and the Rydberg R_H are taken as absolute constants (chapter 8).

All parameters in the post-plasma evolutionary events are estimated based on present values: $\rho \sim 10^{-27}\text{kg/m}^3$, $R \sim 10^{25}\text{m}$ (see section on Unobservable Growth of the Universe in Radius and Particle Number), r_q $\sim 10^{-18}\text{m}$, $T \sim 3\text{K}$, $c \sim 10^8\text{m/s}$, $\sqrt{G} \sim 10^{-5}\text{N·m/Kg}$, $\sqrt{\varepsilon_0} \sim 10^{-6}C^2/Nm^2$, $\sqrt{\mu_0}$ $\sim 10^{-3}Tm/A$ and $k_1 \sim 10^{-10}\text{Nm}^2/\text{C}^2$ and $k_2 \sim 10^{-6}\text{Tm/A}$. The timeline values of these estimates result from working back in time from the present (N = 294) to the initial state (N=0). Thus the particle energy-density ρ_p of the Universe in its initial state would be $\sim 10^{-27} \times 2^{266+28} \cong 10^{61}\text{kgm}^{-3}$ and about 10^{-19}kgm^{-3}, as the quantum era ended with a total of 10^{80} matter particles. Note also that the product εc in the denominator of the fine structure constant requires it to be an absolute universal constant.

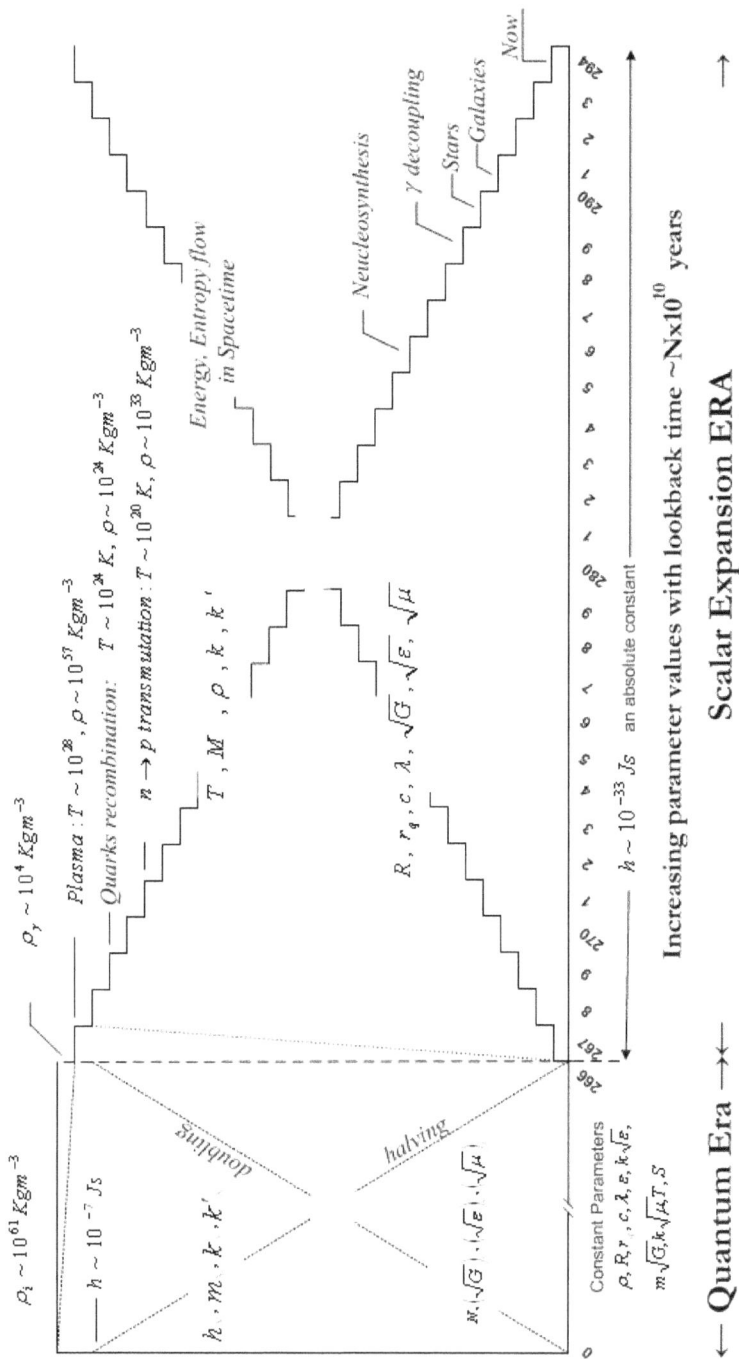

Scalar Expansion ERA

$\rho_\gamma \sim 10^4\ Kgm^{-3}$

Plasma : $T \sim 10^{28}, \rho \sim 10^{57}\ Kgm^{-3}$

Quarks recombination: $T \sim 10^{24}\ K, \rho \sim 10^{24}\ Kgm^{-3}$

$n \to p\ transmutation : T \sim 10^{20}\ K, \rho \sim 10^{33}\ Kgm^{-3}$

$T,\ M\ ,\ \rho\ ,\ k\ ,\ k\ '$

Energy, Entropy flow in Spacetime

Neucleosynthesis

γ decoupling

Stars

Galaxies

Now

$R,\ r_q,\ c,\ \lambda,\ \sqrt{G},\ \sqrt{\varepsilon},\ \sqrt{\mu}$

Increasing parameter values with lookback time $\sim N{\times}10^{10}$ years

$\rho_i \sim 10^{61}\ Kgm^{-3}$

$h \sim 10^{-7}\ Js$

doubling

halving

$h \sim 10^{-33}\ Js$ an absolute constant

Constant Parameters
$\rho, R, r, c, \lambda, \varepsilon, k\sqrt{\varepsilon},$
$m\sqrt{G}, k\sqrt{\mu}, T, S$

$h\ ,\ m\ ,\ k\ ,\ k\ '$

$N, \sqrt{G}, (\sqrt{\varepsilon}), (\sqrt{\mu})$

294   3   2   1   290   9   8   7   6   5   4   3   2   1   280   9   8   7   6   5   4   3   2   1   270   9   8   267   266

0

⟵ Quantum Era ⟶

Fig. 9-1

Fig. 9-1 Timeline: Evolution of the Universe, showing the values of k_N and m_N (particle mass-energy). k_N (particle Coulomb charge), k_N (particle magnetic charge) halve with each doubling of the number of quanta N in the Quantum Era, in which the initial values of ρ, R, r_q, c, λ, T, β and the electric, magnetic and gravitational forces are nearly constant throughout the down-conversion process (note: dissipation energy η). Termination of the down-conversion process led to an epoch-long (10^{1-} seconds) extreme energy-density Plasma State, marking a transition from quantum number expansion to physical expansion of the Universe in which R, r_q, c, λ, $\sqrt{g}$, $\sqrt{\mu}$ and $\sqrt{G}$ incur a 10^1 increase in value in each of 28 epochs to the present time while the 10^{-1} decline of T, ρ, M, k and k' resulted progressively in Quark Recombination, $n \rightarrow p$ transmutation, Neucleosynthesis of the light elements and the decoupling of radiation (γ) and electrons, as shown. From the Plasma State to the Present, the fine-structure constant is absolute and universal, but h reverts to Planck's absolute universal constant in the physical expansion era. This is seen in Table 9-1 as the product of the values of R, c and M in all epochs being equal to 1×10^{-33}Js, where M is the mass equivalent of the total energy of the Universe equal (10^{61}kgm$10^{-3}/10^{-5}$m). All events and parameter variations, in the quantum and physical expansion eras, are driven by quantum indeterminacy and the Nonequilibrium Thermodynamic Imperative (NTI) toward minimizing the rate of entropy increase in the Universe.

Table 9-1 Parameter values of the evolving Universe

	ρ Kgm^{-3}	R m	r_q m	c m s	$\sqrt{G}$ Nm Kg	M Kg	k_i Nm² C²	$\sqrt{g}$ C²Nm²	k_n Tm A	$\sqrt{\mu}$ Tm A	T K	N
Initial state	10^{61}	10^{-5}	10^{-48}	10^{-22}	10^{-35}	10^{-6}	10^{-40}	10^{-24}	10^{-36}	10^{-26}	10^{-2}	0
Plasma state	10^{57}	10^{-3}	10^{-46}	10^{-20}	10^{-33}	10^{-10}	10^{-38}	10^{-22}	10^{-34}	10^{-24}	10^{-23}	266
Quark recom.	10^{45}	10^{1}	10^{-42}	10^{-16}	10^{-29}	10^{-18}	10^{-34}	10^{-18}	10^{-30}	10^{-20}	10^{-24}	270
n→p trans.	10^{33}	10^{5}	10^{-38}	10^{-12}	10^{-25}	10^{-26}	10^{-30}	10^{-14}	10^{-26}	10^{-16}	10^{-20}	274
Neucleosyn.	10^{0}	10^{16}	10^{-27}	10^{-1}	10^{-14}	10^{-48}	10^{-19}	10^{3}	10^{-15}	10^{5}	10^{9}	285
γ decoupling	10^{-9}	10^{19}	10^{-24}	10^{-2}	10^{-11}	10^{-54}	10^{-16}	10^{0}	10^{-12}	10^{2}	10^{6}	288
Present state	10^{-27}	10^{25}	10^{-13}	10^{8}	10^{-5}	10^{-66}	10^{-10}	10^{-6}	10^{-6}	10^{-4}	10^{0}	294

In comparing the initial particle energy-density of the Universe $\sim 10^{61} \text{kgm}^{-3}$ with that in the plasma event $\sim 10^{57} \text{kgm}^{-3}$, the small difference $\sim 10^{-4} \text{kgm}^{-3}$ owes largely to the conversion of some particle energy in the annihilation event into radiation (dissipation). The energy-density per particle declines by ten in each of the 28 epochs of T decline in the scalar expansion era. The values of R_i, r_{qi}, c_i, $\sqrt{G_i}$ m_i ε_i and $\sqrt{\mu_i}$ would be given by the inverse of the factor $(10^{61}/10^{-27}=10^{88})^{-3} \cong 2\times 10^{29}$, i.e., $\sim 10^{-30}$ times their present values, such that $R_i \sim 10^{-5}\text{m}$, $r_{qi} \sim 10^{-48}\text{m}$ and $c_i \sim 10^{-22}\text{m/s}$, $\sqrt{G_i}$ $\cong \sim 10^{-35}\text{Nm/kg}$, $m_i \sim 10^{3}\text{Kg}$, $\sqrt{\varepsilon_i} \sim 10^{-36} C^2/Nm^2$, $\sqrt{\mu_i} \sim 10^{-33} Tm/A$ and k_i $\sim 10^{-42} C^2 Nm^2/C^2$ and $k'_1 \sim 10^{23}\text{Tm/A}$. $\hbar_i$ would vary inversely with N as $\hbar/2^{80/3} \cong 10^{-7}/10^{-26} \cong 10^{-33 \text{ Js}}$, when N=0 (looking back from the present (N=294).

However, each stress-induced down-conversion event is accompanied by the emission of bosons of energy dissipation, thereby marginally reducing the total energy locked up in the matter and antimatter particles. Alternatively we might say the bosons' function is to marginally decrease ρ_N in the quantum Universe, keeping the total energy constant. It is this small dissipation that is reflected in the estimated variance of parameters in the transition from the initial to the plasma quantum states, as shown in Fig. 9-1 and Table 9-1. (Such expansion and decrease in energy-density is driven by the fluctuations associated with $\Delta\varepsilon\Delta t \geq \hbar_i$ and the progressive decline of $\hbar$ from epoch to epoch.) This would necessarily result in some small value of T (near 0K) prior to the annihilation-plasma event. Nevertheless it should be noted that the radii and velocity-of-light parameters remain very nearly constant throughout the quantum era, making the sum of matter particle energy-density in the plasma epoch very nearly equal to that in the initial quantum state. The expansion of the Universe proceeds in the quantum era as an expansion in the number of quanta of progressively lower energy; physical (radial) expansion commenced only after the plasma epoch.

The above prompts several observations about the Universe as an evolving, quantum, nonequilibrium thermodynamic system: (1) The thermodynamic driving force in the quantum era is nearly constant (slowly declining from epoch to epoch) and is closely related to quantum fluctuations, the indeterminacy relations and the epochal decline of the

Planck parameter. (2) While the rate of entropy production in that era increases, from epoch to epoch, the flow-specific entropy production declines, as energy becomes locked up in subsystems (particles), always in a state very close to equilibrium. That is to say all nonlinearities are local in scale and time, as the Universe evolves adiabatically. (3) The annihilation (plasma event) shifted much of the energy of the Universe, formerly stored in the quiescent states of matter and antimatter particles, into free high-energy bosons in the plasma state (extreme temperature), making very little demands to extend the entropy of the Universe from a near-zero state, since temperature is in the denominator of the entropy equation.

As noted in chapter 8, the first major event initiated the down-conversion process, which, after ~265 doubling events, culminated in the annihilation of all the antimatter particles and leaving the quarks from the residue of neutrons in a free-state in a plasma of extreme energy bosons. The values of R, r_q, c, ρ and the entropy S in this quantum state changed relatively little from that prior to the annihilation event (the quantum doubling era), as the value of T rose from near-zero to ~10^{28}K over a period of an epoch. (This Grand Transition was certainly not a Big Bang event in that the interaction of matter and antimatter particles could proceed no faster than the speed of light (~10^{-22}m/s).

Subsequently the value of T declined by a factor of ten, in each of the succeeding 28 epochs, to its present value ~3K. The quantum doubling era was a time in which the gravitational, electric and magnetic forces were absolutely constant and the requirement on null angular momentum ruled over the dynamics of matter and antimatter particles. (The plasma episode, in current theory, is referred to as the GUTs era.) Two further epochs of expansion and cooling led to another transformation heralding recombination of the matter quarks into neutron states of extreme *energy-density*. This transformation left the emerging macro Universe with T ~10^{24}K, R ~10m, r_q ~10^{-42}m, ρ ~10^{45}kg/m³, c ~10^{-16}m/s, $\sqrt{G}$ ~ 10^{-29}N·kg/m, m ~10^{-3}Kg, $\sqrt{\varepsilon}$ ~10^{-30}C²/Nm², k ~10^{34}Nm²/C², $\sqrt{\mu}$ ~10^{-27}Tm/A, k' ~10^{17}Tm/A and the entropy of the Universe S still near zero.

According to current cosmology, this was a time when the Strong, Weak and Electromagnetic forces were equal in coupling strength. However, the force of gravity could not be adapted to the quantum world in the Standard Model. The problem, in present theory, is that G is an absolute constant, which requires the energy of interaction to vary in an unmanageable way, i.e., it may increase without limit, with the increase in energy-density. But if $\sqrt{G}$ varies as ρ^3 and m as ρ^3, then m$\sqrt{G}$ is an absolute constant, and we see that quantum gravity is not only a gauge field, generated by the Lorentz group, but at the same time a quantum force that may be expressed in terms of a phase shift, in a way similar to that associated with the electromagnetic force. Moreover, as we shall see in the section on Emergence of Free Charged Particles, there is no need to speak of a "Weak force" in our scenario. Rather the emergence of leptons (electrons and neutrinos) coincided with the transmutation of neutrons into protons, in accordance with the Nonequilibrium Thermodynamic Imperative (force of the extended *2nd Law*), under conditions of declining average energy-density in the Universe.

The recombination era was followed by a few epochs, during which ρ decreased to the point where the buildup of stress led to the transmutation of some neutrons into protons with the attendant creation of an electron and an electron antineutrino (positrino) to conserve mass-energy, momentum and total angular momentum, as shown in Fig. 9-2. (We use the terms neutrons, protons, electrons and neutrinos, under these conditions, to stand for higher *energy-density* states of these fermions. With the energy of the particle conserved, as the radius of the particle increases, the particle's energy-density must decrease, in accord with that of the Universe.) The parameter values defining this state are: $T \sim 10^{20}$K, $R \sim 10^5$m, $r_q \sim 10^{-38}$m, $\rho \sim 10^{33}$kg/m^3, $c \sim 10^{-12}$m/s, $\sqrt{G} \sim 10^{-25}$N·Kg/m, m $\sim 10^{-7}$Kg, $\sqrt{\varepsilon} \sim 10^{-26}$C^2/Nm^2, $\sqrt{\mu} \sim 10^{-23}Tm/A$ and $k \sim 10^{30}$Nm^2/C^2 and k' $\sim 10^{13}$Tm/A. (See the section below on the Emergence of Free Charged Particles.) We may refer to this time as an epoch of $n{\rightarrow}p$ transmutation in which the incremental increases in the parameters c, $\sqrt{G}$, $\sqrt{\varepsilon}$, r_q and R signaled the adiabatic physical expansion of the macro Universe. As shown in Fig. 13-2, the transmutation entails a reversal of the spin angular momentum of the GSB photon in the two d-quarks in the neutron as a final means of reducing the stress that had built up accompanying the declining energy-density in the Universe. We will see in chapters 11, 14 and 17 that such spin reversal plays a key role in minimizing the rate of entropy production in the expanding Universe, brought about by the NTI (chapter 4).

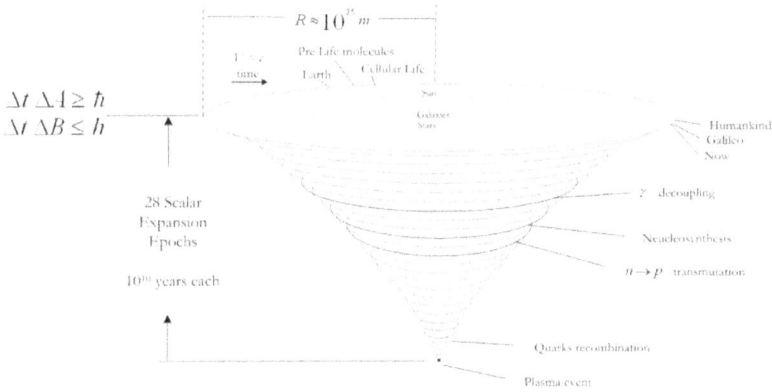

$$\Delta t\,\Delta A \geq \hbar$$
$$\Delta t\,\Delta B \leq \hbar$$

Fig. 9-2   Scalar expansion timeline showing the expansion of 28 circular epochs (appearing in 3-D perspective as ellipses), with each discrete (quantum) adjustment of the group A of increasing parameters satisfying the inequality $\Delta t\Delta A \geq$ h, where A includes t, c, R, r_q, G and the group B of decreasing parameters T ,m,ρ, k and k', which satisfy the inequality $\Delta t\Delta B \leq$ h, in relieving stress buildup in the expansion. These adjustments occur at the left profile. Thus we show, in the present scalar state of the Universe, light traveling uniformly at c m/s over a period ~10^{10} years on a path ~10^{25}m long. Our galaxy, the Milky Way, emerged 4-5 epochs earlier, with the Sun forming an epoch later and planet Earth condensing about four billion years ago in the present epoch. And as little as a million years later, homochiral amino acids were likely created in the hydrothermal vents in the early oceans, followed shortly by primitive cellular systems in the vent environments. Thus, in our scenario, Humankind appeared on planet Earth little more than half-way through the present epoch (2-3 million years ago) and we show the discoveries of Galileo, Maxwell, Einstein and others, in the last three-hundred years, leading to the present.

This transformation opened the Universe, as an adiabatically expanding thermodynamic system, to the eventual creation of atoms, as **enduring open thermodynamic subsystems** at decreasing energy-density levels, as a further means of minimizing entropy production per unit of energy flow and thereby further reducing the rate of entropy increase in the Universe. *The greatly enhanced stability (lifetime) of the proton, relative to that of the neutron derives from the quark structure in the proton, which is* *free of stress buildup with the expansion of the Universe.* This activity commenced, after a dozen epochs of expansion and cooling, with the nucleosynthesis of hydrogen, helium and lithium atoms (see the section below on Nucleosynthesis – Creation of the Light Elements). The parameter values at this time were: $\rho \sim 10^0 \text{kg}/\text{m}^3$, R $\sim 10^{16}\text{m}$, $r_q \sim 10^{-27}\text{m}$, $c \sim 10^{-1}\text{m}/\text{s}$, M$\sim 10^{-48}\text{kg}$, $\sqrt{G} \sim 10^{-14}\,\text{Nm}/\text{kg}$, $k_1 \sim 10^{-19}\text{Nm}^2/\text{C}^2$, $\sqrt{\varepsilon} \sim 10^3\text{C}^2/\text{Nm}^2$, $k_2 \sim 10^{-15}\text{Tm}/\text{A}$, $\sqrt{\mu} \sim 10^5$, and T$\sim 10^9$K.

The next event on our timeline – the decoupling of bosons from charged particles - occurred when the Universe had cooled to $\sim 10^6$K, leaving R $\sim 10^{19}\text{m}$, $r_q \sim 10^{-24}\text{m}$, $\rho \sim 10^{-9}\text{kg}/\text{m}^3$, $c \sim 10^2\text{m}/\text{s}$, $\sqrt{G} \sim 10^{-11}$ N·Kg/m and $\sqrt{\varepsilon} \sim 10^{-12}\text{C}^2/\text{Nm}^2$, M $\sim 10^{-21}\text{kg}$, $\sqrt{\mu} \sim 10^{-9} Tm/A$, $k \sim 10^{16}\text{Nm}^2/\text{C}^2$ and k' $\sim 10^{-1} Tm/A$.

The final six epochs of expansion led to the present parameter values and the evolution of all the large- and small-scale structure (**open thermodynamic subsystems**) we see in the cosmos. (Note that the gravitational coupling strength (gravitational charge) $m\sqrt{G}$ and the electromagnetic coupling strength $k\sqrt{\varepsilon}$ remain absolutely constant throughout both the quantum and scalar expansion eras.) Such structures resulted from fluctuations following the plasma event, and subsequently amplified by the gravitational collapse of regions of relatively high local energy-density, as evidenced in the CMBR. Walls, textures, galaxy clusters, galaxies, star clusters, stars and planetary systems all emerged to light the night in these very few epochs of expansion and cooling, as gravity exerted its occasional dominance in the evolution of the cosmos (see later sections on: Evolution of Large Scale Structure and Evolution of Galaxies and Stars).

Fig. 9-1 and Table 9-1 thus reveal the precision and extent of self-consistency in our model of the origin and evolution of the Universe. But there is more to say in this regard, with reference to Fig. 9-2, which provides a three dimensional representation of the expansion of the Universe from the plasma event to the present time. In this view the Universe is seen to have a minimal angular momentum, the product of the values of R, c and M (equals $1 \cdot 10^{-33}$Js) in each of the epochs in Table 9-1. Thus Planck's Constant is self-consistent in restricting the angular momentum of the Universe in the inequality relation $\Delta t \Delta (R \cdot c \cdot M) \leq \hbar$. In so doing it also conserves the energy of all particles throughout the expansion of the Universe and avoids the critical energy-density Ω that would reverse the flow of energy toward a singularity.

UNOBSERVABLE GROWTH OF THE UNIVERSE IN RADIUS AND PARTICLE NUMBER

In the quantum era, the epoch number N increases and the energy of each particle decreases in the down-conversion process so as to conserve the total energy of the Universe, estimated as being equivalent to $\sim 10^{80}$ atoms. But the epoch by epoch increase in N scarcely increased the radius R of the Universe. However, it was in the succeeding era of physical expansion that the Universe grew rapidly (relative to the down-conversion era) to its present size $R \sim 10^{25}$m. (We note in contrast, that Inflation theory required the Universe to have expanded from a region some twenty-three orders of magnitude smaller than an already small beginning, at which R was $\sim 10^{-2}$m. This value, derived from constant c extrapolation from the present estimated size.) While the numbers are far beyond our ordinary comprehension in either model, the rationale is straightforward in the one and grossly flawed in the other.

Emergence of Free Charged Particles

The end of the down-conversion era and the vanishing of antimatter particles in the plasma era, led to a third symmetry breaking event (actually a fourth if we reflect on the initial symmetry breaking associated with the one-way flow of energy with time), marked by a phase transition and the transmutation of a neutron into a positively charged proton of near-equal mass, accompanied by the creation of two leptons (a negatively charged electron and an antineutrino). This occurred when the local average energy-density in some region of the Universe declined sufficiently, if only ephemerally as a fluctuation, to favor the transmutation as a means of minimizing the flow-specific entropy production in the region (maximizing the energy-density of the baryon) and increasing the potential stability (lifetime) of the reaction products. The antineutrino served to conserve energy-momentum and angular momentum (charge) of the electron. Internally, the transmutation was effected via the conversion of each of the two $-1/3$ charge quarks (d-quarks) into a $+1/6$ charge adding to make a u-quark), thus leaving the baryon with a +1 charge and requiring the electron to have a −1 charge, in order to conserve angular momentum. This came about via a stress-induced reversal and increase in the angular momentum of the d-quarks, resulting in a small decrease in the mass of the baryon, consistent with a large increase in its stability (Fig. 9-3). *This was a major symmetry-breaking event precisely because of the smaller mass of the proton (slightly larger radius of its GSB photon orbit) and because it enabled the creation of much larger and more massive systems* (**atoms as higher energy-density, more stable, open thermodynamic subsystems**), *further minimizing the rate of entropy increase in the Universe.* Thus, the transition was mediated not by force particles, as required in the Standard Model, but by fluctuations resulting in a marginal decline in the local energy-density and temperature of the region and ultimately by the option of further reducing the flow-specific entropy production in the region, with little or no further stress buildup.

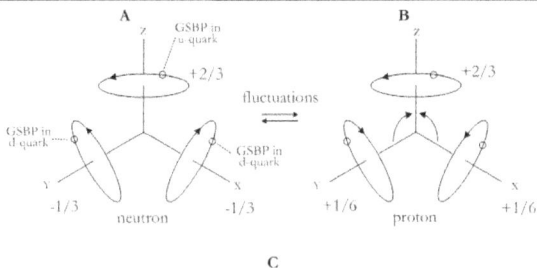

GSBP-Proton, -Positrino dynamics as start of γ-decoupling epoch when an electron was in the rest mass state and a positrino orbited the electron, conserving angular momentum and readily enabled a reversed p-n transmutation, as fluctuations might require.

Photon in GSBP-electron cycles α^{-1} times as the electron orbits the proton once in the H atom.

Positrino extended fine structure relativistic correction to the Bohr model of the H atom of order of a few parts per billion (perhaps 0.1 part per epochal change of e.

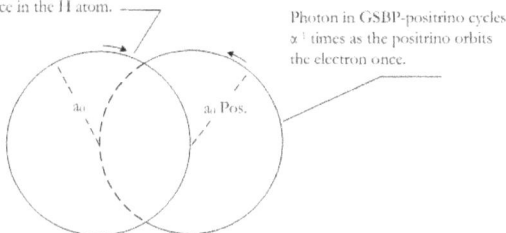

Photon in GSBP-positrino cycles α^{-1} times as the positrino orbits the electron once.

a_0 a_0 Pos.

Fig. 9-3 Passive (stress relief) transmutation of a free neutron into a proton. Panel A shows a GSBP-quark representation of a free neutron in local Euclidean Space in which the angular momentum of a u-quark on the Z axis (+ 2/3 charge) Is exactly cancelled by the angular momentum of the d-quarks on the X and Y axes (-1/3 charge each) in the ground state of the neutron. Fluctuations adjust the positions of the quarks on the Z, X and Y axes so that each GSB photon in each quark cycles α^{-1} times in exact phase closure with a single orbit of the quark about the center of gravity of the neutron. In so doing the GSBP-quarks mediate the dynamics of the nucleus similar to that of the GSBP-electron in its orbit around the nucleus. As the local energy-density ρ_{local} declines (panel B), the angular momentum (charge) of the d-quarks declines, eventually favoring their reversal, each to +1/6 charge, and the creation of proton with +3/3 charge. See Fig 13-1 for a more complete understanding of the transmutation.

But the increased stability of the baryons and leptons were at first necessarily short-lived, as fluctuations could quickly reverse the process. Thus the baryons (neutrons and protons) and leptons have internal composite structure derived from that of quarks after symmetry breaking, i.e., SU(2) symmetry of right- and left-handed gyrating photons, respectively. As we noted in chapter 8, electric charge is seen here as a plus or minus discrete and absolutely constant unit of angular momentum characterizing the various members of the family of fermions. (We leave the question whether, or to what extent, the neutrinos may have charge and mass for later discussion.) We note here a fundamental difference between P-neutrons in the last quiescent quantum state of the Universe and the neutrons referred to above: In the former the GSBP quarks orbit the exact Euclidean axes of the early Universe, while in the latter the GSBP quarks orbit the local Euclidean axes in each of the *free* neutrons, randomly oriented in space, as well as possible difference in energy-density. It is this difference that must be addressed in understanding the quantum gravielectro-symmetry dynamics of the quarks in a nucleon (the "color" force), the "electroweak" force (the residual force between quarks in nearby nuclei) referred to as the "strong" force and the electromagnetic force between protons in a nucleus, as discussed in chapters 11-13. As we shall see later, all such forces are closely associated with the vanishing or minimizing of local energy-density gradients via the emission of bosons acting by recoil impulse on the emitting particle. Thus bosons serve not as force particles *passing between and binding together* two matter particles, but as quanta of thrust, carrying momentum away from each particle so as to increase the energy-density gradients at the surfaces of those particles and acting to bind them together.

All of these transformations, whether involving down-conversions or conversion of neutrons into protons in the creation of new kinds of open thermodynamic subsystems, are characterized by three evolutionary processes: (1) internal stress, resulting from a mismatch in the 3-dimensional structure of the subsystem as it expands (its volume increases) and grows as an inherent prelude to any creative event; (2) stability of the subsystem (particle) is enhanced by any conservative action or transformation that tends to minimize the internal stress and (3) maximization of the energy-density of the subsystem (particle) is equivalent to minimization of its entropy production per unit density as the driving force (**NTI**) of the creation and evolution of the subsystem (particle).

Nucleosynthesis – Creation of the Light Elements

Further expansion and cooling, following after the emergence of free charged particles, led to a gradual increase in the number of protons (electrons and antineutrinos). The neutrons and protons were in thermal equilibrium via the neutrinos, which, at the existing high energy-density, interacted easily with the bosons. But as the energy-density declined, that interaction diminished, eventually enabling the forces of gravity and symmetry, together with the requirement on minimum flow-specific entropy production, to cause a proton and a neutron to combine, if only briefly, into a deuterium nucleus. Further cooling led to the combining of two protons and two neutrons into the helium nucleus ^{4}He, where the process essentially stopped in the creation of the most stable of all nuclei. This process binds all remaining free neutrons into nuclei, preventing them from decaying further and leaving the mass ratio of protons to helium about 0.26, in close agreement with the observed fraction of helium in the Universe. Thus, while the evolution of the Universe to this point is very different from current cosmology, both in process and in time scale, the result is in agreement with an important success of the Standard Model. However nucleosynthesis of the light elements may have lasted many years in the time scale of our scenario, though only ten minutes in the current universally-constant c and G cosmology.

As the temperature declined, the ratio of neutrons to protons declined and the number of electrons (virtually free of interaction with protons) increased. The free electrons interacted strongly with the high-energy bosons, acquiring kinetic energy and leaving them largely immune to the electromagnetic force that would impulse them toward the protons. With further expansion and decline in energy-density, electrons increasingly became bound to the protons and largely free of interaction with the bosons, giving rise to the era of radiation transparency and the Cosmic Microwave Background Radiation we observe today.

It is important to note here that all of these epochs of expansion are accompanied by a corresponding expansion of the radius r_q of gyrating photons in quarks and a comparable increase in the radii of gyrating photons in leptons, as well as the decrease of their intrinsic (rest) *energy-density*. (In this we see the close structural similarity of the quarks and leptons.) All the while, the values of c and $\sqrt{G}$ were increasing episodically, leaving the gravitational *force* and electromagnetic *force* independent of variations in ρ, as the physical size of particles continued to increase as ρ^{-3}.

All of the above is of course in the distant past and so it is instructive to consider how and at what cost we might recreate those conditions in the laboratory. When we try to drive the transition in the opposite direction, e.g., to transmute protons into neutrons, we accelerate particles to high relativistic velocity, such that their collision with fixed targets (or oppositely moving particles) results in an extreme local energy-density, thereby increasing the probability of coupling an electron and an electron antineutrino to a proton and creating a neutron for a brief time. But this of course does not reverse the process; the various high energy particles found in the detector never existed in a fluctuation-induced reversed natural event, as that would have violated energy-momentum conservation. That is, the above strategy greatly increases the energy as well as the energy-density of the reaction. There is in fact no way that we can increase *just* the local energy-density of a free neutron or proton in the laboratory to effect the reverse transformation. That follows from the mandate of the *2nd Law*. Such efforts must be at the expense of a very large rate of local energy flow and entropy production, in gross violation of our assumption that adiabatic processes prevail in all such transformations. We can view such efforts as ill-founded attempts to impose current particle QFT energy-based theory on a radically different strictly-particle theory in which energy-density is the fundamental parameter.

Emergence, Evolution of Cosmic Structure

With radiation transparency, following the decoupling of bosons and charged particles, gravity began to compete with quantum uncertainty, in league with the *2nd Law,* to govern the expansion or contraction of local regions of relatively low or high energy-density, as the Universe continued to expand marginally from epoch to epoch. The outcome in every instance was extremely sensitive to variation in the local energy-density, with the probability slightly favoring expansion overall. If that is so, we should expect to find the total area of "cold" regions in the Wilkinson Microwave Anisotropy Probe (WMAP) mapping of the CMBR or in the Sloan Digital Sky Survey to be slightly greater than that for the "hot" regions. Moreover, the difference in temperature between these regions should scale with Planck's Constant as in the quantum model proposed in chapter 7 and in chapter 8 in which Planck's Constant takes the form of a declining evolving parameter. With these recent technological advances, it is already clear that various regions of these maps define local flat spaces of different energy-density and temperature, seeded perhaps in different earlier epochs. The regions of higher temperature (energy-density) are regions that have already succumbed to the initial stages of gravitational collapse. In current cosmology such regions of extended flatness (over relatively large resolved angles) could be accounted for only in terms of some form of dark matter. We shall see in what follows that there is a far less bizarre answer to this problem, which has become one of the most challenging concerns in current cosmology.

The walls, texture and galaxy clusters, the hot regions of the Universe, are distinguished from the cold regions in that the former define the boundaries of growing open thermodynamic subsystems, created and maintained by the radial flow of energy in the expanding Universe. All such subsystems, undergoing marginal gravitational collapse, are expanding from epoch to epoch, but at a rate significantly lower than that of the Universe itself. The same applies to galaxies, star clusters and the stars, seen as open thermodynamic subsystems at lower hierarchical levels.

Dark Energy and Dark Matter: Phantoms of the Standard Model

It is noteworthy that cosmologists are now comfortable with the idea that the expansion of the Universe is accelerating, less than a decade after evidence to that effect was first revealed. This follows in part from the results of the Sloan Digital Sky Survey (SDSS)[37] and the Wilkinson Microwave Anisotropy Probe (WMAP) that have corroborated the initial supernovae studies. Following the initial shock, cosmologists and astrophysicists turned their attention to exploring various theoretical ways to energize the acceleration, given the demands of the Standard Model. More recently they seek to determine when, how and why the expansion of the Universe changed from positive to negative curvature, perhaps in only the past several billions of years. (In our scenario the curvature has been marginally negative from the beginning and would continue so indefinitely.)

With the temporal evolution of c, $\sqrt{G}$ and ρ and the concerted prodding of quantum determinacy, the *2nd Law* and requirements of the field equations, we see that the need for *dark energy* to account for the marginally accelerating expansion of the Universe is a nonproblem, i.e., a phantom concern, as argued in chapter 6. What is important here is that we should expect these same evolutionary processes and parameters to be sensitively associated with *local* variations of energy-density. Just as the Universe expanded and evolved incrementally, from one flat epoch to another, we should expect that local regions might expand or *contract* incrementally as spatially *flat* features from one epoch to another, as a result of fluctuations in energy-density in earlier epochs. To include *contraction* in such events is simply to acknowledge that local fluctuations may marginally favor gravity over the concerted forces of expansion in the Universe as a whole. Given the extremely small rate of overall expansion of the Universe, this may only mean some regions *expand* less rapidly than others; but it may also mean that other regions may physically contract, such that the total mass-energy of the region is conserved as its *local* energy-density increases. The latter class will likely, over time, suffer relatively strong gravitational collapse, depending ultimately on quantum forces and limitations. The implication in the forgoing wording is that we should expect cosmic structure to be stratified, i.e., organized in a hierarchy of relatively stable, dynamic systems, each such system evolving within the demands of a larger system, while at the same time enabling and constraining the evolution of smaller systems within its boundary. Just as a "flat" Universe, in any epoch, implies virtual freedom from entropy production (such being associated with stress buildup and relief accompanying the discrete expansion separating one epoch from another), so "flat" local regions of the Universe could be considered in a similar light. This applies whether a region is undergoing marginal expansion or gravitational collapse, as exemplified in the CMBR representation of the Universe, ten epochs past. Data obtained by the (WMAP) reveal regions that are relatively cold, cool, warm and hot, for which the cool and warm regions may be taken to demarcate between regions of relatively weak expansion (acceleration) and contraction (collapse) and the cold and hot regions as respectively further evolved states. (Similar results have been obtained by (SDSS).) But our principal concern here is to note that, within the angular resolution of the data, each such region appears flat, as if separated by a membrane from its surroundings.

In all of this, *h* likely plays a very important role comparable in the macro Universe to that in the quantum world, i.e., in characterizing the fine structure of the expanding Universe. This would be particularly important in local regions of accelerating gravitational collapse, as for example in the creation of neutron stars and black holes. Just as the electromagnetic fine structure constant $\alpha = e^2/\hbar c$ characterizes the relativistic velocity of the electron in the stable lowest energy (highest energy-density) state of the hydrogen atom, a *gravitational fine structure constant* $A \equiv (m\sqrt{G})^2/\hbar c$ (where $m\sqrt{G}$ is the gravitational charge or coupling constant) might characterize the relativistic velocity of stars and gas in stable spiral galaxies and other cosmic bodies.

To say that a region of an ideal gas is spatially flat is to say that the gravitational force between any two particles or bodies, separated a given distance, is the same everywhere, on average, analogous to a stationary thermodynamic state. It is also to say that such regions may behave as rigid bodies, with regard to any translational or rotational motion. If such particles or bodies were taken as an ideal gas in a stationary state, the forgoing would be readily accepted, given zero net mass-energy flow through a virtual "membrane" in any epochal timeframe. In the case of the WMAP-SDSS data, the membrane may be taken as the discrete energy-density boundary separating a region in one epoch from that of another in the evolution of the Universe.

But what about regions or objects, spiral galaxies for example, in which the *local* energy-density varies orders of magnitude from that in the core of the galaxy to that at its farthest reaches? How can gravity bind two particles (a distance d apart near the core of the galaxy) with the same force as similar particles (a distance $d(\Delta\varepsilon)^{-3}$ apart), as the long-term stability of galaxies demands? Is there a straightforward way to account for the stability (flatness) of such galaxies? No! Not in the framework of current cosmology, in which some form of dark matter, unspecified and undetected, has been actively pursued in recent years to resolve the puzzle at considerable intellectual and experimental costs. However if the gravitational coupling *force* $m\sqrt{G}$ is an absolute constant, as postulated in our scenario, then the *force* between the two particles in the galaxy would not vary, as ρ varies over time. That is, the *force* is constant, not only over the range of galaxy space in any given epoch, but also as the galaxy evolves (uniformly expands) from epoch to epoch in a stationary state. This follows, as stated before, since $\sqrt{G} \propto \rho_{local}^{-3}$ and $m \propto \rho_{local}^{3}$ and the same holds when ρ_{local} is replaced by ρ in the evolution of the Universe. But also $c \propto \rho_{local}^{-3}$ and that is also a necessary consideration in analyzing the structural form and dynamics of spiral galaxies, in that the relativistic velocity of matter particles and bodies must vary proportionally.

Fig. 9-4 is a reproduction of Begeman's rotation curve for spiral galaxy NGC 3198 taken from Ken Freeman's paper on the hunt for dark matter in galaxies. Shown are the observed rotation curve (points) and the gravitational contribution to the rotation curve for the gas and stars (calculated), at increasing radius from the galaxy core. Freeman[38] reasons that "a dark halo is needed to provide the extra acceleration, V_c, the rotational velocity of the galaxy." (Though not explicitly stated, that argument assumes that G and c are absolute universal constants.) He concludes: "Combining the latest weak lensing data for dark halos with results from the WMAP cosmic microwave background mission and the 2dF Galaxy Redshift Survey yields a census of the different forms of mass in the universe. Stars and cold gas provide just 0.4% of the mass of the universe. When gas in all forms is included, this percentage increases to 4% for all baryons. Dark halos alone contribute about 11%. All types of matter (dark matter plus baryons) add up to 27%. The remaining 73% comes in the form of a currently poorly understood 'dark energy.'" We have already shown here and in chapters 7 and 8 that there is no need for dark energy to square observations of the Universe with theory in our scenario. Here we now need to make a convincing argument that the discrepancy between observed and calculated rotation curves vanishes when c and $\sqrt{G}$ vary as ρ_{local}^{-3}. In our scenario the gravitational force on a star or molecule $F_G \propto (m\sqrt{G})^2$, where $m \propto R^{-1}$ and $\sqrt{G} \propto R$. Consequently F_G is constant for all $R \geq R_{Core}$, in keeping with Begeman's observations. Similarly the centrifugal force on a bound star or molecule $F_C \propto mv^2/R$, where v is the rotational velocity of the star or molecule, which in turn is $\propto c$. But c doubles as R doubles with the result that F_C is constant for all $R \geq R_{Core}$ such that $F_G = -F_C$ in the outer reaches of the galaxy, as long-term stability requires. Since c and $\sqrt{G}$ each vary as ρ^3, the above holds as the Universe expands, making spiral galaxies stable under expansion, translation and rotation, as rigid bodies when gravitationally unperturbed, as depicted in Fig. 9-5. Thus there is no need for dark matter to account for the stability of spiral galaxies and the flatness of the Begeman rotational curve is a reflection of the flatness of the inherent dynamic structure of spiral galaxies. The same reasoning applies to elliptic galaxies and indeed to galaxy clusters, **all of which emerge and evolve as subsystems progeny of the NTI.** As such they are quasi-rigid bodies, marginally stable from one epoch to the next. Close to home, there is new evidence of stars in Andromeda's halo as far as 500,000 light-years from its core, one-fifth of the distance to the Milky Way.

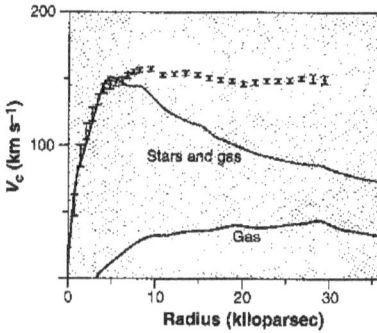

Fig. 9-4 is a reproduction of Begeman's rotation curve for spiral galaxy NGC 3198 taken from Ken Freeman's paper on The Hunt for Dark Matter in Galaxies

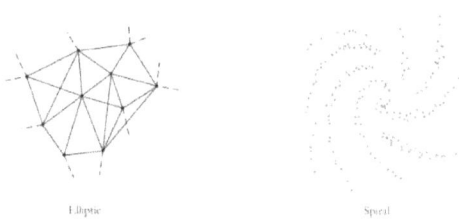

Fig. 9-5 Dynamic stability of galaxies. All distances – between bodies are constant, as rigid systems and increase/decrease uniformly with variation of ε, ε_{local}. The "rotation curve" for spiral galaxies – v_c= constant for all R≥core – and constancy of gravitational force (stability) derive from the variation of c and G with ε_{local} (see text).

All of this would be especially important in understanding the turbulent behavior of collisional systems, globular clusters of stars in galaxies and of galaxies in still-larger structures, as well as the evolution of disk systems at various hierarchical levels.[39] [40] [41] [42] [43] In sum, the phantoms of Dark Matter and Dark Energy, constructs of the Standard Model, appear to vanish in our scenario, as a direct consequence of energy-density-induced variations in the parameters c and $\sqrt{G}$.

The above need not deny and indeed should accommodate the existence of ordinary baryonic *dark* or *relatively-dark* matter in the form of cold or hot gases, whose local mass density (pressure) is very small in the outer reaches of the galaxy but whose total mass is nonetheless an order of magnitude greater than that associated with the observable matter in the galaxy. There is a sense then that we can speak of *dark matter* but only because gravity prevails over ranges far in excess of that possible in models where c and G are taken as absolute constants. It is this dispersed invisible baryonic matter that accounts for the weak gravitational lensing of light from distant sources to and around galaxies (galaxy clusters, etc) and to our eyes. The above also holds for elliptic galaxies as shown in fig. 9-5.

This brings us to NASA's 1E 0657-56 press release, 21 Aug 06, regarding "direct proof of Dark Matter," which has been met with great excitement and relief by cosmologists searching many years for *physical* evidence of the existence of Dark Matter. This study focuses on two clusters of galaxies on a collision course 3.4 billion light-years from Earth accelerated to relativistic speeds, which we show in grayscale representation (Fig. 9-6 panel C). In order to account for the behavior of the system, *in current theory*, the baryonic matter in one cluster is assumed to interact with that of the other causing each to slow down, while the *dark matter*, which normally shrouds each cluster (panel A), is *believed* to pass through the baryonic matter without interaction. *This lack of interaction is a principal postulate distinguishing the one form of matter from the other.* The result, revealed in NASA's photo album of galaxy cluster 1E 0657-56, is interpreted to show the baryonic matter in each cluster *"after the collision,"* in which *dark matter* of each cluster is *assumed* to have passed through the baryonic matter of the other, because it does not interact with the baryonic matter. However, in our analysis, the collision had only begun when the photo was made. As we show, the two clusters are on a collision course far apart, at time A, in which cold gas and dispersed (unobservable) baryonic matter symmetrically shrouds the dense (observable) baryonic matter, due to the *far-reaching* gravitational force when $\sqrt{G}$ and c each vary as $\rho_{local}{}^{-3}$. At time B each cluster is becoming gravitationally aware of the other and structurally adjusting to it.

Fig. 9-6 Grayscale depiction of the collision of galaxy clusters 1E 0657-56. In time A, when they were far apart and time B, when each were just gravitationally aware of the other. Panel C is a grayscale representation of NASA's photo of the collision. See text for comparison of the collision dynamics assumed in current constant c and G theory and in our scenario in which c and $\sqrt{G}$ vary as $\rho_{local}{}^{-3}$.

As the cold, dispersed and unobservable gas and baryonic matter in the separate shrouds approach each other, they form an increasingly high-energy (energy-density) core, further accelerating the baryonic matter in each cluster toward that in the other. As this happens, the baryonic matter in each cluster nearest the emerging core, accelerates faster than that farther away, as does the dispersed gas; the system is unstable under such conditions, with portions of both the baryonic matter and gas being condensed in the leading areas and dispersed or dragged in the trailing areas. The gas is excited to emit X-rays by high-energy bosons emitted by the baryonic matter in each cluster, directed toward its gas companion, thereby demonstrating recoil impulse mediation of gravitational attraction, as discussed in chapters 10 and 11. The emerging high energy-density core will then lead to lower values of c and $\sqrt{G}$ and eventually to the creation of a supercluster in which invisible matter will again symmetrically shroud the baryonic matter as a whole.

119

In short the fundamental issue in all of this is not about the dynamic details of the collision but rather how to account for the far-reaching gravitational and kinetic forces in each of the well-separated clusters, consistent with large variations in the local energy-density in each cluster and what causes the dark shroud to become X-ray visible, just prior to collision. The extraordinary observations of NASA's team, thus beautifully and simply support the above arguments, relating to the role of dispersed invisible or weakly visible baryonic matter in the dynamics of cosmic systems. In so doing they also support our scenario which is based fundamentally on the parameters c and $\sqrt{G}$ as inverse functions of ρ, in the large, and ρ_{local}. In our interpretation they serve also to support our recoil-impulse mediation of the force of gravity and to lift us from the burden and distraction of searching for yet another mysterious fundamental particle or force in order to support an inadequate theory.

Evolution of Large-Scale Structure in the Universe

In speaking of large-scale structure, we need to remind ourselves that in our scenario we can only look back, at best, over a single epoch (a miniscule portion) of the expanding Universe. Thus when we refer to "our sky" or a "whole sky" survey, we are in fact surveying only a very small region of the expanding Universe. An earlier epoch awaits our survey, recorded in a sense in the bosons yet to be detected. Other terminology may require rethinking in our scenario, as for example the notion of a "light-year". The term would have little utility beyond that at which we are just able to detect and verify a departure from the accepted value of the constant c, at which point we start looking into a previous epoch of the Universe. We will have more to say on this later.

The large-scale structure that we map today is the result of four factors at work in the evolutionary process: (1) the early seeding of flat, relatively high energy-density regions, deriving from quantum or classical fluctuations at earlier times; (2) the extremely long time span allowing such regions of enhanced energy-density to be gravitationally amplified; (3) the evolution of c and $\sqrt{G}$ preserving the flatness of those regions as they expand from epoch to epoch; and (4) the Nonequilibrium Thermodynamic Imperative that demands and supports the creation of myriad forms and numbers of relatively stable structures and dynamic systems, all countering the classical thermal imperative to move toward a state of equilibrium.

At various levels and regions of the expanding and evolving Universe, the above factors are at work at the same time, giving rise to the creation and evolution of "large-scale" structure and dynamic systems at all lower levels and locations. We now see the incipient large-scale structure at the cosmic level in the CMBR and the later consequence of that in the images presented by X-ray, optical, radio and infrared telescopes. And the same instruments are revealing the structure and dynamics of the walls and texture, the galaxy clusters, the galaxies themselves, the star clusters in galaxies, the activity near black holes, believed to be at the center of many if not all spiral galaxies, of the Sun itself and of planet Earth. We should expect some major surprises in these pursuits, deriving from the weakly nonlinear evolution of the fundamental parameters characterizing this late period of expansion. *Such surprises may be imminent to the extent that our cosmology directs what we look for and what we look with.* The capability to significantly extend the precision in the determination of energy-density-related variations in the values of c and $\sqrt{G}$ should have a high priority in this regard. In the above we have journeyed through the 2-D timeline of Fig. 9-1, but there is more to be discerned in the complementary 3-D timeline of Fig. 9-2, which focuses our attention on the role of Planck's Constant in the precise scalar expansion of the Universe as did Planck's parameter in the precise quanta *number* expansion in the earlier quantum era. The latter required the Planck *parameter* to halve 265 times to precisely account for 10^{80} quanta now, while the former required the radius of the Universe in the plasma state $\sim 10^{-5}$m times Planck's Constant to double 28 times to account precisely for its current radius of 10^{25}m, a factor of exactly ten for each doubling. Were Planck's Constant different in the 8th decimal place, the above numbers would be grossly self-inconsistent.

Thus Planck's Constant is seen to play a role far more important than simply accounting precisely for the spectral radiation of black bodies, with the one being fundamentally related to the other as is empirically demonstrated in CMBR radiation. Ultimately it is in the above quanta-cosmic relations that we see underlying the origin of Planck's Constant, not simply its statistical derivation. And it is in the same relations on which the most fundamental precise equation of modern physics rests:

$$E = h\nu$$

Systems Evolution of Galaxies, Stars, the Heavy Elements and Planetary Systems

Taking account of variations in the local energy-density and values of c and $\sqrt{G}$ should yield significant further insight into the formation, dynamics and stability of the smaller cosmic bodies and systems, about which there are still puzzling problems. We can do no more here than suggest some areas of inquiry that may benefit from such new insight.

Some have suggested, in current theory, that the first stars, were formed of hydrogen and helium in the first 100-200 million years in an epic event that fundamentally changed the Universe and its evolution. Such stars were likely very massive and luminous in ultraviolet radiation, capable of ionizing the neutral hydrogen and helium gas in their surroundings. But in our scenario, these events may be viewed in a very different light and time scale; black holes could have emerged only an epoch after Recombination of quarks into neutrons followed by proto-galaxies, at the behest of the NTI and driven by fluctuations in accord with Heisenberg's indeterminacy relations. We should expect some major surprises in reexamining these events, both as to their origin and evolution and in their contribution to the evolution of the cosmos.

In our scenario a black hole might have nucleated and evolved as a local fluctuation-induced increase in neutron energy-density which, in a later time, grew via gravitational collapse as it fed on gas, dust and the light elements in its environment. In terms of nonequilibrium thermodynamics, the black hole may be seen as a strongly-nonlinear open subsystem evolving (expanding its energy-density) energized by the greater dissipative flow of energy from the emerging larger galactic system. In so doing it might function as a feedback control to ensure the gravitational flatness of that galaxy structure from one epoch to the next, consistent with general relativity. The hole's "Horizon" is a cosmic example of the "membrane" we referred to in Chapter 4, whose radius varies or otherwise controls how particles compete in the evolution of the hole and its larger galactic (or other large scale) system. Thus, black holes may be viewed as local regions of excessive neutron density, leading to the creation of large scale structures: texture, clusters of galaxies, all the galaxies and star clusters we now see, driven ultimately by the NTI. In effect many of the 10^{80} neutrons emerging from the Recombination time may have played key roles in the organizational evolution of the cosmos, via cosmological and quantum gravitational-collapse, feeding off the nearby gas and stars. If so, it all happened adiabatically in early scalar time when the values of c and $M\sqrt{G}$ were very small and in which M was correspondingly large.

The evolution and dynamics of disks in spiral galaxies and in planetary and sub-planetary systems all emerge in similar ways, though the rate and nature of each must be strongly dependent on the local values of c, $\sqrt{G}$ and ρ_{local}. Much remains to be learned about these key processes in the evolution of cosmic structures, in particular the role of variations of c and $\sqrt{G}$ in the creation and maintenance of magnetic fields in accretion disks and their dynamic implications.

The genesis of the heavy elements is another concern in which we should welcome new insight, as the local energy-density and variations of c and $\sqrt{G}$ in stars must play a pronounced role in their origin, dynamics and destruction and in the evolution and seeding of young stars in later times. Just as the hydrogen atom came into being as an open thermodynamic subsystem which served to minimize entropy production in the expansion of the Universe, so the creation of the heavier elements in massive stars served to greatly diminish the rate at which entropy increased in the Universe, by locking up matter in stable, high energy-density building blocks. The latter later served to further decrease the rate of entropy increase in the evolution of molecular systems.

Spectrographic studies have revealed rapid orbital motions of stars in the central bulge of galaxies and providing firm estimates of the mass of black holes in their cores. But cosmologists reason that those stars are too distant to feel the gravitational pull of the black hole. This is strongly reminiscent of the need for dark mass in the outer regions of galaxies, which, as we have seen, becomes a nonproblem when the constants c and G are replaced by parameters whose values vary with the local energy-density. There is strong correlation between the orbital speeds of these far-flung stars in the galactic bulges and the masses of the black holes at their centers. The more massive the hole, the faster the stars throughout the bulge moved. Moreover, the study raises another puzzle: Each black hole is about $1/500^{th}$ as massive as the bulge of stars it inhabits. No one knows why that figure is so consistent from one galaxy to the next, though it is viewed as another indication that the growths of galaxies and black holes are intimately linked. If as we have suggested the black hole functions as a feedback control to ensure the gravitational flatness of the galaxy structure, the above puzzle has a rational solution.

Summary Thoughts on Cosmology and Relativity

Chapter 7 has sought to directly link the demands of nonequilibrium thermodynamics and quantum and relativity theory to the origin and early evolution (adiabatic expansion) of the Universe in a strictly self-consistent scenario, based on fundamental assumptions relating the permittivity, permeability and velocity of light to the average energy-density of the Universe. The initial radius R_i of the Universe is calculated to be $\sim 10^{-5}$m, when h, was $\sim 10^{-7}$Js in the initial state. That large value was shown to be self-consistent with the requirements of the down-conversion process, energy conservation and the relative forces in the uncertainty relations in the quiescent quantum era. Thereafter h is an absolute universal constant, having the currently accepted value.

The fundamental particles in the Universe were seen, in chapter 8, as primordial matter and antimatter neutrons composed of primordial quarks and primordial antiquarks, in which nearly all of the mass-energy of the Universe is invested. The quarks and antiquarks were simply Gravitational- and Symmetry-Bound Photons (GSBP), traveling in circular radii $r_i \propto R_i$. The number of such quarks and neutrons was postulated to double from one epoch to the next as the mass-energy of each and the value of h halved thereby conserving total mass-energy in a process of energy down-conversion. This epochal sequence terminated after 265 epochs of antineutron doubling and 266 epochs of neutron doubling, resulting in the *annihilation of all the antineutrons and leaving the quarks in the residue of neutrons free in an extreme plasma state*, such that further cooling would result in $\sim 2 \times 10^{10}$ free high energy bosons for every proton.

All of the events corresponding to evolving parameter values, as presented in the above evolutionary timeline (Fig. 9-1) and Table 9-1 thus follow self-consistently from the initial assumptions and arguments in chapters 7 and 8, that is to say, making no requirement on fine-tuning of the numbers. Such self-consistency, in cosmological-relativity perspective, has larger implications: the more *ad hoc* adjustment of parameters required to square with a theory, the more one is prompted to question the theory and perhaps eventually to seek other than natural physical explanations. And the more complex the system (the greater the number of parameters to adequately describe its behavior), the more some may abandon physics and science altogether and rely on metaphysical arguments[44]. Such arguments prompted Nobel laureate Burton Richter to write a book review blasting the anthropic approach as sterile and unscientific. Its proponents "have given up."[45] This is a matter of urgent and growing concern, whose scientific roots lie in the wave-particle duality conundrum, the unquestioned acceptance of the absolute constancy of the speed of light c, and Newton's gravitational constant G and numerous bizarre demands in quantum mechanics and relativistic quantum field theory. More recently cosmologists and astrophysicists have created a cosmology on this shaky foundation that flagrantly denies or circumvents the laws of thermodynamics in the creation of inflation theory. The result has contributed to the estrangement of common-sense reason within the world of physics and indeed of science and from that in other spheres of inquiry and learning. If so it threatens a new era and new kind of dark ages. We will return to this later.

PART III

GRAVITY <u>IN</u>
RELATIVISTIC QUANTUM MECHANICS

What has been said thus far directly challenges the foundation of contemporary quantum theory and requires us to reexamine the fundamental assumptions and representations of nature at the quantum level. We begin in chapter 10 by turning, once again, to long-debated questions relating to the preeminence of particles, waves and fields, in light of the proposed cosmic evolution of $c, \sqrt{G}$ and the principle of minimum flow-specific entropy production, and then to determine whether and in what way a reassessment of the Copenhagen interpretation of quantum mechanics (QM) may now be called for. Our aim in all of this is to focus our attention on the critical assumptions that led us to the current theory and to articulate any seeds of doubt as to the interpretation or adequacy of the relevant theoretical frameworks. Where that doubt emerges, we need to suggest some alternative explanation or theory and explore its potential and problems. If physical particles are to be taken as the most fundamental ingredients of the Universe, it will be necessary, in particular, to account for diffraction and interference phenomena without recourse to waves or wave theory.

Chapter 11 defines an internal structure for all massive "fundamental particles" (quarks and leptons) based on the concept of Gravito-Symmetry-Bound-Photons (GSBP), as sketched in chapter 8 for proto-quarks (p-quarks) and p-antiquarks. "Massless" integer-spin bosons also are shown to have fundamental internal GSBP structure, requiring principally the redirection of the axis of photon spin orientation and resulting in quantized boson velocities very near the present value of c. The gravitational force acting on quarks and leptons, as on neutral massive particles or bodies, is defined in terms of the differential energy-density gradients ρV at their surfaces and is mediated by "massless" spin-1 vector bosons that act through recoil-impulse to maximize the differential of the energy-density gradients $\rho_e V$ at the surfaces of the particle or aggregate, as shown in Fig. 11-3. The electric force acting on quarks and leptons is defined in terms of the differential electric energy-density gradients $\rho e V$ at their GSBP *spin surfaces* and similarly the force is mediated by relatively low energy bosons, which act through recoil-impulse to minimize (same particle charges) or maximize (opposite charges) the differential gradient at their *surfaces*. All such forces act locally and timely on the boson-emitting particle and conserve some fundamental property such as charge or momentum, etc. and hence act as gauge forces.

Based on these propositions, chapter 12 develops a relativistic particle ontology, as an alternative to the current assumption that quantum fields, filling all space, constitute the most fundamental substance in the Universe. The difference in perspective vis-à-vis current theory is thus profound – the one closely keyed to relativistic, physical (quasi-classical) models and the other to highly formalistic, mathematical, non-intuitive models, having no connection with general relativity. And this ontology retains the simple and powerful formalism of the Schröedinger equation to take into account both general and special relativity in the strictly-particle theory.

Chapter 13 sketches a Quantum Gravi-ElectroDynamics (QGED) understanding of the dynamics of quarks in the nucleons that make up the nucleus, which, in principle, should enable a precise relativistic-particle account of the dynamics and energy spectrum of the nucleus of at least the simpler atoms.

Chapter 14 presents a strictly *particle* representation of Relativistic Quantum ElectroDynamics (RQED) and the hydrogen Atom, which presents the fine structure "constant" in a new light and the "Lamb shift" and "anomalous magnetic moment" of the electron as vestigial products of a limited quantum mechanics and equally limited relativistic quantum field theory.

Chapter 15 argues that all the "weird," "spooky," counterintuitive phenomena, characterizing the Copenhagen interpretation of QM and QED, vanish in the proposed relativistic-particle theory of matter. The evolution of quantum states in multi-component systems must proceed, at least statistically, through processes of *thermodynamic selection*, offering another feature in defining the transition between the quantum and macroscopic worlds. And chapter 16 outlines a first-principle theory of the evolution of small molecules and molecular systems based on the principle of maximization of energy-density (minimization of density-specific entropy production) in such systems, seen as open thermodynamic subsystems.

Chapter 16 explores the possibilities for applying the GSBP-electron structure and dynamics in modeling the high-Z atoms and in creating a first-principles theory of the NTI evolution of small- and macro-molecules.

CHAPTER 10

PARTICLES, SPREADING WAVES, FIELDS-REVISITED

Given the extraordinary success of quantum mechanics (QM), quantum field theory (QFT) and quantum electrodynamics (QED), it may appear foolish to ask at this late stage: Is it possible that serious flaws exist in quantum physical theory? Is the Copenhagen interpretation of QM in error? Have we invested too much reality in QFT and therefore in QED? Could such errors derive from still more fundamental errors-of-assumption in the remote past? For example: Could particles, after all, be the fundamental ingredients of the Universe; could the physical world be described and comprehended best (with the fewest *ad hoc* assumptions and inconsistencies) in terms of fundamental particles and fundamental forces acting on them? Do the "fundamental particles" we refer to today have internal structure and if so what is their physical makeup and size? Are some of these particles more fundamental or elemental than others? What, if anything, occupies the space between such particles and how might those particles have evolved as the Universe evolved? Is there any need to invoke the concepts of spreading waves and fields in representations of particles and their interactions? Does the physical vacuum, on which quantum field theory is predicated, really exist? If the vacuum is not "seething" with energy and various fields do not "fill all space", engendering the creation and annihilation of necessary "virtual particles," what alternative is there to square quantum mechanics with special relativity? With general relativity? If a relativistic particle theory could provide answers to these questions, how could it account for diffraction and interference phenomena?

Emergence of Non-Relativistic Quantum Mechanics

We begin by asking: How did we come to our present understanding of non-relativistic quantum theory? What were the major turning points and critical assumptions? We will skip through this history, touching on only the principal contributions and assumptions, as it has all been reviewed many times.

The notion of *electric* properties of matter, associated with the Greek word for amber, existed long before the early debate in the 17th century over the nature of light. But it was a century later before those properties were associated with physical, albeit "point" particles, assumed to have quantized rest mass and (later) quantized intrinsic "spin." Yet, to this day, we have no physical model to support these concepts. We will come back to this later.

Newton espoused a corpuscular theory of light, on the premise that light travels in straight lines and in the vacuum, thus requiring no medium in which to move. This was the beginning of scientific conjecture about the nature of quanta. Huygens argued that light must travel as wavelets, since light is diffracted (spread out) at the edges of physical barriers, including the edges of a small hole or slit, through which light may pass. His principle argued that every point on a wave front may be considered as a new source of waves, and all the points on any wave front may be used, together with the original source, to predict any later wave front. This amounted to a simple algorithm allowing light to travel from one point to another through one or more openings in barriers, with no thought of conservation of energy (not surprising since the notions of energy and energy conservation were yet to be formulated), and his principle set the stage for the later emergence of the wave-particle duality concept. About a century later, Thomas Young provided further evidence in favor of the wave theory by demonstrating interference phenomena in light from the Sun[46]. Later still, Faraday's Law of Induction could be interpreted simply as a relation between electric and magnetic *fields*. Maxwell's equations, melding the principal features of both fields, lent further theoretical support to the preeminence of fields and waves and predicted the propagation of electromagnetic radiation at the speed of light c, though there is nothing in his equations requiring *an* electromagnetic wave to spread out, i.e., disperse in space; nor is there any evidence that it does when considered to be a fundamental particle.

However, with Planck's resolution of the black body radiation problem (deriving from his fundamental constant h) and Einstein's demonstration that light must be absorbed and emitted as discrete quanta while traveling as a spreading wave, the characterization of light and other forms of radiant energy led to the development of early quantum theory, with all of its strange demands. Unfortunately Huygen's spreading-wave concept became deeply embedded in the culture of the new physics, as evidenced in this argument taken from Bohm's Quantum Theory.[47] "...an ordinary plane wave of definite wavelength λ is spread over *all* space and therefore, cannot be used to describe the motion of a pulse, which is localized in a comparatively narrow region." In the same volume he says: "It would be impossible ... to explain interference if one assumed that light was made up of localized particles."[48] This view is echoed today by Mills[49] when he asserts: "a photon isn't really a particle ... it's completely spread out over the whole volume ... and thus bears little resemblance to the particlelike photons..." In discussing Heisenberg's gamma-ray microscope *Gedanken* experiment, Wick[50] concludes: "How, you may ask, did the little billiard ball called 'the photon' know to turn into a wave just in time to pass through the microscope? This is a good question — perhaps it is *the* question of 20th-century physics." Such reasoning was the beginning, a century ago, of a time to the present, when the foundations of physical theory became progressively more abstract and counterintuitive.

With a growing focus to understand the structure and behavior of the hydrogen atom, it soon became apparent that an accurate characterization of its structure and dynamics, based on the nature of its radiation spectrum, required a radically new formalism, a Hilbert space. It represented a radical departure from classical formalism in that it allowed, at best, a non-classical probabilistic representation of the dynamics of quanta. And its linear structure facilitated an open or *ad hoc* reformulation or extended formulation of a system's dynamics as additional empirical evidence might demand. This was exactly what was needed to describe the behavior of multi-particle systems or single-particle systems in a simultaneous "superposition" of two or more states.

The rapid development and representation of quantum theory in Schröedinger's wave mechanics tended to further confirm the wave interpretation of the quantum world, while leading many to conclude that wave-particle duality was simply a fact of quantum life, of Bohr's complementarity principle, to be accepted in order to move on. Only with Born's statistical interpretation of the wave function did we begin to question the conventional view of spreading waves. According to Wick[51], Born's probabilistic interpretation of quantum mechanics "most importantly ... preserved the notion of point particle; in retrospect,

Schröedinger's hope that particle tracks could be explained as little wave packets moving without dispersing was an illusion. Computations soon proved that even a tiny wave packet could grow to a size obviously larger than any particle track – yet the particle did not grow with it. As shown by collisions with other particles or by measuring devices such as photographic plates, it retained its 'particlehood' intact."

Moving on meant simply accepting, in a similar way, Heisenberg's uncertainty relations, the counterintuitive demands of superposition theory (*a* wave-particle "interfering with itself") and the notions of "wave function collapse" (the "measurement problem" implying acausality and nonlocality), the "entanglement" of quanta, and a probability theory peculiar to the quantum world. In regard to the latter, Hughes[52] notes "The principle of superposition tells us something about the set of admissible states, the uncertainty principle something about the set of observables encountered in the theory. Any theory which includes either of these principles is, we may say, *inherently probabilistic*; that is, each principle entails that there are pure states which assign to the outcomes of certain experiments probabilities other than one or zero." (We will return to these issues from a different perspective in chapter 15.)

The Copenhagen doctrine argues that a measurement causes the wave function to collapse to one of the many or infinite superposition states. That idea is not articulated in the mathematics of quantum mechanics; it serves as a postulate external to the formalism. The measuring apparatus can do no more than select, from the many (or infinite) possible states, that one or relatively few state(s), that conform(s) to the pre-established physical constraints, *as time scrolls through the various eigenstates. Nor can we even know or precisely control when that selection process is actualized.* We have to ask: How did this doctrine of superposition of states and their "collapse," brought on by the act of measurement, come into being? Why was it so readily and universally accepted, despite the "measurement problem" which it entailed and the nonlocal conflict with Special Relativity? The formalism of QM provides a probabilistic description of "entangled" states of a system and its evolution under given constraints. A measurement interrupts that evolutionary process at some point in time. *The "collapse" presumption is that <u>that</u> time is under the control of the observer (measuring apparatus). In classical mechanics a description of a system may be characterized by two or more solutions of a polynomial or set of equations, but we do* *not assume that the system is in both or all states until we do a measurement. Why did Bohr and others make and propagate this assumption in the early development of QM?* The problem, still with us, is not with Schröedinger's mathematical formalism of quantum mechanics but with the interpretation bestowed on it by Bohr and others.

Bohm[53] said: "The paradox of Einstein, Rosen and Podolsky was resolved by Neils Bohr in a way that retained the notion of indeterminism in quantum theory as a kind of irreducible lawlessness in nature. ... He argued that in the quantum domain the procedure by which we analyze classical systems into interacting parts breaks down, for whenever two entities combine to form a single system (even if only for a limited period of time) the process by which they do this is not divisible. We are therefore faced with a breakdown of our customary ideas about the indefinite analyzability of each process into various parts, located in definite regions of space and time. Only in the classical limit, where many quanta are involved, can the effects of this indivisibility be neglected; and only there can we correctly apply the customary concepts of *detailed* analyzability of a physical process. ... Now comes the essential point. In order to give the classical laws a real experimental content, we must be able to determine the momenta and positions of all the relevant parts of the system of interest."

Milburn[54] addresses these issues in the contexts of "reality" and randomness in behavior. "The quantum principle is this: physical reality is irreducibly random, but random in a way we could never have expected. Reality exhibits a randomness so constrained, *by an as yet undiscovered principle,* that the odds with which it deals confound anyone who studies it." And he adds: "Photons do behave in an irreducibly random way at a beam-splitter. But the rules for calculating the odds for multiple encounters with beam-splitters are radically different from ordinary probability. ... Why should it be the case that an irreducibly random event be constrained by just these rules? Why cannot multiple events be described by ordinary probability? *No-one knows the answer to that question.*"[55]

Omnès[56] focuses on another quandary: "*The most important consequence of the uncertainty relations for interpretation is their incompatibility with an intuitive representation of a particle as being a point in space. The idea of a space trajectory is also excluded because it would mean simultaneously precise values for position and velocity. The 'concept' of particle becomes obviously much poorer."* (emphasis added) As for the measurement problem, He notes Bohr's acceptance of it: "What can be the meaning of an interpretation that is violating the theory it purports to interpret? Bohr was fully aware of the difficulty. No doubt, he said, the reduction rule is only one of its kind in the whole realm of physics. It has no analogue anywhere because its role is to bridge the gap between two different physics, one of them classical, obvious, and the other symbolic, mathematical. ... Bohr did not waver at considering reduction as being a real physical effect. *The violation of Schröedinger's linear dynamics was thereby ratified, accomplished through an external effect, which occurs only when there is a measurement."*[57] (emphasis added)

The arguments of Bohm and Milburn imply an unquestioned assumption that *one and only one particle* is involved in the complex behavior of a system that would perhaps be analyzable in a multi-particle classical system, while Omnès underscores the folly of giving serious thought to the notion of any real massive particle occupying no physical space.

Auyang's observation regarding the measurement problem epitomizes the "acceptance" philosophy: "I do not speculate on what really occurs in quantum measurement because no satisfactory theory exists and as Feynman said, the physics we are studying is not twisted up with the measurement process. The measurement problem is important, but *at issue is some physical process that can only be illuminated by future physics."*[58] (emphasis added) And as regards the acceptance of dimensionless particles, she remarks: "Spin is a genuine angular momentum that couples as such to the orbital angular momentum. It stands with mass and charge as the fundamental kinematic or intrinsic properties of elementary particles, *which they characterize in the absence of dynamical interactions. In the intrinsic angular momentum we have a case of an 'absolute motion' of a free point particle."*[59] (emphasis added) Wick[60] carries this acceptance mindset a step further in a prescient comment: "Here is the modern physicist's understanding of spin: The point electron exhibits an intrinsic circular motion of less than atomic dimensions **at the speed of light**."(Wick's emphasis, p128) The above inconsistencies also closely relate to the measurement problem which Bell[61] sums up thus: "The continuing dispute about quantum measurement theory is not between people who disagree on the results of simple mathematical manipulations. Nor is it between people with different ideas about the actual practicality of measuring arbitrarily complicated observables. It is between people who view with different degrees of concern or *complacency* the following fact: ***so long as the reduction is an essential component, and so long as we do not know exactly when and how it takes over from the Schröedinger equation, we do not have an exact unambiguous formulation of our most fundamental physical theory.*** (emphasis added)

The notion (frequent in the literature on interference phenomena) that *a* particle or wave-particle can interfere with itself rests on what may be called the *"single particle assumption,"* that when the intensity of a particle (boson, electron, etc.) source is sufficiently reduced, we can be assured that only one particle is involved in the experiment. How could we demonstrate the validity of that assumption, beyond doubt, given the uncertainty relations, and especially since we often go to great pains to create many same-state or nearly same-state particles as particle sources (e.g., lasers) for interference experiments? Wick[62] quotes Schröedinger as saying: *"We never experiment with just one electron or atom…any more than we can raise Ichthyosauria in the zoo."*

In all such concerns, "accepting" has meant: the conceptual framework and its mathematical formulation have worked with great predictive and explanatory power, so why linger over the accompanying unanswerable or meaningless philosophical questions. Indeed one could argue that we soon came to rejoice in the easy and confident dismissal of such questions, as being irrelevant, so that we could get on with the pragmatic application of the new physics. Whitaker[63] refers to Bell on the matter: "Bell regarded himself as a follower of Einstein, and was not in general much in favour of complementarity, or of much of Bohr's efforts to understand quantum theory. *He accused Bohr of taking satisfaction in ambiguity rather than being disturbed by it, and of reveling in contradictions."* (emphasis added)

Creation of Relativistic Quantum Mechanics

But in the late 1920's, moving on also meant reformulating quantum mechanics to the demands of special relativity theory. Early efforts in this direction led to approximate corrections to Bohr's semi-classical model of the hydrogen atom relating to the "reduced mass" of the electron, interaction of the electrons spin with its orbital motion and relativistic adjustment of the electron's momentum. All such corrections were necessarily imposed externally on the electron's dynamics, as there was no internal structure on which to base them. These efforts were soon eclipsed by Dirac's equation, an algorithm involving four new variables in the wave equation whose solutions yielded the *exact* energy levels of the hydrogen atom (when the "reduced mass" of the electron was taken into account) and forecast the existence of antiparticles. But a *physical* basis for the timely interaction of charged particles, required by special relativity, remained undefined, and our attention became focused once again on the possible roles of "particles", "waves" and "fields" in these interactions. *(It should be underscored here that the added variables in Dirac's wave equation offered no connection with general relativity, in which c may be taken as a parameter.)* At the same time, there was growing evidence that there was a small energy difference between the $2S_{1/2}$ and $2P_{1/2}$ levels of the hydrogen atom, thought to be degenerate and in contradiction to solutions of the Dirac equation. The precise measurement of this energy level difference, the "Lamb shift," led to concerted efforts to divine the source of the additional energy in the $2S_{1/2}$ level. That effort focused on the only available hypothetical source, the energy in the vacuum, the existence of which was presumed to derive from Heisenberg's indeterminacy relations. There was need then to tap such energy for the creation of virtual photons as force quanta mediating the interaction of electrons, etc., about which more will be said later. This led to the development of relativistic quantum field theory and this led in turn to the development of quantum electrodynamics, aided and abetted by the development of "renormalization" theory. All of this was marked by a further departure from classical or intuitive reasoning and linked closely with the need to correct not only for the electron's mass but also its charge, the equivalent of gravitational charge in modeling the dynamics of the earth's orbiting the Sun. But this is precisely what Dirac's equation was supposed to yield. To what then must we attribute the Lamb shift and the discrepancies with the Dirac solutions?

The answer again may lie with a questionable fundamental assumption, in this case with the assumption that the measured frequency (or wavelength) of a boson emitted in a transition between a metastable state and its degenerate companion determines exactly the energy difference between those states. But metastability implies some measure of residence in a state over time or some probability that a particle may be found in that state in accordance with the indeterminacy relations. How else would we know or could we determine that it is a valid quantum state? The physical embodiment of a metastable state is the existence of a potential energy well, requiring a certain addition of energy or fluctuation, above that of the lowest level of the well, in order to escape. If this view is reasonable, then what Lamb revealed from his spectral measurements, was not the energy of the $2S_{1/2}$ level, but that value plus the additional energy required to surmount the energy hill or tunnel through it, as illustrated in Fig. 14-1. In the hydrogen atom that additional energy is associated with a required increase in spin angular momentum, which is associated with the *anomalous* magnetic moment emerging in the QED calculations. We will return to these issues in greater depth in chapter 14 from a very different perspective, that is, in the context of a strictly particle development of quantum electrodynamics and representation of the hydrogen atom.

As implied above the electromagnetic force acting on charged quanta was mediated in current theory by photons. In principle, one electron could emit a photon, with a certain energy-momentum, toward another, thereby causing both electrons to gain momentum in moving apart, the first from recoil (while losing some internal energy-momentum) and the second by gaining internal energy and momentum from the absorbed photon, thereby conserving both momentum and energy. This was the basis of Feynman's graphic depiction of QED. (But where was the photon's energy stored in the electron and how might the photon be directed just so to interact with another electron, especially if they are both moving at relativistic speed on different trajectories?) Unfortunately, electrons were (and still are) believed to be "point particles," i.e., dimensionless entities, while having mass, charge and spin angular momentum, but lacking any internal store of these properties. Thus, the creation of photons to mediate the electromagnetic force necessarily called for some unknown, off-the-books, local source of energy. Not surprisingly, this proved to be a major undertaking, requiring the reification of energy in the vacuum (the "physical vacuum"), a process enabling the creation and annihilation of "virtual particles" (virtual photons, electrons, etc.) as needed, and the development (over many years) of mathematical procedures for renormalizing, i.e., discarding or subtracting, unwanted infinities, in the calculation of interacting forces involving relativistic particles. (The physical vacuum, was said to be everywhere filled with fluctuations of various kinds of fields, a necessary consequence of Heisenberg's uncertainty principle, and the energy of these fields could be borrowed and quickly returned to the vacuum to create the short-lived "virtual particles" without being in conflict with the demands of energy conservation.) The resulting quantum field theory has enjoyed extraordinary success, apparently validating the existence of the physical vacuum; yet it leaves us again, as with the wave/particle duality problem and other issues in quantum theory, with an abstract algorithm and procedures that work, often exceeding everyone's expectations, but providing no mechanistic or intuitive model of the observable physical world. So, as Omnès[64] says "The main problem with relativistic particles is to decide first what is real: the particles or the fields. The ghost of complementarity, which was believed to be killed, is popping up again." Steven Weinberg's answer is: "the electron is a particle, described by a relativistically invariant version of the Schröedinger wave equation, and the electromagnetic field is a field, even though it also behaves like particles."[65] He goes on to say "In its mature form, the idea of quantum field theory is that quantum fields are the basic ingredients of the Universe, and particles are just bundles of energy and momentum of the fields. In a relativistic theory the wave function is a functional of these fields, not a function of particle coordinates." But QFT also leaves us with its offspring QED, assigning an anomalous magnetic moment to the electron as mentioned above, again in contradiction to Dirac's theory.

The foregoing is deliberately a highly-simplified account of the evolution of quantum theory over the past century in order to emphasize the extraordinary extent to which that theory has become increasingly complex (conceptually and mathematically), farther and farther removed from intuitive understanding and, most importantly, the product of a few questionable fundamental assumptions – all without connection with general relativity theory.

Critical Challenges and Turning Points

Let us summarize first the principal challenges to conventional thinking that were called for in the creation of non-relativistic quantum theory, challenges which lie at the heart of what has been called the Copenhagen interpretation of quantum mechanics:

- All quanta are said to possess both particle and wave attributes in a "superposition" of concepts or attributes, only one or the other of which can be evident in any given experiment in any particular action. They were in Bohr's view "complementary" attributes. But why did Bohr and others insist that quanta must have complementary attributes? The answer appears to be that our long-standing *understanding of waves in material media, as spreading over all space, was carried over to include electromagnetic waves, even in a vacuum, and we have been unable to incorporate these two very disparate attributes (waves spreading over all space and particles confined to no space) in any rational physical model.* Was this a failure even to contemplate the need for any internal structure in the quanta, to give them some spatial and dynamic representation? ("The idea of looking for a structure of the electron was given up because,' as Dirac[66] suggested, 'the electron is too simple a thing for the question of the laws governing its structure to arise." But if an electron, for example, displays wave properties, should we not expect it to have some form of internal clock capable of defining its phase in spacetime relative to that of another electron with which it may interact, as is the case for the photon? The latter has a precise clock - its intrinsic energy-momentum, stable over its lifetime. But the photon, like the electron, has never been given any *internal* physical embodiment, structure or size, incorporating the attributes of both field and particle. The problem with both the electron and the photon, in strict particle-trajectory terms, is that self-interference appears to require each *single* particle (electron or photon) to be capable of traveling on two or more distinct trajectories simultaneously, in order to comply with the notion of spreading waves. And this brings into question the conservation of energy-momentum in the travel mode as well as in its interaction with a detector, the "measurement/wave-function collapse" problem. How did that notion become so firmly fixed in our thinking? Faced with two conflicting empirical "facts:" the overwhelming evidence of interference phenomena, on the one hand, and the lack of any evidence that two or more photons (electrons, etc.), with identical properties, could be emitted from a source (even at very low intensity, on the other hand, Bohr and others opted for the single-particle assumption. Heisenberg's matrix mechanics quickly supported the assumption, as did Schröedinger's wave mechanics. What is most surprising in this history is that long after the physical and theoretical development of lasers, few if any have thought to reexamine the issue. Perhaps those that did were dissuaded from further investigation because of its disturbing implications and the doubtful benefit of doing so, given the great success of the Schröedinger algorithm. But now the notion of superposition itself appears to lie at the heart of the counterintuitive nature of much of quantum behavior. If a sense of reality, causality and locality is to be restored in quantum theory it will likely hinge on the denial of any validity of superposition theory.

- Quanta or quantum systems can only be described in some suitable Hilbert (vector) space in which the precise values of complementary variables cannot be observed or measured simultaneously in any given experiment. This behavior derives from the indeterminacy principle and constitutes therefore a major departure from classical theory. As Omnès[67] states "The superposition principle is reflected by the fact that a Hilbert space is a vector space and therefore linear. *One can associate with every isolated physical system a definite Hilbert space. Every physical concept entering in the description or analysis of the system should be expressible in a mathematical language into which only notions involving this Hilbert space can enter.*"(Omnès' emphasis)

- The "superposition" concept and formalism, as Dirac labeled it, rests on what may be called the *"single particle assumption": that when the intensity of a source of particles (waves) is sufficiently reduced we can be assured that only one such particle (wave) is involved in the experiment. Has there been any concern or evidence to fully justify that assumption experimentally or theoretically? This is a particularly important question, since it may well be that it is improbable in macroscopic (classical) systems, e.g., lasers, to emit only one quantum at a time, even at an extremely low level of intensity.* As we shall show later, *the time resolved diffraction and interference studies of Garcia[68], et al and interference studies with single photons of Austin and Page[69] are directly relevant to this issue and provide direct evidence that interference phenomena are dependent on the existence of more than one particle in the system. This should prompt us to ask: Are there multi-particle dynamic processes in the quantum world, yet to be explored, which result in correlated emission of such particles, based on the principle of minimum flow-specific entropy production?*

• Thus, Bohr's interpretation of non-relativistic quantum theory hinged at the outset (and remains so today) on the lack of any conceivable way to incorporate the localized (spacetime-bounded) nature of *a* particle, *a* photon for example, conserving intrinsic energy-momentum on *a* correspondingly bounded trajectory, so as to account for the peculiar phenomena of interference. Here the emphasis on the singular article *"a"* informed the early and present-day theorists that we must reject out of hand any possibility that two or more such particles (photons) could be emitted from the same source and travel on different trajectories, so as to converge and interact, constructively or destructively, in the same or nearly-same space of their demise (exchange of energy-momentum). Were that possible such particles would travel as *particles* on *distinct trajectories*, subject to probabilistic momentum change, and *interact* or be *absorbed* as *particles*, without any requirement for superposition of spreading waves subject to "wave function collapse" and their *localized* field and particle natures would remain *complementary* in the sense that we could only precisely measure one or the other property at any given time in any given experiment. There is in this idea no suggestion of a wave packet, nor could there be in the case of a photon traveling uninhibited for billions of light years with stretching of its wavelength, but *without* dispersion. This same underlying, unchallenged belief in the notion of spreading waves (fields filling all space) then required us to accept uncritically the image of a single particle traveling on two or more trajectories simultaneously. But this raises difficult questions: How are we to interpret Maxwell's equations in such a boson's (strictly particle) structure?[70] What could be the structure of charged massive particles displaying similar features? What is charge and how does it differ from one type of particle to another? How do we account for diffraction and interference phenomena? We will explore these and other relevant questions in what follows and in the remainder of Part III.

We turn our attention now to a summary of the principal contributions and assumptions that were associated with the reformulation of quantum mechanics so as to be consistent with the demands of special relativity (SR).

- Dirac's equation joined special relativity with quantum mechanics and yielded the *exact* energy levels of the hydrogen atom (when the "reduced mass" of the electron was taken into account) and forecast the existence of antiparticles. Yet the physical means accounting for the attraction or repulsion of charged particles, in accord with SR, remained obscure.

- The development of quantum field theory was undertaken to (a) account for the discrepancy between the energy levels predicted by the Dirac equation and those derived from precise measurements of the radiation spectrum of the hydrogen atom, including the Lamb shift and the anomalous magnetic moment of the electron (b) characterize the relativistic interaction of charged particles and resolve the wave-particle dilemma in terms of an ontology of fields. All of this was marked by a further departure from classical or intuitive reasoning and linked closely with the need somehow to correct not only the electron's mass but also its charge in relativistic terms, i.e., to endow the electron in some sense with dimension and structure. Is the Dirac equation exactly correct or did he miss something in its formulation?

- The answer again appears to lie with a fundamental assumption, specifically with the assumption that the measured frequency (or wavelength) of a photon emitted in a transition between a metastable state and its degenerate companion determines exactly the energy difference between those states. If this assumption is false, as implied above, then there need be no disagreement with the Dirac equation, nor any need to account for an anomalous magnetic moment, which itself raises questions relating to the long term stability of the hydrogen atom.

- The accepted "point particle" concept of charged particles, in particular the electron, lacking any internal structure or dynamics, lies at the heart of the difficulties associated with the conjoining of general and special relativity with quantum mechanics.

Where Do We Stand?

The failures to challenge the above fundamental assumptions, accepted for the better part of a century, have left us, despite a justified pride in the accomplishments of the past century, with widespread discontent, as the following commentary by those closely associated with the effort attests:

"The shell game that we play to find n and j is technically called "renormalization." But no matter how clever the word, it is what I would call a dippy process! Having to resort to such hocus-pocus has prevented us from proving that the theory of quantum electrodynamics is mathematically self-consistent. It's surprising that the theory still hasn't been proved self-consistent one way or the other by now; I suspect that renormalization is not mathematically legitimate. What is certain is that we do not have a good mathematical way to describe the theory of quantum electrodynamics: such a bunch of words to describe the connection between n and j and m and e is not good mathematics." Feynman[71]

"The resulting [QED] theory is an ugly and incomplete one, and cannot be considered as a satisfactory solution of the problem of the electron." Dirac[72] goes on to say "One can conclude that the fundamental ideas of the existing theory are wrong. A new mathematical basis is needed." 1984 As Schweber[73] says "Almost all the proposals to eliminate divergences that were made during the 1930s ended in failure. ... The pessimism of the leaders of the discipline—Bohr, Pauli, Heisenberg, Dirac—was partly responsible for the lack of progress. ... They had brought about the quantum-mechanical revolution and they were convinced that only further *conceptual* revolutions would solve the divergence problem in quantum field theory."

"We [Dyson and Oppenheimer] are both agreed that the existing methods of field theory are not satisfactory, and must ultimately be scrapped in favour of a theory which is physically more intelligible and less arbitrary. We also agree that the final theory should explain why there are the various types of particles which we see and no more. However, Oppenheimer believes that the nature of the nuclear forces will itself give us enough information on which to build the new theory; in other words, the nuclear forces will not be describable at all except in terms of a theory which explains the existence of elementary particles. ...*If I [Dyson] am right, the discovery of a finally satisfactory theory of elementary particles will be a much deeper problem than those we are tackling at present, and may very well not be achieved within the framework of microscopic physics.*"[74] (emphasis added)

"...the multiple ways of [dealing with the high-energy cutoff in QFT] are not in conflict provided that we understand them as approximations to the structure of some deeper, as yet unknown theory ... If there is one underlying theme to the approach to QFT advocated in this paper, it is this: *the sort of information which we are interested in getting from physical theories is structural information.*" (emphasis added), D. Wallace[75]

"*In spite of [a deep similarity], it has not proved possible to unite relativity and quantum theory in a coherent way. One of the main reasons is that there is not consistent means of introducing extended structure in relativity, so that particles have to be treated as extensionless points.* This has led to infinite results in quantum field-theoretical calculations. By means of various formal algorithms (e.g., renormalization, S matrices, etc.) certain finite and essentially correct results have been abstracted from the theory. However, at bottom, the theory remains generally unsatisfactory, not only because it contains what at least appear to be some serious contradictions, but also because it certainly has a number of arbitrary features which are capable of indefinite adaptation to the facts, somewhat reminiscent of the way in which the Ptolemaic epicycles could be made to accommodate almost any observational data that might arise in the application of such a descriptive framework (e.g., in renormalization, the vacuum-state wave function has an infinite number of arbitrary features)." (emphasis added) D. Bohm[76].

"I admit to some discomfort in working all my life in a theoretical framework that no one fully understands. And we really do need to understand quantum mechanics better in quantum cosmology, the application of quantum mechanics to the whole Universe, where no outside observer is even imaginable." S. Weinberg[77]

"theories in which the most basic entities are points — quantum field theories as they are known — possess unpleasant mathematical properties. They lead to mathematical infinities that must be ignored in the process of calculating observable quantities. This can usually be done by following a systematic recipe which amounts to ignoring the infinite part of any answer, but the procedure is aesthetically rather unappealing. It has only been tolerated in practice because the finite parts that remain in these calculations after the infinite parts have been removed produce predictions of observed quantities that are correct to fantastic precision. There is clearly a deep truth somewhere close to the heart of this picture." Barrow[78]

Eventually it became clear, even to Schwinger, that he was approaching a practical limit to the formalistic demands of QED, as he put it: "'I had gone through … gyrations of these successive canonical transformations, [and] it was clear I couldn't have done this to the next order…"Schweber[79]

"It is on … short distant scales that we encounter the fundamental incompatibility between general relativity and quantum mechanics. *The notion of a smooth spatial geometry, the central principle of general relativity, is destroyed by the violent fluctuations of the quantum world on short distance scales.* On ultramicroscopic scales, the central feature of quantum mechanics—is in direct conflict with the central feature of general relativity — the smooth geometrical model of space (and spacetime)…. The equations of general relativity cannot handle the roiling frenzy of quantum foam…. There are those physicists who are willing to note the problem, but happily go about using quantum mechanics and general relativity for problems whose typical lengths far exceed the Planck length, as their research requires. There are other physicists, however, who are deeply unsettled by the fact that the two foundational pillars of physics as we know it are at their core fundamentally incompatible, regardless of the ultramicroscopic distances that must be probed to expose the problem. The incompatibility, they argue, points to an essential flaw in our understanding of the physical Universe." Greene[80] And on page 87 he writes: "might it be that through historical accident physicists have constructed an extremely awkward formulation of quantum mechanics that, although quantitatively successful, obfuscates the true nature of reality?"[81]

Thermodynamics and the Coevolution of ρ, c and $\sqrt{G}$ in Quantum Theory

ρ, the average energy-density of the vacuum is not a familiar parameter in the conventional literature of quantum theory, though this may be changing as studies of decoherence and the transition from quantum to macro states focuses increasingly on thermodynamic issues. In our scenario, c and $\sqrt{G}$ are taken as inverse functions of ρ in the evolution of cosmological epochs. In accelerator target reactions, where the local energy-density ρ_{local} may be orders of magnitude larger (if only ephemerally), we should expect c and $\sqrt{G}$ to vary accordingly. In chapter 6 we argued that inflation theory, wholly dependent on the reality of the physical vacuum, was invalid on various counts and most importantly because it grossly violated the laws of thermodynamics. Here we seek again to understand in what sense, if any, energy in the vacuum may be considered real and if so how it may function legitimately to support QFT/QED in quantum physics.

The Physical Vacuum

Whether or not the physical vacuum exists as represented in quantum field theory, that is, as all-pervasive fields of various kinds subject to arbitrary energy cutoff constraints, there is a sense of energy in the vacuum in the context of cosmology, i.e., in the form of average energy-density of matter and radiation in the vacuum of outer space. In our region of spacetime, energy-densities in the form of matter and radiation are about 10^{-30} and 10^{-33} g/cm^3 respectively. Such energy-density may be viewed as a pressure against the outer surface of any container in which some of the particles, normally accounting for that energy-density, cannot exist in the container by virtue of its limiting dimensions relative to particle wavelengths. In regard to the Casimir effect, the container may consist of two parallel metal plates separated by a very small distance, such that the pressure on the inner surfaces of the plates is said to be less than that on the outer surfaces, because there are fewer virtual particles, having low energy (large wavelength) transferring momentum to those inner surfaces.

But the issue is not whether there is real energy in the vacuum of outer space or virtual energy, but whether it can be converted into work in any form, as for example the spontaneous mutually-directed acceleration of the two plates or the creation of real or virtual photons whose momentum may be exchanged between charged "point particles," *resulting in some change in their energy-momentum states*. Energy can be converted into work, accompanied by some entropy production, only via the flow over time of mass-energy between systems or regions of different energy or energy-density states, and this is not possible by definition, since ρ is uniformly constant in the vacuum of our spacetime region. The same argument would apply to the all-pervasive fields constituting the physical vacuum. No physical change in spacetime can accrue in any time period except that the work associated with it is accompanied by some measure of energy dissipation and entropy production. *Thus the notion that energy borrowed from the physical vacuum to effect some physical change (do work) without dissipation (returning the full measure of the energy borrowed, however rapidly) is in fundamental conflict with the laws of thermodynamics.*

We are left then to ponder the following questions: If, as will be argued later, Dirac was right after all that his equation *exactly* defines the energy levels of the hydrogen atom in relativistic QM (taking into account the reduced mass of the electron), what did we gain from decades of extraordinary effort in the development of QFT/QED? What parts of the puzzle, for which there were no answers, prompted this persevering effort by several of the greatest theoretical physicists of the past century?

Some have implied that Dirac simply tried this and that equation in his effort to square QM with special relativity; indeed he said as much in his 1963 interview with T.W. Kuhn: "A great deal of my work is just playing with equations and seeing what they give."[82] He surely understood that "reducing" the mass of the electron in the Coulomb field did not yield the same full correction for the non-infinite mass of the proton that was achieved in reducing the mass of a planet in the gravitational field of the non-infinite mass of the Sun. In the latter case the reduced mass also proportionately reduced the gravitational charge $M\sqrt{G}$, the coupling constant, but in the Coulomb field the reduced mass leaves the effective charge of the electron unchanged. This realization must have guided him in some way to incorporate that correction in his equation, even though the correction changed nothing about the nature of the "point electron," which was the other part of the puzzle for which there was no answer. The equation was simply an algorithm that worked. But that success was not sufficiently convincing to dampen the growing belief that relativity could only be understood in some form of quantum field ontology, and especially as there were indications, as laboratory technology improved, that the hydrogen spectrum did not agree exactly with the Dirac solutions.

As to the origin of Dirac's equation, the closest reasoning may be found in Dirac's Nobel Lecture: "It is found that an electron which seems to us to be moving slowly, must actually have a very high frequency oscillatory motion of small amplitude *superposed* on the regular motion which appears to us. As a result of this oscillatory motion, the velocity of the electron at any time equals the velocity of light. This is a prediction which cannot be directly verified by experiment, since the frequency of the oscillatory motion is so high and its amplitude is so small. But one must believe in this consequence of the theory, since other consequences of the theory which are inseparably bound up with this one, such as the law of scattering of light by an electron, are confirmed by experiment."[83] There is much in this passage that sheds light on his thinking. Our emphasis on "superposed" indicates that he was already wholly committed to a field theory conceptualization of quantum mechanics.

In retrospect we conclude, as explored in greater depth in the remainder of part III, that QFT/QED was a misguided, costly departure from what may be called intuitive, semi-classical quantum physics, while leaving deeper fundamental concerns at issue, as implied above and addressed in chapter 14.

The Problem of Diffraction, Interference

The above forces us to return to questions relating to the preeminence of particles, waves and fields and in particular to the diffraction argument that led Huygens to espouse the wavelet concept of light. If there were no need to speak of spreading waves and all phenomena could be explained strictly in terms of particles, how could we account, for example, for the diffraction of light, i.e., bending of the straight-line trajectory of a light particle (boson), as it passes very close to the edge of a physical light barrier? And in particular, how could we account for the phenomena of interference associated with such bending?

Long before the creation of quantum mechanics, Thomas Young[84] demonstrated, to the astonishment of members of the Royal Society, the phenomenon of light interference, and he did that with light from the Sun and little more than a "slip of card." Only in recent times, with the benefit of advanced technology, has this strange behavior been demonstrated in experiments involving only a "single photon." And with this, our *commitment* to Dirac's principle of "superposition" and the underlying concept of wave-particle duality was acknowledged, for most, beyond doubt. Such experiments, by themselves, were sufficient for Feynman and others to admit the impossibility of truly understanding quantum mechanics.

But we need to ask: How can we be so certain that only *one* photon is actually involved in such experiments, especially since we now go to so much effort to generate the photon with a laser, a source noted for its capability of creating enormous numbers of photons, nearly identical in energy-momentum and phase, in a narrow collimated beam? (Note here the fundamental distinction we make at the end of this chapter between *gravitationally-bound photons* and *free bosons*. In referring to other authors' use of the term *photon*, we caution the reader to understand that it would be necessary in our scenario to speak of *bosons*.) We might instead ask: Can a laser emit a single boson or can we select only one of a few bosons from a laser or other source? It is remarkable that the "single particle" assumption quickly became a statement of fact, without generating critical attention until the 200th anniversary of Young's interference experiments. In their invited article, "Time-resolved diffraction and interference: Young's interference with photons of different energy as revealed by time resolution," in the special theme issue, Interferences, in the Philosophical Transactions of the Royal Society. N. Garcia[85], I.G.Saveliev and M.Sharonov say: "We try to study, using femtosecond pulses of photons the wave function collapse time in QM by measuring patterns of photons diffracted by a slit. These patterns are obtained by accumulating many few-photon events (*we have no reason to claim one photon events*) in order to know how many photons are needed to describe the typical diffraction pattern. Here we introduce a new ingredient by using a streak camera to measure the time of flight of pulses from the slit to detectors located at different positions. As our streak camera has a 2ps resolution we can establish that this is an upper limit of the collapsing time because we do not find any sign of a collapsing time. ... All these experiments showed within the 2ps accuracy of our streak camera resolution that the time delays measured correspond to straight line 'bullet' (impact of a wave packet) trajectories from the slit to the detectors. ... We find it difficult to explain these experiments, if we cannot know the photon trajectory and if before wave collapse on the detector the diffraction picture is one wave. ***We believe that this is a manifestation of the particle character of the photons during their flight from source to detector describing straight lines.***" (emphasis added) In short the researchers were forced to conclude that the Fraunhofer interference pattern, visible in the streak camera, was created one or a few photons at a time in each separate passage from the source, through *the* slit and to the camera, notwithstanding the lack of any apparent real wave or wave function collapse, which conventionally would be the signature of wave behavior. How then, we may ask, can each individual photon follow two distinct trajectories simultaneously from source to detector, as Fraunhofer interference would seem to require (see chapter 11), unless two or more nearly-identical photons pass through the slit at the same time, each following a distinct trajectory to the camera? Is there any evidence or argument to support this conclusion? As quoted above, the authors acknowledge the possibility that two or more photons might be involved in "events," yet they say: "In the counting mode, each photon gives a brilliant spot that can be assigned a position (horizontal axis) and a time of flight (vertical axis)." But there is no mention of the numerous non-brilliant spots (distinct though of uniformly lower intensity) in the frames. This would suggest that the equally distinct, brilliant spots are the result of four photons (two pairs) passing through the slit and adding constructively at the detector. Given the relatively wide slits (150 μm), two pairs of photons (particles) may readily travel through a slit at the same time. This argument should have raised another very important issue left unmentioned by the authors, namely, given their conviction that the photons travel as particles, one at a time from the source to the detector, how are such particle trajectories altered in going through the slit so as to give rise to single slit Fraunhofer diffraction? We address this question later in this chapter.

In regard to double-slit interference experiments, the authors state: "The description of particle interference phenomena in QM is based on the probability of the particle being in one of several different indistinguishable alternatives ..." and they quote Dirac in 1958 saying: "*Each photon then interferes only with itself. Interference between two different photons never occurs.*" After citing several papers supporting the QM argument of indistinguishability, they state: "Here however we present Young's double slit experiments that show interference between different photons of different energies as measured by the output of a spectrometer and *knowing which photon colour* [photon energy] *goes to each slit*." Their experimental results are thus a direct contradiction of Dirac's assertion. Moreover their experiments "clearly show that different photons (with different energies) produce interference fringes and the patterns depend strongly on the magnitude of the energy differences between the photons. We believe that each slit is illuminated by definitely different photons and there is no alternative for the photons to be either in one way or another. Although [it] is also true that we do not know which photon is which after the slits, **we do know them before the slits!** (authors' emphasis) In this case we cannot apply the QM explanation (Mandel 1965; Hanbury-Brown 1956) similar to the one given for the Hanbury-Brown and Twis experiments by Purcell (1956) in which interference between different photons can occur due to the fact that it cannot be told which source has emitted individual photons. On the other hand photons are not in an 'entangled state' as occurs in interference experiments with photons created by down-conversion in a nonlinear crystal. In our experiment we can provide 'which-path' information (at least we know that different energy photons, as measured from the spectrometer, are impinging each one of the slits) using both space and time-resolved measurements." And they argue: "**with no overlap of the pulses no interference can be observed no matter how good resolution we have, be the photons of the same energy or different. The fact that the overlap is needed may indicate that indistinguishability of the photons is necessary at the detector but certainly not at the slits. ... the key requirement for interference is to have overlap of the pulses impinging the photo-cathode of the streak camera**" (emphasis added). Here the key word is "*pulses*," indicating that the creation of femtosecond pulses (a subsystem of energy flow in the experimental system) was essential for the accumulation of "many few-photon events" in order to give rise to interference patterns – a lower entropy ordered evanescent structure.

We may conclude that the results of these experiments with high spatial and time resolution argue strongly for: (1) a strictly particle theory of diffraction and interference, (2) the realization that all such single- or multiple-slit interference behavior necessarily involves more than one particle, whether they are entangled with one another at the source or are independent yet sufficiently similar in energy and overlap adequately in time and space at the detector and (3) superposition plays no role in interference phenomena nor likely in the theoretical or empirical investigation of quantum systems. The Garcia *et al.* experimental study is the latest and among the most important in the more than 200-year history of the investigation of light. Yet the authors' intermingling of wave and particle terminology suggests, not surprisingly, a difficulty in confronting the deeper implications of their results (the failure of wave-particle duality and superposition theory), given the largely unchallenged wide-acceptance of conventional teaching over the past century.

In an article published on the Internet entitled "Interference of a Single Photon," Robert Austin[86] and Lyman Page, (Princeton) illustrate further the extent to which we take for granted the "single particle" assumption. They illuminated a triple slit with a laser beam that was attenuated with a neutral density filter to one part in 10^{11}, such that the mean free path between photons (bosons) was about 2 kilometers. They "show that despite this large distance, a diffraction pattern obtains – the photon has 'interfered' with itself." With the aid of a photon counting camera and a CCD frame-grabber card, controlled by computer, a 1/30 second record reveals a total of three recorded events, in good agreement with the expected number. They assume each event to be the signature of a single photon. However all three events in the record have the same relatively-low intensity and appear under magnification to have the same sharply defined rectangular shape, with the long axis in the same (transverse) direction as the length of the slits. It appears that each event is one pixel wide by three pixels long, implying the excitation of three adjacent CCD cells. How could we explain this, assuming a single photon or wave passing through two or possibly three slits? However, if we assume that a triplet of photons (in side-by-side alignment with the length of the slits) pass through one slit simultaneously with that of another identical triplet through an adjacent slit, each being diffracted independently by energy-density gradients at the edges of the slits, as discussed in chapter 11, we should expect such events to result from constructive interference.

A second run, averaging for 1 second, reveals a total of perhaps 100 events, in an emerging pattern of five interference fringes. Even more remarkable, many of these events appear as sharply defined, low intensity square images, two pixels on a side (2:2 events), while others are either 1:2 events or superposed *images* of rectangles and squares, randomly offset in one direction or another. (Curiously there are no 1:3 events in this higher illumination run.) In the overlap cells the intensity is much brighter. It is predominantly in the high-probability fringe regions that such images appear. Here it becomes obvious that there must be, on occasion, more than one photon passing through each slit in the same transverse region at the same time, in order to generate such images. But this is to be expected with bosons, for sufficiently long runs. A third run, averaging for 100 seconds, clearly demonstrates the interference pattern of five fringes, with the distance between fringes corresponding to about a hundred pixel lengths. The central regions of the three principle fringes are very bright as would be expected, but there are several sharply defined black rectangular or 2:2 square areas in such regions, clearly indicating the statistical nature of the detection process and the likelihood of destructive interference. (Such interference could result from the third slit.) From the above we draw the following conclusions: (1) The low-intensity 1:3 (rectangular) events are the result of three photons on adjacent parallel trajectories passing through one of the three slits at the same time and that a second similar pair pass through the closest of the other two slits, such that the separate diffraction of each pair constructively adds at the cite of an event, (2) the low-intensity 2:2 (square) events are the result of eight photons impinging on the detector array simultaneously, as two adjacent 1:2 images, with four photons going through each slit and (3) the events with one, two or three bright pixels in a 2:2 image are the result of partial overlapping of two independent 2:2 low intensity events. While all of this remains to be substantiated with further experiments, the results of this experiment, that was designed to demonstrate the validity of the "single particle argument" appears unexpectedly to deny that validity and to support instead an argument for a strictly particle theory of multiple-photon dynamics. Since there is no evidence of a 1:1 event in these results, there is reason to believe, in answer to an earlier question, that it is highly unlikely for a laser to emit a single photon. (*While the laser was created as a means (subsystem) for creating many correlated photons, the above strongly indicates that such photons are produced in pairs, not unlike those emitted from nonlinear down-conversion crystals, as a means of minimizing entropy production per unit energy flow.*) If so we should henceforth be justified in referring to the emission of pairs or triplets, etc., of entangled photons from lasers. The foregoing also adds support to the concept of dynamic internal structure of photons, as discussed in chapter 11 and as implied in the seminal experimental work of Foster, *et al.*

While the conclusion rendered from the results of the above experimental programs strongly argues for the near-simultaneous release of more than one boson from a source in order to give rise to interference phenomena, we should perhaps turn the argument around and inquire whether there might exist a fundamental requirement for the creation of more than one such quasi-entangled bosons. In the Garcia *et al.* experiment, the creation of femtosecond pulses involves the stimulated buildup of only a few bosons, which must be sufficiently similar in energy and momentum in order that the lase process could accelerate during the femtosecond time limitation. In the Austin and Lyman work, few-boson events are achieved randomly in time by observing only those bosons which are not selected out by the neutral density filter. *Nevertheless the stimulated emission of a newly created boson, that managed to escape through the filter, must have been accompanied by the boson that stimulated its emission. The simultaneous creation of such boson pairs may be seen as the stress-induced creation of an open thermodynamic subsystem in a strongly nonlinear system, which serves thereby to minimize the flow-specific entropy production in the system and maximize the energy-density in the subsystem (boson pair).*

But none of this accounts for Young's original demonstration of interference, in which he sought to prove the wave nature of light, relying on bosons generated by the Sun as a source of light. For a convincing demonstration of interference of two or more bosons, in Young's "slip of card" experiment today, we should have to account for (1) the relatively high correlation of bosons (in energy, phase and momentum) created at the Sun's surface and (2) how the trajectories of bosons, passing as particles on each side of the card, come to be appropriately altered. We have already referred to the latter issue in chapter 9, in a manner closely relating to the bending of boson trajectories passing near to galaxies, i.e., to what is now referred to as gravitational lensing. As regards the former issue, we suggest: (a) Some form of cyclic undulation (thermal?) in the Sun's corona periodically excites electrons over small macroscopic regions (centimeters, meters?), whereupon they subsequently spontaneously decay in unison with the emission of many highly correlated bosons, on well-defined radial directed trajectories. (b) A necessary condition for such periodic emissions to give rise to observable interference is that the intensity of non-correlated bosons traveling on the same trajectory not mask that of the correlated bosons. (This of course requires in addition a means for bending the boson's trajectory at Young's slip of card, as referred to above.) Fiber optics and streak camera technology could open the door to new experimental studies of the Sun's processes of radiation. At the very least Young's initial experiment should long ago have prompted us to question the physical processes involved in diffraction and interference in that experiment and the nature of boson emission at the Sun.

Since the above relates only to diffraction and interference of bosons, there is a need to make convincing arguments accounting for similar behavior relating to fermions and in particular electrons. Any demonstration of interference with fermions nevertheless must be ascribed to minimally-entangled particles in the source, as Garcia, *et al.*, demonstrated beyond doubt: that it is necessary only that the particles (in this case electrons) have sufficiently close energy, momentum and trajectory (and therefore wavelength, phase and trajectory) to enable them to overlap constructively in space and time at a detector. (We conjecture here that the energy-momentum of two electrons cohere in spacetime as bosons (Cooper pairs?) with twice the mass (energy) of one electron, resulting in constructive interference at one location and destructive interference between fringes.) Such interference has been demonstrated in experiments with electron microscopes, as well as, more recently, with other more massive fermions in interferometer experiments[87]. We seek a first-principles theory of the diffraction and interference of particles in the following chapter.

CHAPTER 11
THE FUNDAMENTAL PARTICLES AND FORCES

Our aim now is to articulate the nature and role of "fundamental particles" and "forces," pointing toward the development of a relativistic *particle* mechanics.

INTERNAL STRUCTURE AND DYNAMICS OF THE "FUNDAMENTAL MASSIVE PARTICLES"

The general form of the quarks and leptons and their antimatter counterparts were sketched in chapters 8 and 9 as gravitational and symmetry bound photon (GSBP) composite particles, which we now broadly refer to as fermions. The extreme energy-density $\sim 10^{29} \text{kg/m}^3$ of the bound photon in the electron in our local ρ, for example, endows the electron with a highly stable rest mass of $\sim 10^{-30}$kg, occupying a disk-like space of $\sim 10^{-59} \text{m}^3$. (We assume the gravitationally bound photon to have a wave*length* of $\sim 10^{-12}$m and a width $\sim 10^{-33}$m equal to Planck's Constant. Similarly the energy-density of the lighter quarks is found to be about 10^9 kg/mm^3 in the local ρ_l of an accelerator, which, given the 10^9 increase in energy-density of the local nucleon space of quarks, would compare marginally with that of the electron and that would suggest that quarks are simply electrons in their natural realm. The muon and tau electron and their antimatter counterparts would be taken here simply as higher energy-density representations of the electron, existing naturally only in a time when the average or local energy-density of the Universe was much greater. The same arguments would apply to quarks (antiquarks) and to neutrinos (antineutrinos). We assume the latter particles to be distinguished only by their very low energy-density, perhaps three orders of magnitude less than that of the electron, with their radius of photon gyration being proportionately larger. *Thus we make no reference to three "Generations" or "families" of quarks and leptons as in present theory* (see below and chapter 13). (In so saying we draw an important distinction between what may have existed in an earlier time and energy-density and what we create in high energy experiments: the n→p transmutation leads to the creation of a positrino, as ϱ declines, yet finds no need for the creation of other neutrinos or W particles.)

Given some small momentum along the axis of gyration of the photon, the electron "disk" would move along that axis at a certain velocity v_e such that $\sqrt{(v_e^2 + v_{GSBP}^2)} = c$, where v_{GSBP} is the velocity of the gyrating photon in the electron and c is the present value of the speed of light (Fig. 11-1). But this relation could hold only if the radius r_{GSBP} of photon gyration increased inversely with a decrease in v_{GSBP} in order to conserve exact phase-closure of the photon and the angular momentum (charge) of the electron. This would result in an increase in the energy-density of the electron and a self-consistent increase in its relativistic mass. Accordingly, the electron's added kinetic energy results in an increase in its inertial and electromagnetic mass (constraint on momentum). As the electron's velocity approaches the velocity of light c, the energy required to accelerate it increases without limit, thus setting an upper limit on the speed of the electron and similarly for all fermions. In consequence, the linking of the velocity of light with the internal structure of the electron simply and straightforwardly makes the electron a relativistic *composite particle* in the context of both special and general relativity, the latter since c and $\sqrt{G}$ vary inversely as the third power of the average or local energy-density. The same holds for all leptons and quarks, though the rest mass of quarks would be more difficult to define.

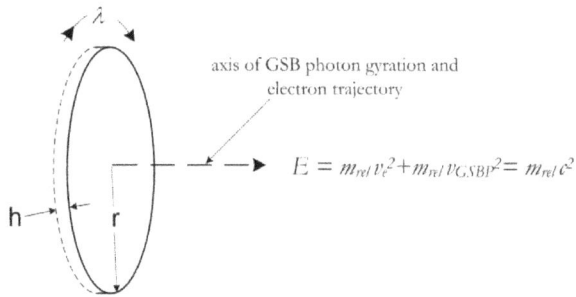

$$E = m_{rel}v_e^2 + m_{rel}v_{GSBP}^2 = m_{rel}c^2$$

Fig. 11-1 Energy-density and relativistic dynamics of the GSB photon in an electron, showing the radius r and axis — —▶ of gyration of the bound photon λ. As we show in chapter 12 (Derivation of E=mc²) the total energy of the electron E=E$_{kinetic}$+E$_{intrinsic}$ requires $v_e^2+v_{GSBP}^2=c^2$, where c is the present constant velocity of light. Consequently any positive values of v_e^2 and E$_{kinetic}$ must be exactly accounted for in $m_{rel}v_e^2+m_{rel}v_{GSBP}^2=m_{rel}c^2$ and, since the intrinsic energy of the electron $m_{rel}v_{GSBP}^2$ must be an absolute universal constant, the kinetic energy of the electron = $m_{rel}c^2$. But in our scenario c varies as ϱ^{-3} requiring only the energy-density of the electron to vary as ϱ^{-3} while conserving its intrinsic energy.

HIERARCHICAL STRUCTURE OF THE "FUNDAMENTAL PARTICLES"

In sharp contrast to current theory, the three quarks in our scenario are finite in dimension and volume, have real angular momentum (electric charge) and spin angular momentum and may be distinguished, in a nucleon, by their confinement to the X, Y or Z axes of the *local* quantum Euclidean space of a nucleon and by the value of their electric charge (See Figs. 8-1, 8-2 and 8-4a and b, relevant to P-quarks and P-neutrons). *They are bound together in a nucleon by the force of SU (3) symmetry, the electromagnetic force, the requirement that energy-momentum, angular momentum and spin be conserved in any transformation, and by the 2nd Law requiring minimum entropy production per unit energy flow (maximum energy-density) in such transformation. General relativity is inherent in that state and in the internal structure of the quarks. As constituents of a nucleon in SU (3) confinement, the quarks and their antimatter counterparts appear as first-order GSBP SU (2) composite particles.*

The charge-carrying leptons -- electrons and their earlier higher energy-density states, as well as their antimatter counterparts, positrons -- are composite particles of GSB photons in SU(2) symmetry, whose quantized radius of gyration varies inversely with the cube of the average energy-density of the Universe ρ (subject to local macro and micro variations), thereby conserving spin angular momentum. The orientation of the spin is determined by the relative helicity of the bound photons in the electrons and positrons, which move on trajectories that are locally defined by the axis of angular momentum of the gyrating photon, as depicted in Fig. 11-1. The same holds for the spin angular momentum of the quarks in the nucleus, though those axes would be the X, Y and Z axes of local nuclear Euclidean space. Like the quarks, the leptons are 1st-order GSBP SU (2) symmetry composite particles, though characterized by lower energy (energy-density) and stability. Also shown in Fig. 11-1 is the calculation of the energy-density of the electron revealing the absolute universal constancy of its intrinsic energy-density (independent of ρ or ρ_l variations), since the energy-density lost accompanying the expansion of the Universe is exactly compensated in the form of kinetic energy-density gain as c increases in the expansion. All of this applies more generally to all leptons and bosons. This is demonstrated in the many epochs of the quantum era in which space is constant and only $\sqrt{G_N}$, $\nu\varepsilon_N$, and $\sqrt{\mu_N}$ double with each increase in N as well as in the increase in the velocity and space of the GSB photons in the fermions and bosons, as N increases in the scalar era. In the latter case we see that it is only the space between the particles that expands.

The "neutral" leptons (electron-, muon- and tau-neutrinos and their "neutral" antimatter counterparts), inferred in high energy experiments, are SU(2) symmetry composite-particles comprised of GSB photons, having very low energy-momentum and spin angular momentum (charge density), but otherwise similar in structure to the charged leptons. As such, the radius of photon gyration of these particles would be several orders of magnitude greater than that of the corresponding "charged" leptons. If this is true neutrinos must have electric charge as well as mass, and this suggests that they may play a more significant role in ensuring the conservation of energy and angular momentum. More will be said on these matters in chapter 13. Likewise we defer discussion of hadron composite particles and their interaction in the nucleus of atoms to chapter 13. These "neutral" leptons should then be classed with the charged leptons, though of lower mass-energy (energy-density) and stability.

The above characterizes various kinds and energy levels of composite massive particles, the fermions. The nucleons - neutrons and protons and their antimatter counterparts - constitute second-order GSBP composite particles characterized by higher levels of mass-energy, but lower energy-density and stability in the present low energy-density of the Universe. The nucleus of atoms may then be referred to as third-order GSBP composite massive particles, and the atoms themselves as fourth-order GSBP composite particles of still greater mass, but lower energy-density. Higher orders of GSBP composite particles and structures characterize the simple molecules and the much greater mass-energy (lower energy-density and stability) of organic macromolecules. (See Table 11-2)

Fund. quanta	1st Order	2nd Order	3rd Order	4th Order	5th Order	6th Order	7th Order	8th Order	9th Order	10th Order	11th Order	12th Order	13th Order
Bosons	Fermions	Nuclei	Atoms	Small Molecules	Inorganic molecules	Organic molecules	Macro molecules	Protocells	Multicell aggregates	Social organisms	Ecosystem	Human systems	Global systems
	Quarks Leptons	Neutrons Protons	Nucleons Electrons	Diatomic molecules	Heavier atoms	Prelife molecules	Protolife molecules e.g. amino acids	Intra-cellular systems	Primitive interaction of cells	Radiation of species, Social Systems of species	Inter-dependence of species	Exploitation of local resources —devel of complex social systems	Global exploitation of resources —'systemic, technological and intellectual systems

Nearly Infinite
Energy   -------- Progressive decrease in energy-density and stability from one order to the next ----------------→
Density   -------- Progressive increase in system complexity, displacement for equilibrium ----------------------→
Range   -------- Progressive dependence at highest structural levels, on stability at the lowest levels,
                    facilitates multiple, varied and adaptable weak bonding at the highest structural levels -------→

Table 11-2

ENERGY-DENSITY ORDERING OF COMPOSITE
PARTICLES, ORGANIZED SUBSYSTEMS

As the only kind of fundamental particle, Bosons have a nearly-infinite range of energy-density. The energy-density of quarks and leptons is determined principally by the *wave*lengths of its constituent GSB photon. In turn, the energy-density of a neutron or proton is somewhat greater than that of its three constituent quarks owing to the compactness of their GSB confinement in Euclidean space. The same holds for atomic nuclei in which the spacial arrangement of one or more neutrons and protons adjusts so as to maximize the energy-density (minimize the entropy production per unit energy-density). The 4th order of composite particles, atoms, emerged as neutral, relatively high energy-density, stable (long lifetime in low energy-density environments) particles, depending on the stability of their constituent quarks and electrons. The large increase in mass-energy of high-Z atoms (with little change in volume) directly demonstrated the principle of maximal energy-density (minimal entropy production per unit of energy-density) that drives the evolution of all of these composite subsystems. Small molecules (order 4) and macromolecules (order5) form as bond lengths and angles adjust to maximally increase energy-densities. The same holds for organic macromolecules involving a much greater range of energy-densities (order 6), leading eventually to the emergence of living protocells (order 7). (It is important to note that for all such composite structures the stationary-state energy-density decreases with the expansion of the Universe and the decline of average energy-density ρ or local energy-density ρ_{local}). All of this, as implied above, assumes that the same principles apply fundamentally in the emergence and evolution of higher-level living systems, though perhaps less readily quantifiable. Stated differently, we assume that the NTI (nonequilibrium thermodynamic imperitive) serves as a fundmental physical proposition on which to base Darwinian theory.

The GSB photon in fermions is thus seen in this view to be the most elemental of all particles, defined by energy, momentum and spin angular momentum *and dynamically constrained for the life of the fermion by its own gravitational force. It is this quantized bounded-energy that sharply distinguishes the bound photon from free massless "vector spin-1 bosons" or simply "bosons."*

FUNDAMENTAL FORCES

Do vector bosons also have internal dynamic structure and a means for the precise storage of their energy, momentum and angular momentum, also deriving from the gyration of a GSB photon? Yes, but that would require a fundamental difference in the orientation of the spin angular momentum of the GSB photon. If the spin angular momentum of the gyrating photon were orthogonal to the direction of travel of the boson, that would deny virtually any assignment of mass to the boson and require the boson to travel always at, and only at, the speed of light *c*, thereby conserving energy and spin angular momentum over the life of the boson, but it would lack any physical means to account for momentum. Thus, we see that the fundamental difference is at root a difference in the internal symmetry and dynamics of fermions and bosons.

How would the bound photon travel relative to the forward motion of the composite vector boson? In fermions, as we have seen the photon defines a path, whose radius and pitch depend on the relativistic radius of photon gyration and the fermion's kinetic energy. In the case of bosons, the bound photon must gyrate in a plane co-defining the polarization of the electric vector and trajectory and *some small physical mass*, moving near the speed of light, if it can be said to have momentum. If so the gyrating photon would appear to travel like a nearly massless wheel on a straight flat track with the gyrating photon advancing a certain distance λ along the track with each exact rotation of the wheel, at the local speed of light $v_{GSBP} <$ c, defined by the energy of the photon. This of course would be possible only if the system were virtually free of inertial drag (Fig. 11-2). (We note here that it would also have to be free of inertial drag in the larger context of general relativity, i.e., in terms of epoch to epoch expansion, since otherwise energy conservation would be violated. And we note further in this analogy that the wheel's surface should be coded with information defining the sinusoidal variations of the electric and magnetic vectors, enabling a measure of the relative phase of the boson.) The larger the wavelength λ of the photon (the circumference of the wheel) the smaller is its energy, local energy-density, momentum and angular momentum. Thus, the gyrating photon leaves the speed of boson travel determined solely by ρ or ρ_l *of the emitting source* and serves in effect as a frictionless flywheel conserving the energy, momentum, angular momentum (charge) and polarization of the boson over the life of the particle, unless perturbed. The photon's spin orientation and the boson's trajectory may be perturbed by quantum gravitation and electromagnetic forces via the emission of relatively low energy vector bosons γ_1, γ_2, ... from the gyrating photon in the boson so as to result in some momentum transfer in a direction orthogonal to the bosons trajectory, via impulse-recoil, while conserving total energy E of the boson, as discussed in the section below on diffraction and interference.

But we see in Fig. 11-2, panels B and C, that the boson, in any conservative system, can travel only at a velocity v_{GSBP} marginally less than *c*, depending on the *displacement* $r_{rel} < r_o$ (r_o being the radius of GSB photon gyration in which the electron has zero potential and kinetic energy relative to the nucleus of the atom), as the energy of the emitted gyrating photon increases corresponding to the rate of decay of the electron's energy in the hydrogen atom. It is this exact *displacement* that ultimately accounts for the energy and momentum of the boson and limits the broadening of emission lines, to values far less than our present measurement capability.

A

Pitch
d

GSB photon

r_0 r_{rel}

$V_e = \dfrac{dc}{2\pi r_{rel}}$

GSB fermion (electron)

$V_{er} = \left[c^2 - V_{GSBP}^2 \right]^{\frac{1}{2}}$

where r_0 refers to m_r of electron

$r_{rel} = \dfrac{cr_0}{V_{GSBP}}$

at which

$r_{er} = 0 = R^{-1} = KE = PE$

R being the Rydberg Constant

B

c

r_0 V_{GSBP}

r_{rel} h

V_{GSBP} →

boson

V_{GSBP}

h

$V_{GSBP} < c$ →

$\lambda = \dfrac{hc}{E} = 2\pi r_{rel} = V_{GSBP}\,\tau$

τ : boson period

C

r_0 V_{GSBP}

r_{rel}

$E = hc(r_0 - r_{rel})/\lambda$

$m_{rel} = m_e h(r_0 - r_{rel})$

Boson momentum $p = m_{rel} \times GSBP$

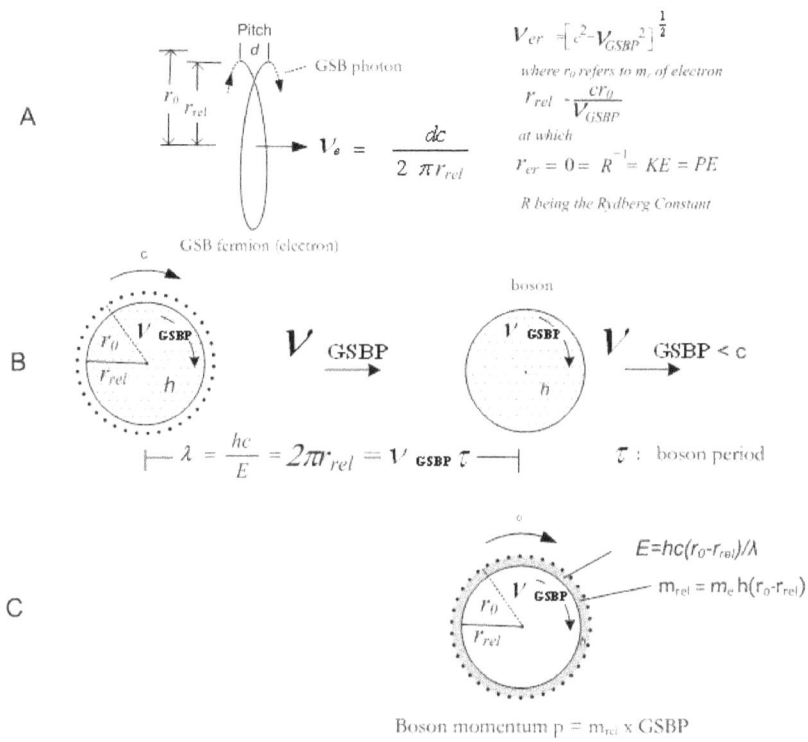

Fig. 11-2 Comparing the internal relativistic structure of fermions and bosons. Panel A depicts the GSB photon (not to scale) in an electron, traveling down a spiral trajectory, such that the velocity of the electron $v_e = dcr_{rel}/r_e = v_{GSBP}$, where d is the spiral pitch, c is the velocity of light, r_{rel} is the relativistic radius of gyration of the photon in the GSBP-electron and r_0 is that radius of gyration at which $v_e = 0 =$ the electron's kinetic and potential energy (relative to the proton in the hydrogen atom); in one gyration the GSBP-electron advances a distance d on the trajectory. From this we can obtain the period and wavelength of the photon, r_{rel} and the momentum p of the electron knowing only the rest mass of the electron and Planck's Constant h. In the conservative system of the hydrogen atom, the decay of the electron from one allowed state to another would be accompanied by the emission of a GSBP-boson depicted as a GSB photon gyrating in a plane that defines in part the trajectory of boson travel at velocity v_{GSBP} (<c) so as to conserve the mass-energy and angular momentum lost by the electron as parameters of the boson, as shown (not to scale) in panel B. In this comparison the velocity of the electron is largely restricted to a small fraction of the speed of light, while any boson emitted by the decay of the electron is constrained to a small

163

discrete range of velocities less than but very near the speed of light. In panel C we see that the momentum p of such bosons is given by the product of a very small relativistic mass $m_{rel} = hm_e (r_0 - r_{rel})$ times the boson's velocity v_{GSBP}, with h defining the angular momentum of the boson and setting a limit on its volume and energy density. This is depicted as the transfer of h mass, from the area within to the region bounded by r_{rel} (to that of the wheel's rim), thereby greatly increasing the energy-density and stability of the boson. Finally we see a beautiful symmetry revealed in the inverse relativistic relations governing the dynamics of the electrons and bosons.

In so characterizing the radiation of bosons we highlight Maxwell's postulate of *displacement current* in deriving his field equations and *we understand that the equations are wholly consistent with our strictly-particle theory. And we understand also that bosons must carry a small mass m consistent with their marginally lower speed of radiation (dependent on their energy or energy-density) such that their momentum p = m·v_{GSBP} = h/λ, all of which holds in general relativity.*

The boson is a precise record of the state transition of the fermion that emitted it. It is a record of the electron's transition in energy-momentum, angular momentum, spin angular momentum (charge) and local energy-density, leading to the emission of the boson, self-consistent with the requirements on conservation of these properties. It is defined also by its phase in spin relative to that of any other similar boson(s) in spacetime relation. In addition it is defined, though less precisely, by its polarization, relative to that of other bosons on a similar trajectory. As such it is a discrete, strictly-localized, quantum of electromagnetic energy – a vector boson subject to gauge invariance - which briefly modulates the local EM energy-density of the vacuum (dissipation free) as it passes any point (free of massive particles) on its trajectory. In passing such point, it carries momentum h/λ and momentum density $h/\pi a_0^2 \lambda^2$, where a_0 is the radius of the ground state of the atom that emitted the boson.

The dynamics of a boson emitted from a hydrogen atom, initially in the $2S_{1/2}$ metastable state are as follows: The angular momentum of the electron, is taken to be zero prior to a fluctuation sufficient to allow the electron to climb over or tunnel through the energy hill separating it from the $2P_{1/2}$ state. Thus the decay to the $2P_{1/2}$ state is characterized by an initial stage of angular momentum acceleration and decline back to zero enabling its escape from the metastable energy well. This initiates the cycle

164

of boson emission, with orbital closing contributing to the final freeing of the boson from the atom, as depicted in Fig. 14-1. In so doing it imparts a small recoil momentum to the nucleus, which then immediately allows the electron to decay (spiral down) to the ground state. Quickly the angular momentum accelerates and declines in the opposite direction, as the electron approaches the ground state. In the latter process, the electron makes one complete orbit of the proton, while the GSB photon in the electron completes 137 gyrations, such that both cyclic closures occur simultaneously in a nearly dissipation-free process. This results in the emission of a boson at nearly the speed of light, carrying detailed information about the particulars of its creation in the atom.

If we could anticipate the travel of a single boson on a particular trajectory and plane of polarization and measure its electric vector in passing, the detector would reveal the absorption of the boson's energy-momentum as a sinusoidal variation in the electric potential over a period $\tau = \lambda/c$. Such radiation would be characteristic of all bosons emitted from conservative systems, in which energy-momentum is conserved, but it would certainly not be characteristic of the propagation of a wave packet or a wave spreading throughout all space. *In this view there is no need for wave-particle duality, yet there remains a local EM field/particle duality in the sense of countable numbers of* bosons, *having the same or similar trajectory, λ, phase, spin and polarization, at any position affording measurement in spacetime. This is in fact what we do when we measure the field strength of a radio wave at a particular point in space; the result is the vector summation of all such* bosons *at that point in spacetime. The invariance of the photon's energy E in the* boson *and momentum $m_{rel}v_{GSBP}$, over the life of the* boson, *serves as a universal clock and measure of phase correlation among similar* bosons *in the same energy-density region or of their relative velocities in different regions.*

As shown in Fig. 12-2, two phase-coherent bosons (particles characterized by *local* EM fields), on sufficiently close parallel trajectories, appear, each to the other, as oppositely charged particles tending to propel the two together via recoil impulse mediation of the Coulomb force. This could lead probabalistically to the merging of the two bosons into one of total energy-momentum equal to twice that of one boson. The reverse would account for few if any bosons in regions where the bosons are out of phase. It is this phase-dependent Coulomb force that accounts for the intense illumination, in interference experiments, in one locale of a screen and little response in an adjacent region. (See below for a first-principles explanation of Fraunhofer single-slit, far-field interference.)

Only in such special - accidental or contrived - circumstances do bosons directly interact. In our scenario, bosons are emitted from leptons and quarks for the principle purpose of moving the massive particles that emitted them either away or toward one another, depending on whether their charges are similar or opposite in sign; and they do so by timely, direct impulse-recoil reaction, thereby minimizing or maximizing the *local* electric charge density (angular momentum density). Such scattering action is in response to differential charge density gradients at the surfaces of the bound photons, prompting the emission of bosons toward a similarly charged lepton or quark, thereby recoil propelling them apart, or by the opposite process, in the case of oppositely-charged leptons or quarks, leading to their mutual attraction. The latter behaves similarly to that of the relatively much weaker gravitational charge-density gradients, as illustrated in Fig. 11-3, in our accounting for the Casimir force. (Since the energy-momentum of the boson increases as the sixth power of the energy-density within the plate relative to that in the vacuum, we see that the force must increase more rapidly than the second power of the distance between the plates, and indeed it must increase as the fourth power of that distance. c and G, taken as absolute constants and gravitons as agents of attractive force between massive bodies, could never offer a viable alternative explanation for the Casimir force. Most importantly we see that classical gravitational theory fails at the quantum scale.) The assumption in all of this – whether gravitational or electromagnetic force – is that it is the energy-density gradient that engenders the release of energy-momentum in the form of low energy bosons. For similarly charged particles the scattering would be repulsive. The greater the differential charge-density gradients, the greater would be the rate and/or energy of bosons emitted and the resulting attractive or repulsive force.

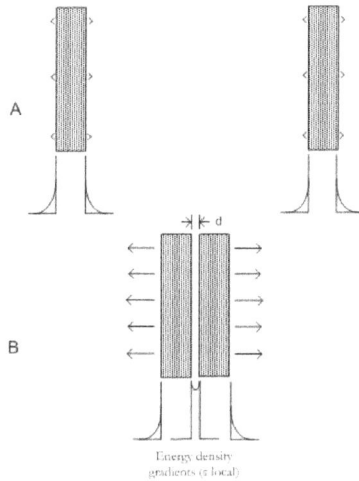

Fig. 11-3 Energy density gradients at inner and outer surfaces of Casimir plates at (A) when the plates are far apart and (B) close together relative to the plate thickness. The gravitational force F_G moving the plates together derives from differential boson emission from the surfaces of the plates and resultant impulse recoil, which acts essentially as a compressive force when plates are far apart (A), but more and more as an attractive force, as the distance d between the plates decreases (B). Since $F_G \propto m\sqrt{G}$, $\sqrt{G}$ varies as ρ_l^{-3} and m varies as ρ_l^{-1}, F_G varies as d^{-4} in agreement with experiments.

Since all massive particles or bodies are composites of quarks and leptons or agglomerates of such composite particles, either with or without net charge (for example neutrons, non-ionized atoms or stars), bosons may be emitted in a direction away from such particles or bodies in response to mass energy-density gradients so as to propel them toward one another in order to increase the local mass energy-density. In this case the electric, magnetic and phase properties of the bosons play no role in mediating the attractive gravitational force, except at very close range as in the extreme energy-density of the nucleus of atoms.

Is there any evidence to substantiate the above? There is, in a *Science* report by E. Goulielmakis[88] *et al.*, Direct Measurement of Light Waves, in which they "directly observed how the field built up and disappeared in a short, few-cycle pulse of visible laser light by probing the variation of the field strength with a 250-attosecond electron burst." The interaction of the electric field of the pulse with the correlated burst of electrons resulted in field-induced momentum change of the electrons, as measured by a detector. The authors' terminology implies that *photons* (we would say bosons), making up the pulse, oscillate as they travel at the speed of light and thereby perpetuates the notion of waves. But there is no physical mechanism to support this idea and indeed "the electron probe needs to be localized not only in time to a tiny fraction of the wave period…but also in space to a tiny fraction of the wavelength…of light to be measured." From this it is evident that the probe scans the electric potential of the pulse from a fixed position, as the pulse (a fixed unchanging pattern) moves past it at the speed of light. This is equivalent to saying that a photon (boson) is a *fixed unchanging pattern of local electric field variation, traveling at the speed of light*. Of course the pattern displayed was not that of a single boson of light, given the laser light source. In our interpretation, the unchanging pattern is the constant cycling of the bound *photon* in the *boson*. It is this internal dynamic structure that is of course lacking in the authors' account. (A simple simulation of the boson would appear as a projected two-dimensional pattern of length λ and varying height on a screen (the latter representing the electric charge displacement from a baseline), moving as a single sinusoid across the screen, at some slow fixed velocity. Covering the field of view with a narrow slit, we would see the *apparent* oscillation of displacement from the baseline, over the length λ and period τ.)

In a *Physical Review Letter* entitled: Quantum State Reduction and Conditional Time Evolution of Wave-Particle Correlations in Cavity QED, G.T. Foster[89] *et al.* "report measurements in a cavity QED of a wave-particle correlation function which records the conditional time evolution of the field of a fraction of a photon. Detection of a photon prepares a state of well-defined phase that evolves back to equilibrium via a damped vacuum Rabi oscillation." In essence they use photons (bosons) from a laser to excite rubidium atoms traversing a high finesse Fabry Perot cavity. Part of the signal escaping the cavity creates a photodetection at a photodiode, which triggers a digital oscilloscope to record in coincidence the photocurrent from a Balanced Homodyne Detector (BHD). "The notable feature of our measurement is that according to the reduction postulate, the detection of a photon (particle aspect of light) conditionally selects a field amplitude (wave aspect of light) oscillating in time with an anomalous phase. Our experiment catches the field fluctuations as they occur, and thus we have been able to observe the anomalously phased oscillation directly. … The recorded correlation function is nonclassical and provides an efficiency independent path to the spectrum of squeezing. Nonclassicality is observed even when the intensity fluctuations are classical." While the title refers to wave-particle correlations, it is the correlation of the strictly local electric field of the particle escaping from the cavity that is measured coincident with the free photon (particle) triggering the recording of the local field fluctuations. Since the photons involved have ~40 MHz frequency, the electronics can easily resolve sub-40 MHz frequencies in the fluctuations, i.e., "a fraction of a photon." The resulting nonclassical fluctuations may be interpreted as revealing the oscillations of the strictly local field of a boson (particle) as it decays in the BHD. If so this experiment, along with the above *Science* report and others relating to photon interference phenomena (chapter 15), contributes greatly to freeing us from the shackles of our two-hundred year old commitment to wave-particle duality and its formalization in superposition theory.

The above should prompt us to ask: How is it that electric *charge* (via the small displacement of many electrons) must propagate a distance of about 1m in an antenna in order to radiate bosons of the same frequency and wavelength as those emitted by an electron in a Rb atom, the dimension of which is about ten orders of magnitude smaller than that of the antenna? This question seems seldom if ever to have been asked, with physicists preferring (or having no recourse except to accept) simply that a "quantum jump" occurs. We will pursue this question in chapter 14 and offer a physical model of the dynamics of an electron decaying from one energy level to another in the process of emitting a boson.

But the above assumed, in keeping with our fundamental particle concept, that *all* such radiation, including that from macroscopic antennae, is in the form of quanta. This assumption could only be valid if what we perceive in the classical case consists in actuality of a very large number of bosons emitted in slightly different directions and frequencies, associated with nonclassical as well as classical fluctuations about the carrier frequency and any modulation of it. What Goulielmakis *et al.* and Foster *et al.* succeeded in doing in their experiments was the display of the pattern of local field-strength variation characterizing a photon [boson] as it gives up its energy-momentum to a detector in a few-cycle stream of photons [bosons], in the one instance, and in the other a fraction of a photon [boson], associated with a wave packet.

Returning to our discussion of the structure of fundamental particles, we see then that gravitationally-bound *photons are the only proper elemental particles in the Universe, with all the energy in the Universe represented either in the form of gravito-symmetry-bound photons (massive particles) or free massless* bosons. Today, disregarding the current great concern to discover the nature of "dark energy" and "dark matter," all but about one part in a thousand of the estimated energy of the Universe is accounted for in the massive particles – GSBP composites of quarks and leptons - with the remaining energy in the form of free bosons. While much of the forgoing discussion focuses on bosons as composite particles, we note here that neutral massive particles, for example atoms or Bose-Einstein condensates, may behave as bosons, which warrants special attention in our theoretical inquiry.

In QCD gravity plays no role in the nucleus, believed to be orders of magnitude too weak, given that G is taken as an absolute constant (see below). The "color force," is said to be mediated by "gluons," which act in binding together in SU-3 symmetry, the triplet of quarks that make up each nucleon. But such postulates leave us with no intuitive understanding or physical model in relation, for example, to the gravitational force, despite the fact that they both act indiscriminately and universally as attractive forces. Moreover, this model makes no connection with general relativity. The gluons are dependent for their existence on the *physical vacuum* and are therefore problematical, as we have argued. But assuming their existence, how are the gluons *directed physically*, one to another, to mediate the binding force so as to preserve "color neutrality of the nucleon?" The color force, mediated by gluons is also said to bind a quark and an antiquark of a given color into mesons, thereby preserving a kind of gray-scale quark neutrality in mesons, as the embodiment of the "strong force" in QCD. Are the mesons real or virtual? Mesons have relatively large mass, yet they are bosons that somehow generate an attractive force in their exchange between nucleons. Do massless bosons really acquire mass with the help of Higgs bosons? How are they directed toward, and exchanged with, a nearby nucleon? How does such interaction occur as a residual force, a kind of van der Waals force? To be sure, these questions were asked in the making of QCD, but the answers again appear as contrived and limited in the algorithm, especially in view of the very limited extent to which renormalization theory, so essential in QED, has been applicable to the extreme environment of the nucleus.

In QCD massive W and Z bosons are assumed to play key roles as *force* particles in "weak" and "electroweak" theory. In our scenario a free neutron does not decay "spontaneously", it is forced to decay, or rather its (local) energy-density is forced to decay (decrease) by fluctuations driven ultimately by the expansion of the Universe. There is thus a profound difference in the two views: the one makes no direct connection with the evolution of ρ, the average energy-density of the Universe, or ρ_{local}, or with gravity (the field equations) or the *2nd Law*, yet postulates an exact detailed accounting of the transformations, while the other derives from the above connections and presumes only to account for the products of transformations in probabilistic terms of self-consistent fundamental assumptions. In the latter view the local energy-density of a free neutron or proton determines the probability of its transformation in one direction or the other, in the same way as the local temperature may determine the direction of a chemical reaction. Indeed the average energy-density of a system or subsystem and its temperature are two sides of the same coin. There is nothing to be gained in QCD or in our schema by postulating W and Z particles *as the mediators* of the transformations, since all we can actually observe are the products of the reactions. The issue is not about the possible reality of the W and Z particles, but about the role they play: in the one case as unique participants in the transformations and in the other as facilitators, bringing energy-momentum to the system or freeing it for other functions. As a facilitator *a decrease or increase in the local energy-density acts, in conjunction with the principle of minimum flow-specific entropy production, as a forcing function.*

The electromagnetic force in QED is mediated by "virtual photons" (virtual bosons) whose energy-momentum, like that of mesons, is on no-interest, short-term loan from the "physical vacuum," in strict accord with the Heisenberg indeterminacy relations. How do the charged particles emit such photons (or enable their reification) and orient their trajectories in frequent interactions with one another, in precise accord with the requirements of the Coulomb force, relativity and the conservation of energy-momentum and charge? In retrospect, and from the perspective of those who were seeking only a working algorithm, many of the above questions would no doubt have appeared unproductive if not foolish. Still, it is remarkable how that algorithm came about and how well it has succeeded in an initial quantum theory, given that there was relatively little empirical evidence at the time to suggest a different course.

The following offers a very different understanding of the nature and dynamics of the fundamental forces acting on all massive particles or aggregates and with regard to the above questions.

In our scenario it is important first to underscore several fundamental distinctions in the way that forces are perceived, created, directed and adjusted. Since the advent of Einstein's field equations, the force of gravity has been perceived as the mass-energy warping of spacetime, though we still felt the need for "gravitons," exchanging *between* particles and bodies, exerting an *attractive* or binding force. Here we extend the warping paradigm to apply in the extreme local mass-energy-density of the orbiting GSB photons in quarks, requiring them to travel in infinitesimal circular orbits, as depicted in Fig. 11-1. But, as we see in chapter 8, the force of gravity is too small to account for the orbiting of the three quarks around the local X,Y and Z Euclidean axes (analogous to the orbiting of electrons about the nucleus of atoms in our particle scenario). In this and all higher-dimensional levels in the nucleus and atoms, it is the local electromagnetic force that dominates. (In quantum field theory the quarks are bound in nucleons by the "color force".) Similarly in our model the "strong force" now appears simply as two quarks, in different nuclei, constrained to move and orient themselves and their individual nuclei in accord with local electromagnetic energy-density adjustments, a residual van der Waals kind of force required ultimately and self-consistently by the NTI.

However the above provides no physical or intuitive model to account for responses to perturbations or fluctuations, which would disturb them from their stable states. The creation of such a model requires another major change in how we should perceive such responses, in order to dispel the notion of "virtual particles" or "all-pervasive fields" acting between fermions to mutually attract or repel them, as in QFT/QED or "gravitons" acting to bind both neutral and charged particles or bodies together in gravitational theory. In its place all such responses would act via the emission of bosons, proportional to differential charge-coupled density-gradients ($\sqrt{\varepsilon}\nabla$), in the one case, or mass-energy-density-gradients ($\sqrt{G}\nabla$), in the other, at the surfaces of each interacting particle or body, giving rise to a timely change in the energy-momentum of the particle by impulse-recoil. *The quantum gravitational force would thus be manifested as the resultants of all such density-gradients on a particle or body, having characteristic internal energy-density, and the emission of* bosons *in a direction that would tend to maximize those resultants.* In a sea of many such massive interacting bodies, each would emit bosons propelling them individually by impulse-recoil toward one another, in accordance with the field equations. *Since the resultants imply the difference between impulses applied to opposite sides of a particle or aggregate, there is also an implied compressive force on the particle and on the aggregate, as gravitational theory requires. The same argument applies to the EM force in the case of oppositely charged particles. That is the EM force would be seen as a impulse-recoil force brought about by the emission of* bosons *from the spin surface of each such particle so as to maximize or minimize the resultant of electric energy-density-gradients* $\sqrt{\varepsilon}\nabla$ *at each charged particle or agglomerate, thus minimizing or maximizing their spatial separation via impulse-recoil, depending on whether the particles have the opposite or same charge.* In the case of particles of like charge, the resulting impulses act to maximize the separation of two such particles or, in a "sea of electrons" for example, to minimize the average electric energy-density of the sea. *It is important to note that all such forces act locally, incrementally and immediately and make no reference to dispersed waves or fields. As we have said they are in essence gauge forces.*

The forces of symmetry that conserve gravitational momentum, angular momentum and charge, are mediated, by the emission of bosons. Those that act to conserve EM energy-momentum, angular momentum and charge are likewise mediated by the emission of bosons, as are those that conserve momentum, angular momentum and quark electric charge in the SU(3) symmetry EM force, binding quarks in a nucleon and between quarks in nuclei, in the 3×3 degrees of freedom in local quantum Euclidean space. In all cases these responses to fluctuations are orchestrated everywhere locally by the system in spacetime as if via the sampling of gradients at the surfaces of such particles, agglomerates or bodies. Such timely corrections would evolve statistically (cf., processes of thermodynamic selection), i.e., as the result of many small adjustments of the system, at the speed of light, in spacetime.

The above is most clearly evident in regard to the acceleration of an electron from its rest mass state. In that state the electron has radius r_0, the radius of gyration of its GSB photon) and very small thickness Δd (orthogonal to the plane of gyration), determined by the uncertainty relation $\Delta p \Delta d \geq \hbar$. In the presence of a moderate EM field, the resultant of EM energy-density gradients at the surfaces of the "electron disk" ("GSBP disk") and the acceleration of the electron (disk) would both be nil initially. Both would increase gradually as the electron momentum increased, so long as the resultant of the local energy-density gradients was non-vanishing. (Similar arguments may be made in regard to gradients on each surface of "quark disks.") This would occur in response to the electron's emission of bosons in a direction opposing the applied EM field, thereby accelerating it in the direction of the field through impulse-recoil. (This is the electromagnetic equivalent of gravitational inertia.) *Suffice it to say here that the behavior of particles or bodies, acted upon by gravitational or electromagnetic fields in classical or quantum theory, can be accounted for in terms of energy-density gradients and impulse-recoil theory and in a way that is consistent with the demands of general and special relativity.* There will be further discussion on these matters in chapter 14.

But the above also suggests that something similar characterizes the nucleus of the hydrogen atom. The quarks (structurally resembling the electron, except for their mutual association in quantum Euclidean space) exist in stable orbital EM ground states about their respective quantum X, Y and Z local Euclidean axes. Just as calculations of the ground state of the hydrogen atom requires various corrections, including the cancellation of the proton and electron spins, so the ground state of a nucleon requires similar corrections and, much more significantly, the correction for net angular momentum of the quarks in the extreme energy-density of the proton. *Whether in the case of gravitationally bound macro or quantum systems (e.g., galaxies, planetary systems or atomic or nuclear systems) or systems subject to scattering of particles or bodies, momentum gradients and angular momentum gradients are involved.* In the case of a planet in a stable orbit around the Sun or the electron orbiting the proton in the hydrogen atom in the ground state, the momentum gradients are almost exactly zero, except for fluctuations. However the momentum gradients for scattered bodies or particles may only be minimized or maximized by the emission of vector bosons. *Thus, in both classical and quantum conservative systems, there is only one kind of local*

174

gradient (momentum) acting on massive bodies or particles and one kind of mediating response acting to minimize or null the gradient. In this scenario, the "gravitational force" (response to mass-energy-density-gradients) acts to minimize the potential energy of a planet, orbiting a star, while maximizing its kinetic energy in the form of orbital angular momentum, resulting in a nearly exact null momentum gradient in the path of the planet and null angular-momentum-gradient for the system. The same holds for the null momentum gradients characterizing the orbits of GSB photons (composites of quarks) in planes orthogonal to the axes of nucleon Euclidean space and the orbiting of quarks in SU(3) symmetry about those axes (the "color force"), all in response to energy-density and angular momentum (spin) gradients. And the same holds also for binding nucleons together in the nucleus of atoms in SU(3) symmetry in response to energy-density and spin gradients (the "strong force), limited only by the Fermi statistics forbidding multiple same-state Fermions occupying the same space. The "electromagnetic force" acts in atoms in the same way, in response to the angular momentum (spin) gradient (with virtually zero mass-energy-density gradient playing a role), to require the electron in the hydrogen atom to follow an orbital path of zero momentum gradient in the ground state.

All forces acting on a massive particle, or aggregates thereof, are responses to local gradients in the spacetime position of the particle or aggregate and are therefore gauge responses. That is, they are timely local corrections that take account of global spacetime variations in the structure of the Universe. In the case of uncharged particles or aggregates, the responses are to local mass-energy-density gradients.

As argued above, the "weak force" appears a misnomer. The transformations associated with the "weak force" may be most simply explained in the context of the local mass-energy-density of a nucleon. The decline of that energy-density in the environment of a neutron would favor its transformation into a proton, as discussed in chapter 9, accompanied by the emission of a high-energy, left-helicity boson. As the boson (traveling at the very low velocity in the high energy-density quark environment) leaves that space, its energy, momentum, and polarization are abruptly conserved in the relatively low-density environment of an accelerator's detector in the form of a massive particle W$^-$. (The low velocity and high momentum of the boson are directly conserved in the low velocity and high momentum of a W massive particle, and the left-

175

helicity spin of the boson gives the W its negative charge.) The W⁻ rapidly decays into a 'muon' (a high energy-density state of the electron), which then decays into an electron and an antielectron neutrino, leaving the nucleon in a lower energy state and a relatively free electron. The reverse transformation, of a proton into a neutron, can be made to occur only by sufficiently increasing the energy-density environment of the proton. That is accomplished, either in nature or in an accelerator, by the interaction of a positive pion π^+ (a massless right-helicity boson, having low-velocity c and high-momentum, seen in a detector as a massive particle with positive charge) and a proton in the nucleus. In our scenario that interaction simply elevates the local energy-density sufficiently to favor the transformation of a proton into a neutron, with the excess energy and momentum leaving in the form of a right-helicity boson, seen in the detector of the accelerator as the massive boson W^+, as described above. The W^+ decays to an antimuon (a high energy-density state positron) and an electron neutrino, followed by the decay of the antimuon into a positron which then annihilates with an electron to yield only radiant energy and a neutrino.

The above makes no call for Higgs particles to give mass to the W or π mesons or a *force* (weak or otherwise) to bring about the transformations. The transformations are dependent not on the *force* of high-energy *massive* bosons, but on the capability of a high-energy boson to carry away excess energy-momentum from a nucleon, in the one case, or raise the local energy-density in the other, sufficient to bring about the transformations. In the reverse process, the π^+ serves only as a means to turn the clock back locally to a time in the Universe when the average energy-density was much higher. This is readily apparent since the reverse transformation in a free neutron in the laboratory occurs unaided, with a *massive* boson emerging to take away excess energy and spin. Given the vast amount of equipment and physical and intellectual energy required to effect the reverse transmutation in an accelerator experiment, it is clear that such process must be accompanied by an enormous excess of entropy production, as the *2nd Law* requires.

The electromagnetic force, like that of gravity, has played a fundamental role in the origin and evolution of the Universe. These are the only forces that are mediated by a relativistic particle, the boson, with specific measurable properties (energy, momentum, spin angular momentum, propagation speed and polarization) as discussed below, though we can only obtain a limited knowledge of them without destroying them. The gyrating photon, which embodies the quarks and leptons may emit a boson in response to a minimal local energy-density gradient set by the indeterminacy relations and the wavelength of the photon. The electromagnetic force acts either to attract or repel charged massive particles in the nucleus and in the atom, depending on the orientation of angular momentum of the bound photons in the quarks and electrons. The attraction between a neutron and a proton in the nucleus derives from opposite spins in a quark in the neutron and in one of the quarks in a proton, resulting in a residual attractive force, while the like spins of two quarks in protons in a nucleus acts to repel them. Similarly, the opposite spin of the quarks in a proton in an atom and that of an electron act attractively, while the same spins of two or more electrons in the atom act to repel one another.

The above argues for a strictly particle representation of the organization and dynamics of the nucleus of atoms, in which the values of the parameters c and $\sqrt{G}$ vary as ρ^3 and the mass of particles varies as ρ^3 over three hierarchical energy-density ranges, as depicted in Table 11-2. The regions of greatest energy-density and stability are occupied by quarks, whose "rest mass" are taken to be about four times m_e, the rest mass of the electron, occupying a volume of $\sim 10^{-6}$ that of the electron and making its energy-density $\sim 10^9$ greater than that of the electron, or $\sim 10^{-21} \text{kg/m}^3$. Whereas current constant c and G theory asserts that the total energies of the quarks are $\sim 10^9 \times m_e$, *our assumption here is that quarks are seen simply as electrons*, whose mass in the region of very high energy-density increased by a factor 10^3, along with the reduced radius of GSB photon gyration and volume decrease by a factor 10^{-6}. (Under earlier conditions of greater local energy-density, the electrons would have had still greater energy-density.) The next to highest level is that of protons, having about the same mass as the quarks but occupying a relatively large space. With the decrease of local energy-density, such regions are characterized by an increasing negative energy-density gradient. It is this gradient that favors a spin reversal in d-quarks, transmuting them into +1/6 u-quarks each and the neutron into a proton of +1 charge, along with the emission of a W⁻ boson, in keeping with the principle of minimum flow-specific entropy production. Constituting the nucleus of the hydrogen atom, the proton defines a region of space in which a large positive gradient would activate the emission of bosons, opposing by impulse-recoil, any fluctuation or external force tending to remove a quark from the proton or neutron (in the case of deuterium). And it is this gradient that accounts for the phenomena of "confinement" of quarks in a nucleon and the "asymptotic freedom" of quarks within and between quarks in nucleons in which the minimal ground-state energy (maximal energy-density) has relatively low curvature. Beyond this range the energy-density of multiple nuclei increase stepwise, as more and more protons and neutrons are enabled to bond tightly with little increase in the size of the nucleus. The boundary of this space is characterized by a very large negative energy-density gradient, separating the nucleus from the electrons in atoms and ultimately the extremely low average mass-energy-density now existing in the Universe. As before, this results in a measure of asymptotic freedom in the dynamics of the nuclei and especially serves to disperse the protons, tending to repel one another, while yet binding them in a stable nucleus.

In current theory (the Standard Model), the strength of the gravitational force was said to equal that of the electromagnetic force in the earliest time of the Universe, in which the temperature $T \geq 10^{30}K$, and declined rapidly thereafter. If so, did the force of gravity really decline, orders of magnitude, relative to the Coulomb force from an earlier time to the present in which $T \sim 3K$? Again, the problem traces back to a very different beginning and evolution of the Universe in our scenario (as compared with the Standard Model), in which gravity plays the dominant role in the creation of all the GSBP quarks and antiquarks in the quiescent Quantum Era, and the EM force governs in binding those quanta in a three-quark Euclidean space. The *distances* over which gravity acts in the creation of the quarks are three orders of magnitude smaller (energy-densities greater) in such bound systems than is the case for the EM force binding the quarks into a local Euclidean space. (The same holds today for the GSBP electron orbiting the proton in the hydrogen atom.) *That is, the two forces never act or dominate in the same local space or energy-density. Neither the magnitude nor relative strength of these forces vary as the average energy-density of the Universe declines with the particle number expansion in the Quantum Quiescent Era or physical expansion in the Scalar Era. Simply stated, in all evolutionary time, the GSB photons, constituting the quarks and antiquarks, are bound in an energy-density space six orders of magnitude greater than that binding the quarks and antiquarks into their Euclidean space. Only the energy-density of the particles declines, as their mass-energy is conserved.*

Why are there six orders of magnitude difference in these energy-density spaces and how can we compute the radius of photon gyration in the up and down quarks in the nucleus of the hydrogen atom, as we do (chapter 14) for the electron? And how might we estimate the "rest-mass" of those quarks? The answer to the first question is: *the radius of photon gyration must be smaller than the radius of quarks orbiting their Euclidean axes by the factor 137 (the inverse fine structure constant α) making the energy-density ratios differ by a factor α^3, $\sim 2 \times 10^6$.*

We must approach the answer to the second question indirectly. Since we cannot assign a "rest mass" value to these quarks, as is possible in the case of the electron, we assume the rest-mass of the proton to be associated largely with the ground state kinetic energies of the three quarks (nearly equal) in their orbits of the Euclidean axes. That is, the kinetic energy of each quark is approximately equal to one-third of the rest-mass of the proton and we take their nearly-equal orbital radii to be the present value of r_q (10^{-18}m) in Table 9-1. The radius of photon gyration in the GSBP quarks should then be $\sim 10^{-18}/137 \cong 7 \times 10^{-21}$m.

Lastly, we can calculate the "rest-mass" of a quark (in analogy with that of the electron) as the proton nucleon mass in MeV/c^2 divided by 3 (number of quarks) divided by 137 (α) $\cong 4.5$MeV/c^2, only about four times the rest-mass of the electron. This is in close agreement with recently reported lattice gauge theory simulations of the u and d masses (a QCD 20-year study).[90]

Just as the GSBP in the electron represented a radical departure from current theory in modeling the hydrogen atom, so the GSBP quark greatly extends and simplifies the modeling of nuclear structure and dynamics.

Particles Diffraction, Interference From First-Principles

There is nothing that more uniquely defines and separates the quantum world from common sense reasoning than the phenomena of diffraction and interference. In order that particle theory may convincingly account for such phenomena, it is necessary to find rational answers to these questions: What forces are capable of altering the trajectories of bosons, fermions or neutral particles and bodies? How are such forces created and implemented, consistent with the requirements of relativity and the conservation of energy, momentum and angular momentum? How do we physically distinguish between photons, bosons and fermions? What adjusts the relative phase of two or more bosons, along with trajectory-bending, to give rise to the phase correlation and anti-correlation that characterizes the phenomena of interference? Answers to some of these questions are provided in chapter 12. Here, we focus on the last question and show that the Coulomb force acts in conjunction with gravity to join or separate two or more GSBP nearly identical bosons (particles) traveling on closely extended trajectories, according to their relative phase, to account for their bunching and antibunching at detectors in Fraunhofer single-slit, far-field interference. Similar action accounts for high-resolution, near-field interference in double- or multiple-slit interference phenomena.

In figure 11-4 we show a pair of phase-correlated bosons passing (vertically in time) through a slit, whose internal energy-density ρ_I is several times that of the vacuum, causing their trajectories to bend toward the walls of the slit via impulse-recoil action. The energy-density would decline exponentially with increasing distance from the walls (edges of the narrow opening in the screen), with the values of c and $\sqrt{G}$ increasing accordingly in the opening. The effect of the energy-density gradients in that space would be to cause the bosons, passing near the edges of the slit, to travel at a slightly reduced velocity, thereby decreasing their forward momentum, while acquiring a component of momentum toward a wall, i.e., in the direction of increasing energy-density. In the process the boson's total momentum would be conserved. All of this follows from our postulates regarding the inverse variations of c and $\sqrt{G}$ with ρ_I. The extent of this change of momentum (angular deviation in trajectory) would depend on the energy-density of the screen (e.g. silicon wafer), the distance of the boson from the edge of the wafer, the thickness of the wafer, the original trajectory of the boson relative to the plane of the wafer's edge and the value of λ of the boson, relative to the thickness of the wafer. (Note that similar diffraction phenomena occur with low energy radio "waves" — long wavelength bosons — bending around the

180

earth's surface and with micro*wave* bosons bending around buildings. All of such bending would now be seen as the gravitational scattering of light).

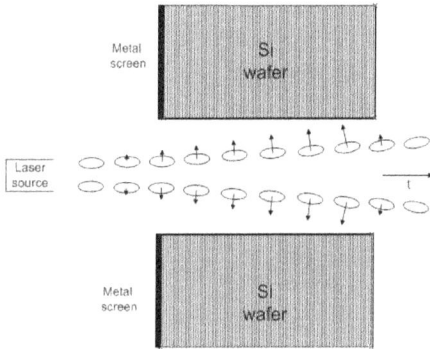

Fig. 11-4 Showing the gravitational force (not to scale) bending the trajectories of a pair of bosons toward the walls of a slit in a silicon wafer optical screen, as they travel through the slit at the speed of light.

Conventional theory argued that such diffraction patterns could *only* be accounted for in terms of Huygens' wavelet theory, though it is problematic on two fundamental issues: How does the original wave from the source give rise to two or more daughter wavelets of the same wavelength (two or more times the total energy) and what instructs the waves just so to conform to Huygens' wave theory? These are fundamental energy-momentum conservation issues. But any strictly particle theory would be equally problematic in a world in which c and G are absolute universal constants in classical interpretation. Here, we argue that it is the Coulomb force that acts between close trajectory GSB photons in the bosons to bunch the bosons together (in-phase) or spread them apart (out of phase) in their relatively long travel toward a detector.

To make this clear, we show in Fig. 11-5B a boson pair, in phase in the central region of the slit, bunching together, while the gravitational energy-density at the walls of the slit contribute to a reduced forward velocity and a bending of the trajectory of bosons to result in low energy-density regions in the far field of the detector. The inner bosons will experience little, or no gravitational bending of their trajectories and only small reduction in their velocity, while their close trajectories and phase will gradually bunch them together, mediated by the Coulomb force. In adjacent regions, where the gravitational bending is somewhat greater and the boson velocities lower, boson antibunching results in relatively few detected events in the far field, with the missing bosons contributing to, or extending the range of in-phase bunching. And those bosons passing closest to the edges of the slit will be bent most and contribute simply to the background gravitational scattering of bosons. The statistical result of many such bosons over time would reveal a Fraunhofer far-field, single-slit interference pattern. Ultimately, these actions derive from the requirement on conservation of angular momentum of the involved GSB photons in the bosons and even more fundamentally from the requirement of the Nonequilibrium Thermodynamic Imperative in minimizing local entropy production. Is this scenario sound? Could impulse-recoil mediation of the electromagnetic and gravitational forces sufficient to support the theory?

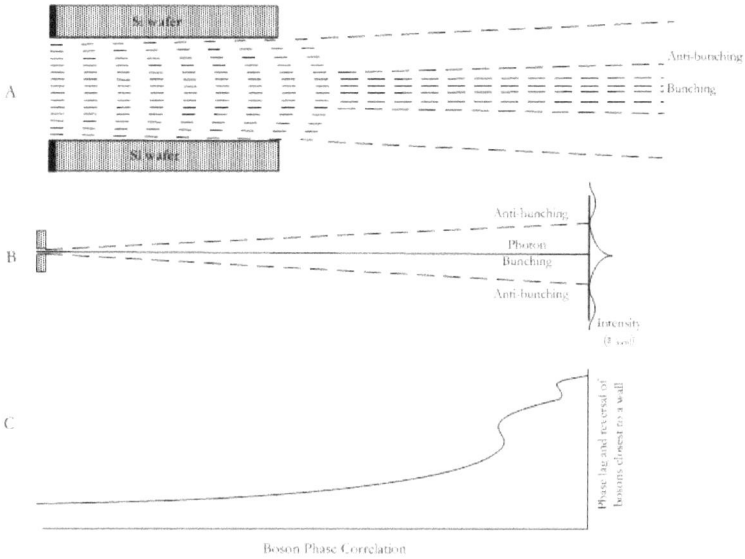

Fig. 11-5 Fraunhofer Diffraction and Interference:

Panel A shows a phalanx of photons passing through a slit in an optical screen characterized by the Coulomb force tending to bunch bosons together and the gravitational force acting principally on those bosons closest to the walls of the slit and resulting in phase related anti-bunching of bosons. This results in a slight curvature of the phalanx emerging from the screen and the bending of photon trajectories. Panel B shows the far-field consequences of such diffraction giving rise to interference phenomena at a viewing screen in which phase-correlated photons in a central group result in high intensity light and low intensity regions on each side. The pattern is aided and abetted by the bunching and anti-bunching of phase correlated and anti-correlated photons. Panel C shows the combined effects of high areal-density photons and high phase correlation in creating the regions of strongly bunched photons (high intensity) and those in which anti-correlated photons have left those regions with low intensity. It is the resulting increase in ρ_{local}, with increasing phase-correlated photon areal density that accounts for the tendency of bosons to group together, driven by the NTI. This phenomenon is shown here in the extreme, when it gives rise to a phase transition in which two photons of energy E may result in a single photon having energy 2E.

Consider two bosons, in the center of the slit on their way to the far field at a slightly reduced speed of light and assume each to have energy E

$\cong 10^{-19}$J (λ=~10^{-7}m) on parallel trajectories ~10^{-8}m apart (d) and in phase correlation. (While this separation is small compared to λ and the width of the slit, it is orders of magnitude larger than the thickness h of the gyrating GSB photons in the bosons, making the bosons very sensitive to Coulomb differential energy-density gradients. The Coulomb force would act to decrease or increase boson separation, over some travel distance D and time t at the speed of light c from the slit to the far-field detector. The Coulomb acceleration would determine t, the travel time of the bosons, according to d $\cong$ ½ at², where a is given by F_c = ma, and F_c is the Coulomb force = $q^2/(\varepsilon_0 d^2)$ = ma, where ε_0 is the dielectric constant and m=10^{-19}ev. (Note here that the acceleration is orthogonal to the trajectory of the boson and therefore we take the mass m simply as the energy of the boson in electron volts.) From this we get a $\cong 10^8$m/s², t $\cong 10^{-8}$s and the distance to the screen D $\cong$ tc, or about 1m. Thus there is a minimum time and distance d in order to demonstrate single-slit Fraunhofer interference. Note also that the demonstration of Fraunhofer interference is more demanding than for double-slit interference, since the latter depends principally on the gravitational scattering of particle trajectories in each slit.

At the inner walls of the slit the gravitational energy-density is nearly equal to the internal energy-density of the matter constituting the screen, such that the GSB gyrating photon in bosons very near the wall would emit low energy bosons, in a direction away from the wall so that the recoil would be sufficient to bend its trajectory toward the wall, as shown. Such trajectory bending could thus be seen as a quantum state example of gravitation lensing.

The foregoing makes a strong case for the gravitational and electromagnetic forces acting on bosons to give rise to the extraordinary phenomena of optical diffraction and interference. But it does so only if bosons also can account for the mediation of these forces via impulse-recoil action. And that can happen only if such bosons have an internal dynamic store of energy, momentum and angular momentum, in the form of an internal gyrating GSB photon, whose axis of gyration is orthogonal to the direction of travel of the boson. And in turn the boson may alter its course only if the GSB photon can emit relatively low-energy bosons (in response to $\sqrt{G}$ or $\sqrt{\varepsilon}$ gradients at the gyrating photon) so as to impulse-recoil the free boson in a direction orthogonal to its travel. Thus we see that these several ifs-and-only-ifs make sense in particle physics because they each self-consistently contribute to the notion of particles interacting with particles in a timely manner. And these arguments should be seen as opportunities for advances in precision experimental physics to verify or deny the validity of the above strictly-particle theory. Fermions and bosons are composite particles because they each derive from the dynamics of an internal GSB photon; however, they are incommensurate because the one characterizes nearly all the mass in the Universe, while the other mediates all of the forces, consistent with their incommensurate GSBP spin orientations.

The above reasoning applies also to double- or multiple-slit diffraction and interference, in which the far-field contributions play little or no role. It applies also in accounting for the diffraction and interference and bunching and antibunching of electrons traveling on closely-parallel trajectories, with similar constraints on energy and phase, when traveling at relativistic velocities. Such kinetic properties partially nullify their fermionic proclivity (in low energy states) to shun one another. Under these conditions, the GSB photon in the electron travels on a helix of radius $\sim 10^{-13}$m about the electron's trajectory, with kinetic energy orders of magnitude greater than the electron's rest mass See also chapter 15. But the bending of fermion trajectories is mediated not by impulse-recoil action but by the torque imposed on the GSB photon in the fermion due to a gravitational energy-density gradient in the space of the GSB-photon, causing a differential in the velocity of the photon on one side or the other of its circular travel. Consequently, we should conclude that all gravity acts at the quantum level. That is to say, gravity and relativity are joined at that level and at all higher levels. Moreover, we could refer to the unique processes of boson (or fermion) antibunching as legitimate instances of anti-Coulomb action.

In contrast with the above, the current reasoning, asks us to accept the notion that a boson, having some quantum of energy-momentum, is emitted from an atom, travels as if it were a spreading wave, whose wave function may collapse at some point on a barrier or perhaps at a hole in the barrier, whereupon, instead of being absorbed, it subsequently spreads out the back side of the barrier, as a new spreading wave, and all somehow conserving energy-momentum in the process, eventually to be absorbed by one atom or molecule (as a particle) at one or a few of an infinite number of possible localities. The same scenario may be represented more abstractly in QFT and with essentially the same predictive results, but at the expense of even greater departure from intuition.

Thus our scenario finds no need for spreading waves or the wave nature of light. The bosons are created as particles, travel to the slit and interact with it as particles and then to a photo-sensitive emulsion or detector as particles, where their energy-momentum is exchanged conservatively with electrons in atoms or molecules or dissipatively with particle aggregates or massive bodies. In the latter regard, we need to explain how uncharged massive particles (e.g., neutrons), unionized atoms or aggregates thereof may interact gravitationally via GSBP-boson emission and impulse-recoil action. From what and how are the bosons emitted?

As uncharged massive particles, neutrons are the simplest of all such composite particles or aggregates thereof to account for their gravitational interaction in terms of GSBP-boson emission and impulse-recoil action. They have rest mass $\sim 1.6 \times 10^{-27}$kgm, on the basis of which we can assign an internal dynamic structure (gyrating photon), as in the case of the electron. The essential difference between the EM and gravitational structures is that the latter, governed by the gravitational charge $m\sqrt{G}$, is immune to negative charge ($\sqrt{-G}\sqrt{G} = G$), apart from differences in the strength of their coupling force and the fact that the mass and energy density of neutral agglomerates are not constant; they evolve in competition with the local forces of expansion.

So saying, we are left with two important questions *open to experimental verification*: First, could we detect and quantify the increase in directed low energy-momentum impulse-recoil bosons in interference experiments and in Casimir experiments and thereby validate the impulse-recoil model of gravitational and electromagnetic force (see Fig. 11-3).. *Second, could similar experiments shed light on the requirements on phase overlap of two or more photons (Garcia et al.) and the transition from photon bunching to antibunching?*

We turn now to another puzzle in quantum physics, though little explored. The wavelengths of electromagnetic radiation λ range from nanometers to many tens of meters and accordingly the radius of the GSB-photon in a boson may vary over the same range, as must its cross-section for interaction with matter (in emission and absorption). How can a boson of wavelength, say 1cm, be created or absorbed by an atom of dimension $\sim 10^{-9}$m, as in various experiments? (Here, it is necessary to make a distinction between radiation emitted and absorbed by quantum systems and that by macro systems, where the dimensions of the latter must closely correspond to the wavelengths of emitted and absorbed radiation. Thus a low frequency boson emitted by a Rydberg atom may be compared with a "wave packet" having the same fundamental frequency and created by a radio transmitter and radiated from an antenna, where the dimension of the antenna is many orders of magnitude larger than the radius of the Rydberg atom. The former (atomic) process of radiation is strictly conservative of energy-momentum, while the latter is a dissipative process. This suggests a very important question scarcely raised in the literature: What is the trajectory and dynamics of an excited electron as it falls to a lower energy state in an atom thereby emitting a boson, whose energy corresponds exactly to the difference of the two electron energy levels? We will pursue this in chapter 14.) The strict conservation of energy-momentum from one epoch to another suggests that the GSB-photon may be the most fundamental or elemental of all particles. We will pursue this in greater detail in the next chapter.

Also, our accounting for the force of gravity on bosons, in terms of boson emission and impulse-recoil reactions makes no demand for "gravitons," since there is no need for a force of *gravitational attraction*. The same holds for the force of gravity on fermions, in terms of the asymmetrical gravitational gradients acting on the gyrating photon in the fermion. We have thus disposed of one fundamental particle essential in the Standard Model that is unneeded in our scenario.

Finally, the various factors listed above, as contributing to the bending of boson trajectories at a barrier, could be thought of as "hidden variables," though pragmatic considerations and the indeterminacy relations would add little or nothing to our predictive power. Bohr, Born, Heisenberg, Pauli, etc., would say that's as it should be, there's been too much time wasted already in trying to discover a more fundamental theory than quantum mechanics. But Einstein, de Broglie, Bohm, Bell, etc., would likely welcome the banishment of the dictums of wave-particle duality and superposition theory and the implied complementarity associated with them. We will consider these and related issues in greater detail in chapter 15.

Summary:

The GSBP internal structure of fermions and bosons is self-consistent with the demands of special and general relativity, providing a simple and straightforward theoretical base for the development of quantum theory.

Quarks, leptons and bosons are shown to be GSBP composite particles in several hierarchical energy-density categories. The GSB Photons in fermions are seen to be the most fundamental and elemental particles in the Universe.

Many of the high-energy particles, found in accelerator experiments and believed to be fundamental particles, are simply products of the extreme energy environment in the targets of accelerators in *efforts to simulate earlier higher energy-density states of the Universe*. The same holds for high energy ("mass") GSB photons and bosons that accompany the creation of the massive particles in these experiments.

The forces – gravitational and electromagnetic – that act between neutral or charged particles or bodies are mediated by timely impulse-recoil action in response to differential energy density gradients at the surfaces of such particles. They are in essence gauge forces. There is hence no need for attractive gravitons.

The internal structure and dynamics of massive particles is conserved as the energy-density of the Universe declines, as does the internal structure and dynamics of the bosons.

The conservative mechanics of atomic radiation is presented in GSBP fermion-boson theory.

Particle diffraction, interference and the bunching and antibunching of bosons and electrons is explained in terms of quantum gravitational and electric differential energy-density gradients mediated by impulse-recoil action.

CHAPTER 12

A GSBP-SCHRÖEDINGER RELATIVISTIC QUANTUM (PARTICLE) MECHANICS

We have argued, in chapter 6, that the inflation scenario, which postulates energy in the vacuum (the "physical vacuum") to energize the superluminal expansion of the universe, is untenable, principally on the grounds that it violates the laws of symmetry and thermodynamics and in particular the principle of flow-specific entropy production and the extended 2nd Law. As shown in chapter 11, the average energy-density of the universe in the vacuum in our region of spacetime would account for the Casimir force, given the variation of c and G with the local energy-density gradients and their dependence on the spacing of the plates. But the "physical vacuum" is not able to drive real physical processes, because the dissipation associated with such processes (e.g., the acceleration of the plates against frictional forces) would violate the arguments on which the notion of the "physical vacuum" was founded. This, together with its counterintuitive nature and its failure to connect with general relativity, makes the formalism of QFT and quantum electrodynamics highly problematic.

What alternative might we have to transform quantum mechanics into a relativistic mechanics in terms of a strictly particle theory? More to the point, what must we demand of such an alternative? The following propositions must be capable of empirical demonstration, in the sense that they facilitate the creation of quantitative models (at the nuclear, atomic, free particle scattering and many-particle systems levels) and consistent with both general and special relativity and the laws of thermodynamics, in particular the extended *2nd Law*:

1. All matter, antimatter "fundamental" particles (quarks, leptons and their antimatter counterparts) must have internal energy and structure, enabling the conservation of quantized intrinsic energy, angular momentum (electric charge), spin in any interaction, such that the requirements of special and general relativity are satisfied in any spacetime epoch, taking into account variations in the average and local energy-density environment.

2. All particles, whether massless or massive, travel as particles, along probabilistically-defined trajectories, and react in relation to other particles as particles, i.e. all wave or field properties are strictly local and the concepts of spreading waves, wave packets or fields that fill all space may on occasion serve as useful statistical or mathematical representations, but otherwise have no physical embodiment or essential utility.

3. No particle can occupy two or more distinct positions simultaneously nor can it travel on two or more distinct trajectories at the same time or, in consequence, "interfere" with itself. All interference phenomena require the "entanglement" of two or more similar or identical particles in a quantum system. This would effectively deny the validity of superposition theory.

4. Two or more particles may be created as subsystems of a nonlinear thermodynamic/quantum system or become entangled in the creation of such systems. Entanglement emerges in the system as the increasing flow of mass-energy through the system leads to increasing stress and ultimately to a local state transformation in which mass-energy, symmetry or other parameters are conserved, thereby relieving the stress. In the process, the entropy of the entangled particles, taken together, is decreased at the expense of a larger increase in the total entropy of the system or its environment. In one form, this may be manifested in the down-conversion of one particle into two or more particles, whose combined energy-momentum is nearly equal to that of the initial particle and whose quantum states are more or less strongly correlated. The degree or strength of entanglement (correlation) varies inversely with the net entropy production of the system. (Compare the near-zero entropy production, associated with the nearly-adiabatic down-conversion of one proto-neutron into two identical ones of half mass-energy in the early Universe, with the relatively large net entropy production in the nonlinear crystal down-conversion of bosons in the laboratory. See the section on Violation of Bell's Inequality Theorem in chapter 15.)

The internal structure of massive fundamental particles, as sketched in chapter 11, derives from the origin of all such particles in the evolution of the early Universe, under conditions of extreme average energy-density

ρ in which the "constants" c and $\sqrt{G}$ have infinitesimal values. That structure embodies the concept of Gravitational and Symmetry Bound Photons or Gravito-Symmetry Bound Photons (GSBP). As the Universe expanded (and is still expanding) in the scalar era, ρ decreased and c and $\sqrt{G}$ increased inversely as the third power of ρ, preserving all the GSBP quarks and composite proto-neutrons and their antiparticles. The gravitational force (coupling strength $m\sqrt{G}$) holding the photons in their very small orbits remained constant throughout the scalar expansion because $m \propto \rho^2$. As the Universe expanded further, the lower energy-density favored the transmutation of increasing numbers of neutrons into protons, accompanied by the creation of leptons, as discussed in chapter 13. Thus we might say that the massive particles we experience today have their origin in a decreasing scalar field ρ, the average energy-density of the Universe.

But the GSBP representation of massive particles does not suffice to tell us how such particles communicate with one another in spacetime, just as Feynman diagrams provide no physical insight as to how photons (real or virtual) are *directed*, so as to account for the interaction of two electrons, thereby conserving energy and momentum in their interaction. As we saw in chapter 11, the interaction of all particles occurs via the emission of bosons from massive particles so as to propel them toward or away from similar particles or aggregates thereof, as the particular situation requires, by impulse-recoil.

The above is the basis for a proposed relativistic strictly-particle mechanics to free us of centuries-old counterintuitive beliefs and enabling us to return to a semblance of intuitive models, even in the remoteness of the early Universe or of the quantum realm. The GSBP particle provides a foundation for the development of a semi-classical theory in that all such particles exist in real physical spacetime as well as state space during their lifetimes, consistent with the lower-order approximations of Newtonian mechanics or non-relativistic quantum mechanics. Such theory would be *semi*-classical in that those states may be specified or measured only within quantum limits. And they would be inherently relativistic, consistent with special relativity in the vacuum of any epoch of the Universe where $\rho_{local} \equiv \rho$ and with general relativity as ρ varies from epoch to epoch or in any region where ρ_{local} differs significantly from ρ. Just as the "fine structure" of general relativity accords with Newtonian dynamics, so the fine structure of relativistic *particle* mechanics should accord with Bohr's less-demanding classical model of the hydrogen atom and in particular with the spin-orbit corrections characterizing the Schröedinger quantum formalism. The latter would lack only exact relativistic (special and general) corrections to constitute a self-consistent quantum (particle) mechanics, save for extraneous (non-formal) assumptions associated with the "Copenhagen interpretation" of quantum mechanics. That is, a strictly-particle theory should have no need or tolerance for wave-particle representations, spreading waves or fields, a particle being in two or more places or on multiple trajectories at the same time or interfering with itself, if it fully accounts for the spin-orbit/relativistic interaction of quanta. *Each attempt toward this end (the Schröedinger, Dirac and QFT formalisms) has failed ultimately because it made no requirement for internal structure of the fermions and thereby offered no rational connection with or dependence on general as well as special relativity.* This is revealed in each case with the tacit acceptance of c and G as absolute, universal constants.

A GSBP- Schröedinger Relativistic Quantum (Particle) Mechanics

A quantum (particle) mechanics can have full relativistic validity only if it refers to physical particles, having mass, internal dynamical structure and occupying physical space. Lacking such *internal* model of the fermions, the founders of quantum mechanics had no alternative but to develop external models – relativistic spin-orbit corrections to "point particle" Schröedinger mechanics. Ultimately, that path lead to the replacement of particles, as the most fundamental representation of matter in quantum theory, with an all encompassing field theory. It has been a marginally successful but costly enterprise. The challenge for us in developing a relativistic particle mechanics is to comply with the above general requirements, while melding the well-established formalism and power of Schröedinger mechanics with the inherently relativistic structure and dynamics of GSBP particles, so as to directly satisfy the demands of general and special relativity. If, in the one-dimensional Schröedinger time-independent equation $-(\hbar^2/2m_r)\nabla^2\psi - [e^2/(4\pi\varepsilon_0 a)]\psi = E\psi$, we rewrite the second term (Coulomb potential) by substituting e_r for e and ε_r for ε_0, where $e_r = e/(1+m_e/m_p)$ as the reduced electric charge of the electron, $\varepsilon_r = \varepsilon_0/(1+m_e/m_p)$ the reduced dielectric constant and *a* is the spatial separation between the nuclear charge e and the electron charge e_r in the ground state of the electron. The Coulomb force is thereby reduced in the same proportion as the mass of the electron. The result of this is to increase the radius of gyration r_0 of the GSB photon in the electron and a corresponding increase in the orbital radius of the electron about the proton in the ground state. Thus corrections for both the mass and charge of the electron enter into the relativistic dynamics of the electron, including compliance with the requirements of general relativity, since *m* varies as ρ^3 and ε as ρ^{-3}. In the above we see that the unit electric charge $k\sqrt{\varepsilon}$ *plays the same role in reducing charge in EM theory as does the gravitational charge* $m\sqrt{G}$ *in reducing mass in gravitational theory. Here we are clearly referring to electrons as physical particles, having dimension, position, momentum, angular momentum and spin angular momentum (charge) in spacetime, in which the notion of waves and fields have meaning only in a strictly local sense and the notions of wave-particle duality, superposition and wave function collapse have no meaning or function.*

The above directly derives from our initial assumptions in chapters 7, 8 and 9, which led to the creation of all of the massive particles, and so we should expect relativistic GSBP-Schröedinger mechanics to significantly extend our understanding and representation of the structure and dynamics of the nucleus. Specifically it should apply, at one level, to the *external* characterization of the nucleus, i.e., how the electrons relate to the mass-charge structure of the nucleus, and, at a more fundamental level, to its internal structure, i.e., how the GSBP quarks in the neutrons and protons interact to define the internal structure and dynamics of the nucleus. We leave the latter much more complex problem area for further investigation. In principle a GSBP-Schröedinger equation could characterize the ground state relativistic dynamics at these two structural levels as well as for the atom as a whole, at least for the lower-Z atoms.

How should we interpret the above relativistic Schröedinger wave equation? Our arguments in chapters 11 and 12 and in the following chapter deny any interpretation as a spreading *wave* or as a superposition of all of the possible solutions of the equation subject to particular boundary conditions, as Bohr and others would have it. Yet those solutions characterize the radiation, at the speed of light, of electromagnetic energy of discrete physical wavelength and so the Schröedinger equation is clearly a wave equation, though one that is unique to the representation of quantum systems. However, to emphasize once again, *all such wave properties are strictly local – to the nearly massless boson as it travels very near the speed of light or to the massive fermions in their travel at lower relativistic velocities.* The illusion of a "spreading wave" derives from the emission of a very large number of bosons from atoms in a localized space in a random distribution of close trajectories (see chapter 11).

In broader perspective we are challenged also to square GSBP-Schröedinger mechanics, self-consistently, with fundamental theoretical issues, as below:

Lorentz Transformation

In our scenario, the velocity of the electron v_e and that of the GSB photon v_{GSBP} are governed by the relation $v_e^2 + v_{GSBP}^2 = c^2$, as required in order to conserve spin angular momentum of the electron, and making it inherently relativistic. Thus the displacement dr of an electron can be described equally well in terms of $dr^2 = dc^2t^2$ or $dr^2 = c^2dt^2$. (The latter appears to have been chosen by Einstein with little question, as the evidence at the time strongly supported the belief that the speed of light was an absolute universal constant.) But in our schema it is the element of time advance dt that is taken as an absolute universal constant and the velocity of light c *as a local or epochal parameter.* Consequently, an observer in frame S would perceive a displacement dr of an electron as $dr^2 - dc^2t^2 = 0$ and an observer in frame S', moving at velocity V relative to S, would see the displacement as $dr'^2 - dc'^2t'^2 = 0$. From this we obtain the equivalent Lorentz transformations $x' = \gamma'(x - Vt)$, $y' = y$, $z' = z$ and $c' = \gamma'c$ where $\gamma' = (1 - V^2/c^2)^{1/2}$, for the reduced velocity of light c' in the higher energy (higher kinetic energy) environment of frame S', with the rate of time advance taken as a universal constant and elapsed time in S' is equal to that in S. (Note: γ' is formulated in direct relation to $(1 - V^2/c^2)^{1/2}$, rather than inversely, as in Einstein's special relativity because the measures of time and velocity are inversely related.) Here, as in conventional theory, the transformation is linear, symmetrical and reduces to the Galilean form in the nonrelativistic limit. It takes account of the underlying dependence of c on the average energy-density of the Universe or local variations thereof, which Einstein could not have foreseen. Indeed, the transformation is grounded fundamentally in the inverse relation between energy-density and the velocity of light (see chapter 7). In consequence the notion of time dilation (in constant c theory) is replaced by that of light-velocity dilation, which may be expressed as $\Delta c' = \gamma'(\Delta c)$ or $\Delta c' = (\Delta c)/\gamma$. In this light we see that the old counterintuitive argument about a person aging less rapidly in a spaceship leaving Earth is turned on its head: The gain in energy-momentum of the spaceship and its occupants (as an NTI-crafted maximal energy-density subsystem of mankind's exploratory systems), at the expense of a small decrease in the local velocity of light and *increase* in the aging of its occupants, comes at the expense of a very large increase in the dissipation and entropy production of the support systems. *It is important to note here that the potential advance of mankind's understanding of the Universe, resulting from such exploratory activity, may later foster multiple advances capable of further reducing the rate of entropy production associated with human activity on Earth as well as in space.* Thus we see that the GSBP model of fermions reveals a more general formulation of the Lorentz transformation, self-consistent with general relativity.

Uniformity, Constancy of Boson Velocity

The above fails to inform us how bosons (particles) are emitted from, for example hydrogen atoms, so as to travel very nearly at the velocity c. What is there about the electron's structural *properties* and *dynamics* that guarantees this in the emission of bosons? In our scenario, the GSBP electron doesn't just jump from an excited state to a lower one, it travels down a potential energy spiral staircase at relativistic velocity, such that the period, corresponding to the wavelength of the emitted boson at velocity v_{GSBP} <c exactly defines the distance traveled by the electron, at relativistic velocity, between the initial and final states of the electron. (Fig. 11-2 (see also chapter 14). The electron's loss of potential energy and gain in kinetic energy (gain in momentum, conserving angular momentum) results in the release (emission) of the boson (particle), having energy hc/λ, momentum h/λ and traveling very near the velocity c. (As we show in Fig. 11-2, all bosons gain in marginal momentum what they lose in velocity.) Thus the emission of bosons in the hydrogen atom occurs as the electron spirals down from one allowed discrete energy level to another, results in energy-momentum loss in the electron exactly equal to that acquired and stored in the emitted boson. The distance traveled by the electron, at relativistic speed, is equal to the wavelength of the boson, thereby minimizing line-broadening of the emission, *allowing only a maximal*

error of ±h, and where h is Planck's Constant. The same process works in reverse in accounting for the absorption of a boson and excitation of the electron to a higher energy state. This same process would hold in any other epoch or local region in which c and $\sqrt{G}$ are uniformly constant and for the emission of bosons from the higher-Z atoms, self-consistent with the requirements of general relativity. According to the above, the direction of the boson's travel would be determined by the orientation of the emitting atom, either randomly in spacetime or under more controlled conditions.

Electron Mass Gain with Velocity Increase

An increase in the electron's velocity Ve would result in a corresponding decrease in v_{GSBP}, which in turn would require a decrease in the radius of gyration r_0 of the GSB photon (in the electron) for exact closure of the GSBP. This could occur if and only if the mass increased as the square of the resulting increase in the electron's energy-density. The same is required also to conserve spin angular momentum.

Relativistic Correction of The Electron's Mass and *Charge*

The relativistic dynamics and spectral radiation of the hydrogen atom can be characterized in GSBP-Schröedinger mechanics, knowing only the mass of the electron and that of the proton, the Rydberg constant for hydrogen, Planck's Constant and the velocity of light c. Schröedinger mechanics lacks only a rational inherent correction for the electron's mass and charge so as to make the kinetic energy of the proton (its wobble) vanish. A similar problem occurs in planetary dynamics; the mutual orbiting of the Sun and the Earth about a fixed point in space is readily corrected by assigning the Earth a reduced mass:

$$m_r = m_E \left(1 - m_E m_S / (m_E + m_S)\right),$$

where m_E and m_S are the masses of the Earth and the Sun. This simple correction works because adjusting the mass of the Earth automatically corrects also the gravitational charge $m\sqrt{G}$ in the model to make the mutual orbiting vanish. But there is no way in Schröedinger mechanics to adjust the charge of the electron simply by reducing its mass, because the electron is taken to be a point particle with absolute values of electric charge and spin angular momentum and offers no logical means for reducing the charge.

This most fundamental problem in quantum mechanics is straightforwardly resolved in GSBP-Schröedinger formalism by reducing both the mass and *charge* (spin angular momentum) of the electron relative to that of the proton, as above for planetary systems. The charge may now be reduced by decreasing the GSB photon's energy (increasing its radius of gyration proportionately to that of the proton's mass) in order to maintain exact relativistic closures of the GSBP and electron orbits, in accordance with the fine structure constant (see chapter 14). The result of this is to increase the ground-state orbital radius of the electron (Fig. 12-1). The complete relativistic/spin-orbit correction is thus inherent in the GSBP-Schröedinger formalism, anchored in general relativity (since c varies as ρ^{-3}) as well as in special relativity. The same applies in defining the ground state of the three quarks in SU(3) symmetry (local Euclidean geometry) in the proton constituting the hydrogen nucleus, in which the local energy-density is orders of magnitude greater than that of the electrons (see chapter 13). (The above is intended to suggest that no insight is gained by referring to waves and indeed, by doing so, we tend to confuse our own thinking and that of others.)

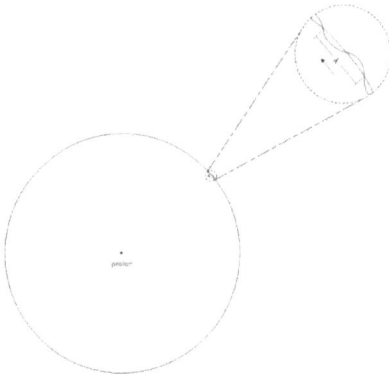

Fig. 12-1 Helical path of the GSB Photon in the electron as it travels in exact closure with one orbit of the electron about the proton.

Relevance of *Waves, Fields, Phase Etc.?*

We have already argued that the terms *field, field-strength, beam*, etc. refer in our scenario to greater or less number-density of similar-state bosons, electrons, etc., as typified in a laser beam of bosons, in which their relative *phase* (entanglement) is of key importance. By that we mean that nearly-all such bosons of a group are emitted at the same instant and advance at the speed of light on closely-parallel, more or less closely packed, trajectories. Thus, the *phase* of a boson has significance only in relation to other such bosons in the group or to some means for detecting or measuring their position in spacetime relative to some event, as in the case of interference phenomena (Fig. 12-2). And to speak of *resonance* is to speak of the efficient transfer of energy-momentum from the gyrating photon in a boson to that in the bound electron, for example, in the ground state of the hydrogen atom just sufficient to excite the electron to a particular discrete higher energy state (Fig. 12-3). Such transfer occurs when the momentum of the boson h/λ is exactly equal to the momentum required to relativistically boost the electron to a higher metastable energy state. As noted in a previous section, the distance the electron must travel up the energy "spiral staircase," at relativistic speed, must be exactly equal to the wavelength of the boson, thereby *allowing only a maximal error (among alternative bosons) of ±h, where h is Planck's Constant.* The *Compton* wave*length* of the electron is related simply to the radius of gyration r_0 of the GSB photon in the rest state: $r_i = \hbar / m_r c$, where r_i would be the *increased* radius corresponding to the *reduced* mass of the electron. Again, we gain nothing by referring to a *wave*, though taken in strictly local terms it provides a connection with conventional reasoning.

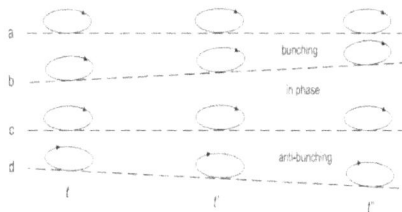

Fig. 12-2 Showing four of a volley of GSBP-bosons emitted from a laser on close trajectories (dotted lines). Bosons a and b are close enough in trajectory and phase to lend to bunching together commencing at time t and progressing in time periods t' and t''. Boson c is in phase with bosons a and b but lacks sufficient coulomb coupling to bunch with them. Boson d has moved out of phase with a, b and c at time t'' sufficient to align with other bosons (not shown) to contribute to anti-bunching as evidenced in interference phenomena. See Fig. 11-4 characterizing GSBP-boson internal structure and dynamics.

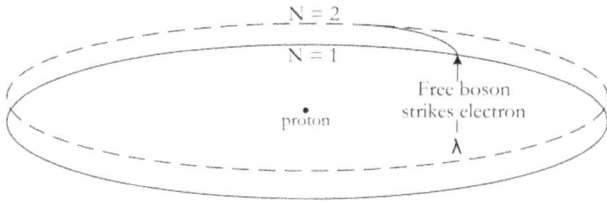

Fig. 12-3 Helical path of the GSBP photon in the electron as it travels from energy level n=1, L=+h to n=2, a^{-1} λ, in exact closure with one orbit of the electron about the proton. This process can be better understood as a resonance transfer from a boson's energy to the electron.

Derivation of E=mc^2

Einstein's most famous equation can be derived straightforwardly from the relation $v_e^2 + v_{GSBP}^2 = c^2$. An increase in the square of the electron's marginal velocity $(\Delta v_e)^2$, would require v_{GSBP}^2 to decrease in the same proportion, i.e., in the ratio $(\Delta v_e)^2 / v_e^2$, and its radius of gyration to decrease in the same ratio in order to preserve exact closure of the gyrating photon. But this would require the mass of the electron to increase in the ratio $((\Delta v_e)^2 / v_e^2))^2$ in order to conserve spin angular momentum and the internal (intrinsic) energy of the electron, the latter, presciently referred to by Einstein as stored energy. Thus, multiplying the above relation everywhere by the relativistic mass m_{rel} we obtain $m_{rel}v_e^2 + m_{rel}v_{GSBP}^2 = m_{rel}c^2$. The first term on the left represents the electron's increase in energy (kinetic energy) and the second term is the electron's conserved intrinsic (stored) energy ($\Delta E_{intrinsic} = 0$), with the result: $m_{rel}v_e^2 = m_{rel}c^2$ or $E_{kinetic} = m_{rel}c^2$. The greater the m_{rel} the smaller is the radius of gyration of the GSB photon and the greater is the local energy-density, thereby conserving the intrinsic energy-density of the electron, as v_{GSBP} declines. The result is to make the kinetic energy $m_{rel}v_e^2 = m_{rel}c^2$, where c is the present constant velocity of light in our corner of the Universe. But there is more to understand in this derivation, as we see that the relationship $v_e^2 + v_{GSBP}^2 = c^2$, where c varies as ρ^{-3} applies, with the only physical variation being that of the energy-density of the electron. Consequently, Einstein's equation, valid in GSBP-Schröedinger mechanics holds in general, as well as, special relativity, i.e., in earlier epochs of scalar expansion and in higher local energy density regions of the Universe. And the above holds for all matter and antimatter fermions, including quarks. It also requires the conservation of their intrinsic spin angular momentum.

Physical Basis of the Complex, Wave-Function Formalism

The GSBP-electron *physically requires* what could only be *algorithmically assigned* to a "point particle:" Like a spinning top, a force (electromagnetic in this case) in a certain direction is manifested as a precession or motion orthogonal to the force. Such action is at the heart of Heisenberg's matrix mechanics and Schröedinger's complex wave-mechanics and the non-commutativity relations, which characterize the quantum world. Here, the electromagnetic force plays the role of gravity and defines the direction of attractive force (the angular momentum of the electron), about which the electron's spin axis (the GSBP angular momentum axis) precesses. What distinguishes conservative quantum systems is that the GSBP internal structure and dynamics of a fermion requires the precise quantization of allowed energy states of the system – a stable lowest ground state, having zero angular momentum, and, in the hydrogen atom, a set of higher metastable states in which the angular momentum may be briefly null. The ground state is defined by the exact closure of $n^2\alpha^{-1}$ cycles of the GSB photon with that of the electron's orbit about the proton, where α is the fine structure constant and $n=1$, while $n=2,3, \ldots$ refers to the higher-order metastable states (see Fig. 14-1). As for the $2S_{1/2}$ state, the higher-order metastable states are shielded somewhat from fluctuations by the requirement to surmount an angular momentum energy-barrier distinguishing the metastable state from its degenerate companion. The stability of all such states depends on the intensity of fluctuations in the atomic environment. This is the physical basis for the characterization of the allowed principal states of the hydrogen atom. Thus, we see that the GSBP model of the electron is self-consistent with the Schröedinger formalism in the investigation of time independent and time dependent *particle* interaction, while greatly enhancing its range of application and relativistic precision.

Indeterminacy Relations

If we think of an electron as a cylinder of radius r_0 (radius of gyration of the GSB photon) and length r_0, we could expect to determine the electron's position in space only over the volume $\sim r_0^3$, when the electron's velocity is small compared with the speed of light. In this case the electron's energy, energy-density and momentum correspond closely with those parameters defined by its rest mass. However, when the electron's velocity approaches the speed of light, its energy and energy-density increase rapidly (volume decreases and mass increases) as does its momentum. The result of this is to increase the range of momentum Δp, and the range of position Δr, such that $\Delta p \Delta r \geq \hbar$.

In our scenario electromagnetic energy-density is an important parameter in the interaction of charged particles (fermions), just as is gravitational energy-density in the interaction of massive particles or bodies in general relativity. In the latter case, a planet, in a stable state of minimum potential energy in its orbit around a star, may be seen to define the star-planet system as being in a state of maximum energy-density. The same reasoning applies to the electron in its orbit around the proton in the hydrogen atom and to the more massive atoms that leave their radii little changed from that of the hydrogen atom.

GSBP-electrons in Deuterium, Helium and higher mass atoms

Since the GSBP-electron is inherently relativistic, it should greatly simplify and enhance precision in the calculation of energy states and optical spectra for the heavier atoms. For example, we cannot take the ground state of an ionized helium atom to be one-half the ground state radius a_0 of the electron in the hydrogen atom, as in the case of simple models, because the velocity of the electron increases by a factor less than two due to the increase in energy-density and the attendant increase in its mass. All such adjustments are straightforwardly accounted for in the relation $v_e^2 + v_{GSBP}^2 = c^2$ (See chapter 14). It would also be necessary to include an adjustment for the electron's reduced mass and charge, in the absence of the ionized electron. All such adjustments become even more significant as Z increases in the heavier atoms.

Reflection, Absorption of Bosons (*Particles*) at Matter Surfaces

What determines whether a boson (particle) is reflected (the direction of its momentum is altered) or its energy-momentum is absorbed at the surface of a particle or body? What makes a radiometer spin? We offer here yet another theory to account for the rotation of a radiometer: A boson's momentum has a high probability of being redirected and conserved at a mirror's surface in a nearly dissipation-free interaction and a high probability of being dissipated at a black body surface. The former can only occur when the boson interacts with electrons in a flat, dense and inertial Fermi sea, as characterize electrically conducting thin films on glass. Where multiple bosons having the same energy, momentum and phase are involved, as in laser beams, the reflection of the beam is accounted for as in classical theory. Absorption of bosons of relatively high energy at black surfaces and the radiation of some portion of that energy by lower-energy bosons can occur involving various dissipation processes, including thermal equilibration, photo-chemical reactions, energy down-conversion in non-linear crystals, partial reflection at liquid surfaces, etc. The emission of such low energy bosons results in impulse-recoil rotation of the radiometer.

Zero-Point Energy

The measure of radiant energy is manifested in one of the simplest and most fundamental equations in all of physics: $E=h\nu$, where E is the energy of a boson, h is Planck's Constant (empirically derived and *assumed* to be an absolute universal constant) and ν is the frequency of the photon. It was the foundation on which quantum mechanics was erected, which soon spawned a new theoretical issue of fundamental importance, the requirement for non-vanishing energy in the vacuum at 0K temperature. As with other interpretive appendages to early quantum theory, the "zero-point energy" *postulate*, on which relativistic quantum field theory was erected, has become an essential element of quantum theory. This is especially disturbing, since a flurry of theoretical and experimental studies by respected scientists in the pre-quantum era failed to make a convincing case, for or against, the postulate. It is as though the need for such energy, the "physical vacuum," in QFT (to square quantum theory with relativity) was so great and urgent as to simply accept the postulate as fact.

Listed below are some of many problems in current quantum theory and cosmology that have been resolved straightforwardly in our strictly-particle scenario, with no demand for the physical vacuum:

- Seamless joining of relativity and quantum theory
- Evolving c and G as parameters $\Rightarrow$ rational cosmology and
- No requirement for dark energy or dark matter
- Creation of all particles in the quantum Universe
- Inherent relativity of all particles
- Accounting for the present lack of antiparticles in Nature
- Gravitational, electromagnetic forces rationalized
- NTI creation of all organized subsystems from galaxy clusters to quantum and living subsystems
- Quantum theory freed from Copenhagen quandaries: Complementarity, Superposition Theory, etc.
- Gravitational explanation for the Casimir force

But demonstrating no demand for energy in the vacuum is not equivalent to demonstrating its nonexistence; here we suggest a first principles approach to the latter possibility. A GSBP electron having zero energy of translation and vibration at 0K still has rotational energy, the GSB photon gyration (spin angular momentum). That could be reckoned null only statistically, either in terms of an ensemble of weakly interacting electrons or a Milliken electron, whose precession about a fixed magnetic field declines to zero, on average, with decline of the magnetic field. This follows from the requirement for minimization of total angular momentum in any isolated system as it evolves toward equilibrium.

Implications for Modeling the Nucleus of Atoms

GSBP-Schröedinger mechanics holds great promise also for modeling the structure and dynamics of quarks in neutrons and protons and their interaction in the nucleus of high-Z atoms. Like electrons the quarks have internal dynamical structure, making them inherently relativistic particles whose mass and electric charge may require corrections leading to the maximization of nuclear energy-density and stability. There is much to be learned in this effort, likely requiring and enabling the development of computer software to facilitate calculations.

Implications for Molecular, Macromolecular Theory

In our scenario energy-density is a key parameter and strongly-nonlinear thermodynamic theory (chapters 3-5), accounts for the creation, under suitable energy-flow constraints, of stable subsystems characterized by minimum density-specific entropy production (maximal energy-density). As noted above, all of the atoms came into being in this way in the earliest history of the Universe (chapter 8, 9) and there is every reason to believe that all molecules, macromolecules and the much larger subsystems of living things came into being in the same way (chapters 16 and 17). Though the calculations, relating to the maximization of energy-density in such stable subsystems, becomes progressively more complicated, as energy-density increases, the GSBP-Schröedinger formalism provides a fundamental relativistic foundation for such calculations, especially as it should enable the relatively straightforward development of computer software toward that end. The same applies to time-dependent calculations (modeling particle scattering, etc.). As in modeling the nucleus, we are only beginning to formulate relativistic theory at these hierarchically lower levels.

The Photon's Role in GSBP-Schröedinger Quantum Mechanics

Clearly, the photon plays a key role in GSBP-Schröedinger mechanics, serving as a precise record, traveling at the speed of light, of the energy-momentum decay of a fermion in a bound subsystem that created the boson. It is the GSB photon in the fermion that requires each of a class, for example electrons, to be identical to all others in the class, and it is in the precise communication of the fermion's energy-momentum decay that informs us about the fermion's local environment, e.g., the electron's state in an atom. Had we discovered electrons not exactly identical or photons incapable of precisely communicating the process by which an electron decays from one energy level to another, there would have been little or no empirical base on which to develop quantum theory.

Are The Born-Derived Quantum Statistics Relativistically Correct?

There is a tacit assumption in all the writings on quantum statistics that Born's probabilistic interpretation of the wave function provides a valid basis for the prediction of observable (statistically measurable) quantum properties. This is most notable in John Bell's theorem, demonstrating the impossibility of improving such predictions (in keeping with classical demands for *locality and causality*) by any "hidden variable" or set of variables. Our aim here is to question the validity of the statistics and suggest that they may be grossly wrong, since they derive from the *non-relativistic* time- independent Schröedinger equation. More importantly, predictions at all stationary-state energy levels are enabled by the relativistic time-independent GSBP-Schröedinger equation, constrained only by reference to Planck's Constant. What then accounts for the much wider statistical spread in current quantum theory? In chapter 14, we argue that the Lamb shift (*apparent* energy difference between the $2S_{1/2}$ metastable state and $2P_{1/2}$ stationary state) is due to the fact that the $2S_{1/2}$ state lies in a small energy well from which it must escape, on its way to the $2P_{1/2}$ state and then quickly to the ground state. Here we argue that it is this small energy hill, out of which the $2S_{1/2}$ state must tunnel-through or climb above, that greatly increases the statistical spread. The same would apply to the higher degenerate energy-level dynamics. If this is true, we can in reality point to a "hidden property" that accords with much sharper statistical predictions – the GSBP internal structure and relativistic dynamics of the electron. This accords with Bell's larger view: "For me … this is the real problem with quantum theory: the apparently essential conflict between any sharp formulation and fundamental relativity. *It may be that a real synthesis of quantum and relativity theories requires not just technical developments but radical conceptual renewal.*"[91] (emphasis added)

GSBP Gauge Theory?

"It still remains a mystery that the geometrical analysis which led to such a deep understanding of gravitation has had no success elsewhere in physics."[92] F. Dyson

Gauge theory is yet a further development of quantum field theory and, in view of our discredit of the latter, might prompt us simply to argue that it would add little if anything to GSBP-Schröedinger theory.

However, the fundamental features of gauge theory underscore the importance of symmetry and symmetry-breaking, the gauge symmetry group, the gauge potential (which defines the connection between particles at different positions in spacetime), and the physical particles (which are the sources of the gauge field) and which interact with each other via the gauge potential. And it is on such foundation that general relativity emerged, suggesting the possibility of close ties with quantum particle theory, as with our GSBP-Schröedinger formulation. We leave this for future investigation, along with other fundamental questions relating to the lack of any rigorous determination of the "free mass" of quarks and a corresponding lack of "Rydberg" reference for the spectra of nuclear emissions.

CHAPTER 13

GSBP-QUARK ELECTRODYNAMICS –
THE ATOMIC NUCLEUS

In this chapter our concern will be to sketch the principal features of Quantum GSBP-Quark ElectroDynamics (GSBP-QED), a physical particle model of the structure and dynamics of the nucleus, and contrasting that with the field-theory algorithm of Quantum ChromoDynamics (QCD). The names assigned to these models reveal a profound conceptual distinction. The *Chromo* in QCD refers to an abstract geometry governing the dynamics of three quarks (mathematical point particles) in a nucleon, whose properties are uniquely designated by one of a tricolor set, while GSBP-QED refers to a real local Euclidean space governing the dynamics of X-, Y- and Z-quarks (GSBP composite particles), in which the force of gravity plays a fundamental role together with the electromagnetic force. It is important to note that the electromagnetic force plays the same fundamental role in governing the dynamics of the GSBP-quarks as that governing the dynamics of the electrons in their orbit(s) of the nucleus – hence the similar heading of the following chapter.

Toward the above end, we seek first to underscore the importance of *energy-density states* in bound GSBP-QED nucleons, as contrasted with *energy states* in QCD theories. The latter refers to *excited* states of particles or *families* of particles found in high-energy accelerator experiments, while the former may refer to the higher *energy-density* state of a particle, e.g., an electron, in the local high *energy-density* time of its evolution in the expanding Universe. The transmutation of quarks, key to such evolution in GSBP-QED, will be seen not as the work of "weak force" particles (W^+, Z^0), but simply as the passive underlying imperative of the *2nd Law*, as the average energy-density of the Universe declined to a critical enabling energy-density level. The action of force particles, bosons, which govern the dynamics of quarks in nucleons and bind nucleons together in the nucleus of atoms, will occupy our attention in much of the remaining part of the chapter. Lastly we address a number of questions, yet unanswered, that have lessened credence in the Standard Model, but which will be seen as non-problems in our scenario.

Evolution of Protons, Electrons and Antineutrinos

Present efforts to understand the structure and dynamics of the nucleus of atoms may be viewed, from our perspective, as attempts in the laboratory to recreate locally (in accelerator experiments) energy-density and temperature conditions that prevailed epochs ago, when the size of the Universe and that of the nucleus was much smaller. This was a time also when the average values of c and $\sqrt{G}$ and the dimensions of quarks and leptons were much smaller and their energy-densities correspondingly greater than they are today. Chapters 7 through 12 have made it clear that the fundamental particles, quarks and leptons, have characteristic intrinsic energy and dimension (radius and thickness) and hence volume and energy-density. In our scenario, the emergence of high energy (high energy-density) protons, electrons and positrinos in the expanding Universe is seen as a major evolutionary event, a second symmetry-breaking event, as stated in chapter 9.

In the Quantum ERA the neutrons dealt with the stress buildup via the down-conversion of each neutron of mass m into two neutrons, each of mass m/2, in which the parameters h_N, m_N, k_N, k'_N halve and N, $\sqrt{G}_N$, $(\nu\varepsilon)_N$ and $(\nu\mu)_N$ double (see Table 9-1) keeping the total mass-energy and angular momentum constant. This was effective because the local stress buildup increases nonlinearly with the decline of ρ_b, leading to a phase transition and a doubling of N and halving of h. That option came to an end as the inherent nonlinearity, giving rise to the stress, and the rate of radial energy flow per unit area became insufficient to support the process.

Subsequently the quarks in the neutrons did the next best thing to alleviate the stress buildup: the d-quarks reduced their mass and transformed their electric charge from $-1/3$ to $+1/6$, thereby transforming the neutron into a proton and leaving the nucleon with a net positive unit charge and greatly increased long term stability. We will go into this in greater detail in the next section. The net loss in mass and gain in angular momentum remained to be accounted for by the emission of bosons from the converted d-quarks, with energy and helicity appropriate to the creation of an electron and a positrino. Thus the mechanism of energy and charge transfer, as a near zero-mass boson, is posited, but it remains to account for the transformation of the bosons into a massive electron and a positrino.

While such bosons have little mass in any local region in which c is constant, they may have considerable momentum and that momentum, as well as a unit of angular momentum (electric charge), must be conserved when they travel into a region in which ρ_{local} may be orders of magnitude smaller than that in which they were created and the value of c correspondingly greater, as in the detector of an accelerator. The bosons must accommodate the change by rapidly accelerating to the greater velocity and in such a way as to conserve their momentum. This could happen if their momentum became gravitationally confined in the form of a GSBP-electron and a neutrino. *We assume that the mass-energy lost in the transmutation of the neutron into a proton is restored largely in the creation of an electron, with the remaining mass associated with the creation of an electron antineutrino (positrino). Likewise, we assume that any excess charge (angular momentum) of the electron is cancelled by the positrino (weakly charged) orbiting the electron. In effect and in a sense literally, one boson pair is gravitationally rolled-up into an electron and the other pair into a positrino, such that one orbit of the positrino about the electron leaves the electron-positrino subsystem in a lowest energy (highest energy-density) ground state and the proton-electron subsystem in its lowest energy (highest energy-density) state.* Like the hydrogen atom, such a system would be relatively stable over a small range of ρ_{local} variation. But the foregoing could be true only if: (1) neutrinos, like electrons, have specific *intrinsic* mass and angular momentum (charge) and became free, stable particles in our low energy-density Universe; (2) the opposite spin of the positrino serves, as we said, to cancel the excess angular momentum of the electron by its orbit about the electron in an earlier time, just as does the electron in its present-day orbit about the proton in the hydrogen atom; (3) the rest mass of the positrino is two to four orders of magnitude smaller than that of the electron and the magnitude of its charge is equal to the excess charge of the electron, thereby binding the newly-created positrino to the electron and the electron to the proton and (4) in such systemic form the positrino would serve to enable a rapid reverse transmutation (p→n), as fluctuations might require. The electron might thus be stable over a small range of potential energy (relative to the proton) and local energy-density, in accord with empirical studies. In this view, the positrino is seen as a positively charged lepton, in which the radius of the GSB photon gyration in the positrino is two to four orders of magnitude greater than that in the electron in the rest state, or about 10^{-10} or 10^{-9} m in the vacuum in which it now travels at near the speed of light, characteristic of the present low energy-density state of the Universe. In this light it is not surprising that neutrinos are now virtually free of interactions, except those associated with extreme energy-density conditions – extreme gravitational or electromagnetic energy-density – referred to in QCD as the "weak or "electroweak" force. Just as the proton and electron emerge as stable stress-free subsystems of a decaying system (neutron), so a positrino evolves to cancel or diminish the local angular momentum (charge) of the electron in the high energy-density environment of its creation.

In the above high energy-density conditions, small fluctuations could readily initiate transformations in the reverse direction. It is important to note here that, while the above transmutation of neutrons into protons has been depicted in our scenario as a cosmic evolutionary event, it is inferred in current theory from tracks in particle detectors, brought about by high energy reactions. In our scenario we see that the radius of photon gyration in the converted d-quarks is necessarily increased, as ρ and the ρ_{local} in the d-quarks declined in an epoch. That is, the *intrinsic* energy of the converted d-quarks *was reduced, as was their energy-density*. It is important also to remember that the transformation proceeded passively via fluctuations and made no specific demand for the creation and exchange of virtual *force particles* (W^{-+}, Z^0) to mediate it, as in QCD. To be sure a real physical force was necessary in that earlier time to statistically mediate the demand of the *2nd Law*. That force was in the form of real bosons emitted by d-quarks in the neutrons, acting by impulse-recoil to minimize the energy-density gradients on the surfaces of the quarks and thereby the momentum and stress in the d-quarks, as they orbited the X and Y local Euclidean axes. The effect of this was to raise the potential energy of the d-quarks (lower their kinetic energy), such that at some point the lower energy-density favored reversal of the electric charge polarity of the d-quarks and thereby the transmutation of the neutron into a proton. (Fig. 13-1) The accompanying violation of parity, in this transmutation, is readily explained in the GSBP model as a polarity inversion of the d-quark's angular momentum, requiring the emission of left helicity bosons and their decay into a GSBP-electron and a positrino.

A B

C

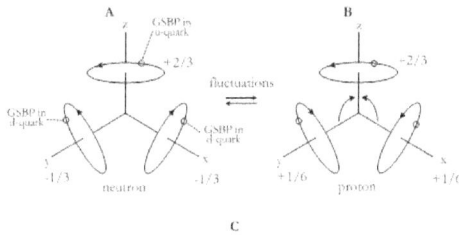

GSBP-Proton, -Positrino dynamics as start of γ decoupling epoch when an electron was in the rest mass state and a positrino orbited the electron, conserving angular momentum and readily enabled a reversed p-n transmutation, as fluctuations might require.

Photon in GSBP-electron cycles α^{-1} times as the electron orbits the proton once in the H atom.

Positrino extended fine structure relativistic correction to the Bohr model of the H atom of order of a few parts per billion (perhaps 0.1 part per epochal change of c.

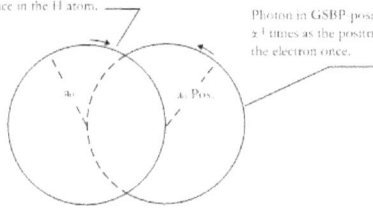

Photon in GSBP positrino cycles α^{-1} times as the positrino orbits the electron once.

Fig. 13-1 (expands on Fig 9-3) shows the passive (fluctuation induced) transmutation (panels A and B) of a neutron into a proton in the early γ-decoupling epoch of the Universe, in which one of the *d* quarks (-1/3 charge) responds to increasing stress in the neutron by reversing its spin and transmuting to a +2/3 charge and resulting in a proton having unit of positive charge. This was a time in which the electron was in a rest mass state and a positron orbited the electron as a means of conserving angular momentum. In so doing a reversed fluctuation could easily result in the passive mutation (p→n) of the proton into a neutron. Such action was driven at root by the NTI requirement to minimize entropy production in the Universe. In panel C we show the GSBP-Proton, GSBP-Electron, and GSBP-Positrino dynamics at the start of the γ-decoupling epoch when an electron was in a rest mass state and the photon in the positron cycled α^{-1} (~137) times as the positrino orbited the electron once in exact closure. Just as the photon in the electron cycled α^{-1} times as the electron orbited the proton once. The result is a positrino extended fine structure relativistic correction to the Bohr model of the Hydrogen atom of order $(\alpha^2)^2 \sim 10^{-9}$.

But what was the purpose of the positrino, apart from conserving mass and charge? Why *antielectron* neutrino? The electron was created and existed in its high energy-density environment as a nearly-free massive particle with negative electric charge. In that environment the electron was nearly free in random orbit about the proton, as was the positrino to the electron, and thereby the angular momentum and spin of all particles was effectively cancelled. But there was much more to this puzzle, as we show in Fig. 13-1. The positrino extends the fine structure corrections to the Bohr model of the hydrogen atom, pointing toward a precision physics of $(\alpha^2)^2$ order of magnitude, or about one part in 10^9. With possible further advances in measurement technology, could we begin to discern the variation in energy density $\Delta\rho$ at which the velocity of light increases from one epoch to another?

The above is necessarily a very different view from that envisaged in the detector of a super-accelerator. There we would have to infer the existence of high energy-density quarks making up an inferred neutron, which would rapidly decay into an observed proton, and then infer the existence of a massive boson, which would decay into an observed electron, and finally infer the existence of an antielectron neutrino. All of these are very high-energy (high energy-density) unstable particles. That is to say, a great deal of energy would have to be expended, with associated entropy production, in order to create such a neutron as would have existed prior to the evolution of protons, electrons and electron antineutrinos. Indeed the energy required to observe the Ω-particle in the spin -3/2 baryon decuplet and its decay to lower states is only a fraction of what would be required in the above super-accelerator scenario. This can be appreciated best in terms of the evolutionary decline of the average-energy-density parameter ρ and the requirement in the laboratory to reverse this decline locally in order to observe these particles in action.

What prevented all neutrons from transmuting into protons? This of course never happened, since otherwise there would be no atoms heavier than hydrogen and the Pauli Exclusion Principle would have prevented any significant agglomeration of matter in the Universe. As the number of free neutrons declined and the number of protons increased, the local energy-density declined, due to the mass difference between the two, relaxing the stress in all quarks in the remaining free neutrons and slowing the transmutation process. With further ρ decline, the stress buildup would accelerate, leading to a repeat of the process, though of more limited extent. However such events would have left us with far too many free neutrons, had it not been possible for neutrons and protons to combine into stable high energy-density nucleons in the nucleus of atoms, as discussed below. *It is important here to note that such combination may be seen as yet another example of the creation and evolution of subsystems, which serve to minimize entropy production in the Universe.* The above reveals a profound distinction between the two approaches toward understanding the nucleus of atoms, the one relying on physical energy-density models to achieve self-consistency, taking into account the demands of both general and special relativity, and the other on black-box algorithms to make up for inconsistencies ("point particles," etc.), unaccountable to general relativity. This distinction needs occasional reminding as it characterizes much that follows and in the remainder of Part III.

GSBP-QUARK ELECTRODYNAMICS (GSBP-QED) OF THE NUCLEUS

Attempts to employ the perturbation methods of QED to account for the dynamics of quarks in the nucleus were met with relatively little success, as it became apparent that the "point particle" description of quarks led to non-renormalizable results, owing to the much larger coupling strengths and their variance over the range of the nuclear energies. Here, the GSBP model of quarks, with their internal spatial representation and dynamics and inherent accord with general as well as special relativity, holds promise for a more precise characterization of the interaction of quarks in nucleons and between nuclei in the nucleus.

In sketching the principal features of GSBP-QED, the first task is to make clear the roles of gravity, quantum Euclidean geometry, fractional-thirds electric charge and symmetry in accounting for the binding of quarks *in* nucleons and to compare that with the exchange of "gluons" carrying color and electric charge in QCD. Similarly we need to understand the roles of Euclidean geometry, fractional-thirds electric charge and symmetry in accounting for the binding of baryons together in the nucleus of atoms, in GSBP-QED, rather than via the exchange of massive bosons (mesons) as in QCD. In QCD the "strong force" refers to the binding of quarks *within* and *between* nucleons, while in GSBP-QED we refer to quarks electromagnetic force in local Euclidean space.

Today's quarks (yet unseen) are direct descendants of our postulated proto-quarks in the early quantum Universe. The extreme energy-density, that warped photons, traveling at $\iota \rightarrow 0\text{m/s}$ into circular spin orbits in the proto-quarks and the electromagnetic forces function similarly in today's neutrons, albeit at energy densities orders of magnitude smaller and orbital radii and velocities of the quarks comparably larger. The EM force serves also to warp the spin of the three quarks in the neutron to null (exactly and in noncommutative order) its total angular momentum, and in the proton to give rise to $\hbar$ total angular momentum with right helicity, as shown Fig. 13-1 panel B).

Quarks thus have mass and occupy real space in our scenario and so we should expect *that* space to be local Euclidean and, like intrinsic "isospin" in QCD, subject to the same SU(3) symmetry, i.e., to non-Abelian group transformations. However there is an important distinction and advantage of the real Euclidean system over the isospin algorithm: the GSBP Euclidean model inherently incorporates or takes into account the requirements of general as well as special relativity, just as does the GSBP model of the electron in the U(1) symmetry of quantum electrodynamics (chapter 14). Indeed, as implied in Fig. 13-2, the GSBP Euclidean model of the nucleus of atoms is characterized by U(1) symmetry. So, instead of the colors red, green and blue, we henceforth refer to the Z-, X- and Y-axes of the local Euclidean geometry and to the individual quarks occupying those subspaces of a nucleon. The quarks are bound in this space not by the exchange of "gluons," carrying color and electric charge, but by the electromagnetic force, acting in conjunction with the forces of gravity and symmetry, to adjust the orbital angular momentum of the quarks in their positions on the Z-, X- and Y-axes, so as to null the angular momentum and charge of the neutron. The same guarantees the appropriate values of these parameters for the proton (with charge +1). The gravitational and electromagnetic forces act together toward these ends, with the latter mediated not by the exchange of virtual gluons *between* quarks but by the emission of bosons so as to directly adjust the momentum of the emitting quark by impulse-recoil.

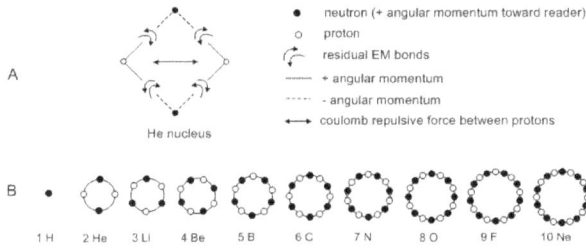

Fig. 13-2   Internal structure of the nucleus of the Helium atom and higher Z atoms. Panel A depicts bonds ↔ that are formed between the +1/6 charges of the X and Y d-quarks in the emerging protons and the residual –1/6 charges of the d-quarks in the newly decayed neutrons. We see also in panel A the Coulomb repulsive force >– < between the protons. The former acts to compact the nucleus, to increase the energy-density, while the latter serves to minimize the repulsive force between the protons and therefore as a limit on the compaction. The same applies (in panel B) as more and more neutrons and protons are added make up the heavier nuclei such that the radius of the nucleus tends to increase as the square root of Z, the atomic number. This nonlinearity is due in part to the dipole moment action of the neutrons between proximate protons tending to increase the Coulomb repulsive force proportionally with that of the compacting force. The result for the size of the atoms is to cause their radii to increase only marginally from that of the Helium atom to the heaviest stable atoms.

For an isolated and relatively stable neutron (in an earlier epoch where $\rho >> \rho_0$), the linear and orbital angular momentum gradients of the quarks are nearly zero and rise marginally with increasing potential energy (increasing orbital radii of the quarks within the neutron radius), as ρ_{local} declines and stress builds in the fixed orbit of the photon in each quark. That is to say, the binding force of the quarks in the neutron varies only marginally, since the local EM energy-density *within* the radius of the neutron varies relatively little. This is the underlying physical mechanism accounting for the counterintuitive "asymptotic freedom," characterizing the relatively-weak bonding within the neutron. However, the local EM energy-density declines steeply (negative energy-density gradient increases rapidly by as much as twenty orders of magnitude) beyond the radius of the neutron and, as a consequence, the EM energy-density rises, increasing the binding force on the quarks via impulse-recoil response. And it is this latter mechanism that accounts for the extreme

215

"containment" of the quarks, preventing them from leaving the neutron. Similar dynamics would hold for the proton.

In the laboratory in which a free neutron decays (transmutes) into a proton plus an electron and a positrino, stress buildup will result in breaking the angular momentum symmetry of the local Euclidean space, requiring the two d-quarks to reduce and ultimately reverse their orbital angular momentum and charge, each by transferring (rotating) a +1/6 charge to the Z axis. This resulted in a 3/3 positive electric charge for the proton and requiring a -1 charge on the electron, thereby conserving mass-momentum, angular momentum, and charge. The resulting +1/6 d-quarks in the proton are thus prepared to bond with other neutrons as we show in Fig. 13-2. However, as a member of the lepton family, the positrino is a spin ½ particle and should therefore have discrete intrinsic energy (mass) and *electric charge*, i.e., it should be a stable GSBP composite particle like the electron, and quarks, despite evidence to support this idea. (Particle physics has been challenged by the inability to observe particles characterized by extremely small energy-density and angular momentum. Recent evidence of the oscillation of neutrinos between energy levels may relate to their sensitivity to local energy-density variations in their travel from their source to detectors.)

The above physical process could of course and did proceed in the reverse direction in an earlier epoch, and it could and does happen in reverse, where the local energy-density of the proton is raised (as in an accelerator) to sufficiently high levels. All of the above is thus seen to be driven by the forces underlying the expansion of the Universe: the *2nd Law*'s requirement to minimize the rate of entropy increase relative to the rate of energy flow in any interaction; the nonlinear behavior of gravity (general relativity); and the requirements of symmetry and symmetry breaking on conservation of momentum, angular momentum and charge.

In chapter 11, it was argued that the direct exchange of force quanta between massive particles was untenable in our scenario because no physical (intuitive) basis could be posited to account for the precise exchange of momentum. (QFT encountered no such problem, either in QED or QCD, as the virtual photons, etc., were simply assumed, as part of the algorithm, to travel from one vertex of a massive particle to another.) Instead such force would be implemented via the emission of real bosons from each such particle, thereby acting directly on the emitting particle through impulse-recoil to alter its momentum, as required in the larger system. The gravitational force (never negative), would be mediated via the emission of spin 1 bosons, as with the electromagnetic force (positive or negative depending on the electric charges of the particles). In either case, the force acts locally on the emitting particle (not on other particles) in response to energy-density gradients at the surface of the particle and in the form of impulse-recoils. Unlike the supposed exact exchange of virtual photons in QED/QCD, the adjustment of the emitting particle's own momentum occurs statistically in multiple impulse adjustments to null or minimize the energy-density gradients on the surface of the particle, due to the presence of nearby particles. Since both the gravitational and electromagnetic forces may act at the same time on a particle, there is a nonzero probability that an electron, for example, could emit bosons, propelling it gravitationally by recoil-action toward another electron, while also emitting bosons, tending to move it, by recoil-action, away from the electron. In practice of course, in the average energy-density conditions of the laboratory, the gravitational impulse is infinitesimal and probability favors complete dominance of the repulsive over the attractive impulse, as explained in chapter 11. The above applies also to the interaction of charged and uncharged particles in the nucleus.

With the emergence of positively charged massive particles (chapter 9), a neutron could bind with a proton, via the emission of bosons, if that reaction resulted in a decrease in the mass-specific entropy production (increase in local energy-density and stability) of the product, together with a decrease in the local rate of flow-specific entropy production in the larger system. This could occur as an electromagnetic-symmetry interaction between the plus half-charges in the d-quarks in the decayed neutron and the minus half-charges in d-quarks in a non-decayed neutron, as residual or van der Waals type bonds of half charges (-1/6) so as to result in a full charge for the proton; such passive transmutation would then leave the proton able to bond together two different neutrons in different nucleons, as we show in Fig. 13-2

Before pursuing the physical nature of this bond, it is necessary to examine the internal structure of the proton. Externally, the proton behaves as a fermion having U(1) symmetry and spin ½, i.e., the mass-energy of the three GSBP quarks, constituting the proton in the rest state, must add up to the known rest mass of the proton, and the net electric charge must be +1. In this model the original u-quark, together with the transmuted d-quarks (into a u-quark), account for the net +1 charge, as depicted in Fig. 13-1 panel B. As sketched, the spin of the converted u-quarks is electrically warped in Euclidean space to add to that of the original u-quark, thereby accounting for a net charge of +1 and revealing the role of symmetry along with that of the EM force in governing the behavior of quarks in a nucleon. However, the above leaves us with no way of knowing how much of the mass-energy of the proton may be ascribed to the "rest masses" of the quarks. Indeed, the rest mass of a quark, unlike that of an electron, is devoid of meaning. Yet, like the electron, we should expect the negatively charged d-quarks to orbit the local X and Y-axes and the positively charged u-quark (spin axis of the proton), evolving toward a lower potential energy while gaining in kinetic energy (orbital angular momentum). All of this – transmutation of a neutron into a proton and its evolution toward the lowest energy state, the ground state of the proton--would come about as a means of minimizing the flow-specific entropy production of the system, under favorable energy-density conditions.

As we will see in chapter 14, the GSBP relativistic model of the electron greatly simplifies its relativistic electrodynamic interaction with the proton in the hydrogen atom. That interaction is driven, at root, by the requirement that the electron, in the ground state, be characterized by null total angular momentum and potential energy gradients. This principle may be applied in modeling the GSBP electro-symmetry dynamics of the protons and neutrons in the nucleus of the helium atom, in which bonds are formed between the residual spin-reversed d-quarks (+1/6) in the decayed neutrons and the residual d-quarks (-1/6) in the protons, as illustrated in Figs. 13-1. However the mass of the two nucleons is very nearly equal, thus resulting in a reduced mass correction for their nearly equal mutual orbital radius, depending on the strength of the residual force. And such correction would likewise entail reduced charges of the d-quarks, thus further weakening the bond. (Interestingly, if our best estimate of the constituent masses of the quarks comes from such interactions, then we might speculate that the true masses of the quarks have been underestimated by a factor of ~2.) Moreover, the neutrons serve also as a dipole agent to reduce the coupling strength between the protons, thereby further increasing the energy-density of the nucleus.

Thus, this situation resembles that in the coupling of atoms in a molecule, via the van der Waals (electromagnetic) force, subject to kinetic constraints. In both cases symmetry plays a critical role. The concern in the nucleus is with mutual attractive motion of the neutron and the proton so as to achieve a ground state in the nucleus, in which the total angular momentum and potential gradients of the neutron and proton are null, subject only to small fluctuations. As in the case of the binding of atoms in a molecule, we should expect excited (resonant) metastable states (corresponding to the greater number and complexity of the spectrum associated with these states) the greater the number of nuclei. The same symmetry options and constraints apply in the makeup of the helium nucleus and that of all atoms of higher atomic mass, when the repulsive force acting to disperse protons in the nucleus is taken into account. Thus we see in the helium nucleus the two protons at maximally separated positions with their axes of angular momentum oriented in the same direction (Fig. 13-2).

It is important to note here that this view of the nucleus, i.e., of a nuclear structure in SU(3) symmetry in Euclidean local space, derives from the initial assumption that the internal space of a nucleon is in fact a local Euclidean space and not simply some "internal," "rotational" or "color" space. And this in turn derives from the existence of quarks as GSB gyrating photons, having dimension and orientation in local Euclidean space and in which much of the energy of the Universe is now locked up. It is important also to note that each higher level of organization, at which particles are bound together with long-term stability, serves the demands of the *2nd Law*, while greatly facilitating the creation and evolution of complex systems at higher and higher (more complex) levels of organization. With the increasing size and stability of such systems, the lower levels of particle organization become less subject to external energy fluctuations, i.e., relatively shielded from or immune to the larger environment, though yet contingent on possible catastrophic events. This holds for quarks in nuclei and nuclei in the nucleus and atoms in molecules, as well as for cells in organisms and (in weaker and more flexible communicative bonds) for organisms of a species in a social system or an ecosystem, all in keeping with the principle of minimum flow-specific entropy production and the *2nd Law*, i.e., the Nonequilibrium Thermodynamic Imperative. We will return to this in part IV.

The Hierarchy and Other Problems with The Standard Model

The rest mass (intrinsic energy) of the fundamental particles we investigate today (quarks and leptons taken as GSBP composite particles) derive ultimately from the evolution (expansion) of the quantum state Universe, as discussed in Part II. The range of energy of these particles extends from $\sim 5 \times 10^5$eV, for the electron, to $\sim 2 \times 10^{11}$eV, for the top quark, or a range of about 4×10^5 times the energy of the electron. Moreover, if the neutrinos are demonstrated to have measurable mass, as proposed above, the range would then be extended by two or more orders of magnitude. This is the "hierarchy problem" that disturbed Weinberg when he said "we need to "understand it (this vast jump from one level to the next in the hierarchy of energy scales) not just by adjusting the constants in our theories to make the ratio come out right but *as a natural consequence of fundamental principles."*[93] (emphasis added)

In the GSBP model of the composite "fundamental" particles, we might expect that the largest-mass particle, the top quark or some predecessor, may have been among the earliest quanta to emerge in the very high energy plasma state of the Universe as it expanded and cooled. The lower mass particles condensed out progressively on further cooling, with the electrically charged fermions emerging, coincident with the transmutation of d-quarks into a u-quark in some neutrons, in the still relatively high energy-density environment of the nucleus. *This marked a profound transition in the state of the Universe in that it opened the way to a much wider range of stable particle energies and systems for decreasing the rate of entropy increase in the Universe. In particular it led to the possibility, in the then-distant future, of great numbers of stable particles (protons, neutrons, electrons and neutrinos), capable of combining to form stable atoms and combinations of atoms in still more complex structures and dynamic systems. The energy hierarchy is thus intimately associated with the storage of energy, at relatively high hierarchical levels and in relatively stable nuclear quanta as well as progressively lower levels in particles that may combine flexibly in organized dynamic systems (e.g., the molecules and organisms of living systems) at still higher hierarchical levels of organization.* The top quark and others that are today very short-lived were, in earlier cosmic epochs, before the u- and d-quarks condensed out, relatively stable in their much higher energy-density environment. There is in this logic no need to invoke supersymmetry or new forces or new massive particles to make sense of the wide range of particles we encounter. *Were it less extended we would not have the hierarchy of stable dynamic systems that characterize the Universe from galaxy clusters to solar systems, quantum systems and living systems of almost infinite variation.* The Standard Model has little to say on these matters. But we see that this pattern of gradual, minimal increase in entropy production (increase in numbers of relatively stable particles) accords with the principle of minimum flow-specific entropy production and the *2nd Law*.

Gordon Kane[94] has listed ten concerns that have no explanation in terms of the standard model:

(1) the "cosmological constant" problem;

(2) the minimal accelerating expansion of the Universe;

(3) the inflation fine-tuning problem;

(4) the matter/antimatter asymmetry problem;

(5) the cold dark matter and dark energy problems;

(6) the Higgs field that gives specific values to the mass (inertia) of particles;

(7) the huge mass of the Higgs boson that leads to huge values for the mass of all particles;

(8) the structural isolation of gravity from the other fundamental forces (theoretical isolation of general relativity);

(9) the masses of quarks and leptons;

(10) the existence of three generations of quarks and leptons without apparent need for more than one. Kane says, the Standard Model "cannot *ever* explain" these puzzles, but he goes on to say "One reason the supersymmetric extension [to the Standard Model] is attractive to many physicists is that it can address all but the second and the last three of these mysteries."

Here, we show that all of the above concerns are non-problems in our scenario. The "non-problem" characterization of items 1–5 derive directly from the fundamental assumptions relating to the coevolution of ρ, c, $\sqrt{G}$ and their inherent compatibility with the field equations. The asymmetry of matter and antimatter, item 4, is accounted for also by the application of adiabatic down-conversion theory, in experimental quantum studies, to the quantum state of the early Universe. The specific values of the mass (inertia) of present-day particles, item 6, derive from the mass-energy assumptions about the proto-quarks constituting the early Universe, as that energy was parceled out via down-conversion processes; one could say that the intrinsic masses of the ultimate progeny were set by the energy-density scalar field ρ. (It is remarkable that the Brans-Dicke[95] scalar field φ, proposed in 1961 to satisfy Mach's principle by allowing G to vary in time and space, and Higgs scalar field φ, proposed in 1964 to give mass to matter particles, were eventually seen in a similar light, while failing to see their possible connection with ρ or ρ_{local}. The inertia of GSBP massive particles derives from their physical dimensions and volume and energy cost of changing that volume. Having no meaning or likely existence in our scenario, the Higgs boson, item 7, is a non-problem. The structural isolation of general relativity, item 8, is surely sufficient by itself to question the validity of the Standard Model. The demands of both general and special relativity are inherent in the GSBP model of the fundamental particles and the gravitational force is therefore structurally related to the other forces. Item 9 is here considered as a particular aspect of item 6 and is therefore seen as a non-problem. The three generations of present-day quarks and leptons are distinguished principally by their relative masses or rather, as stated above, by their relative local energy-densities (energy cannot be gained or lost in the adiabatic expansion of a spatially bounded physical particle; it is only the energy-density that can decline in the decay of particles of fixed intrinsic energy). In our scenario there is only one generation of quarks and one of leptons (item 10), the energy-density of which, in each case, varies as ρ^{-3}, as does the radius of gyration of the GSB photon in the quarks and the leptons, with energy strictly conserved. Similarly, the radius of the neutron and the proton vary as ρ^{-3}, resulting in their decrease in energy-density in spacetime. Thus item 10 is a non-problem.

The above is only a sketch of a GSBP theory of the structure and dynamics of atomic nuclei. In principle it could be further developed in SU(3) space in a manner similar to that presented in the next chapter relating to relativistic quantum electrodynamics of the hydrogen atom. As for the electron, it would be necessary first to determine the radius of

GSB photons in the quarks in the nuclear equivalent of the electron's intrinsic rest mass in terms of their energies $(h\nu)$. In the case of the electron, this refers to a state of zero potential and zero kinetic energy in which the electron is said to be free. But the quarks can be free only in an extreme energy plasma state. Yet experiment informs us that they are in a state of "asymptotic freedom" in the ground state of a nucleon. By analogy we ask: Does asymptotic freedom imply a state of zero potential and kinetic energy, within the constraint binding the quarks in the nucleon, to which we could assign the equivalence of a Rydberg radius on the X-, Y- and Z- local Euclidean axes and thereby their mass-energy values and spin in terms of radii of GSBP gyration and values of $h\nu$? The ground state of the quarks could then be characterized as the radius at which the u- and d-quarks orbit their respective X- and Y- axes in which the system is in a state of null orbital angular momentum. And, as for the electron, such a model would be inherently relativistic, in terms of both general and special relativity. There are some very big "ifs" here, which could perhaps be clarified by more precise experimental studies.

Closely related and even more challenging is the need to quantify the dynamics of the relatively-weak, residual electromagnetic force between a quark in a neutron and one in a proton, binding the two together in the nucleus via the emission of bosons in impulse-recoil formalism. As this force declines rapidly with the distance separating the quarks, a great deal of energy-momentum is required to bring them together in order to initiate such a bond. That possibility may occur in the binding of protons to neutrons to form deuterons or $2H^{++}$ in the extreme energy-density of the early Universe or, more generally, in stars to form the nuclei of higher-mass elements. There is much more to learn in light of the GSBP model.

CHAPTER 14

GSBP-ELECTRONDYNAMICS –
THE HYDROGEN ATOM

Here we discuss the process by which quanta of electromagnetic *thrust* are emitted (in the form of *real* bosons) from GSBP protons, electrons and their related charged matter and antimatter counterparts. We then consider the implications of this strictly particle dynamics for the structure and dynamics of the electron in the hydrogen atom. Here again, we take this opportunity to draw attention to the fact that the hydrogen atom, and all of the heavier elements, evolved as open thermodynamic subsystems, characterized by maximal energy-density, which serve to minimize entropy production in the Universe by locking up much of the energy in the Universe in long-term stable composite particles.

GSBP Model of the Electromagnetic Force, Electron Dynamics

Since the electron, in conventional theory, is considered a point particle, i.e., lacking in internal structure or energy store, it cannot, as a free particle, emit real bosons and interact thereby with nearby electrons, in such a manner as to minimize their mutual separation in a hypothetical electron sea, as energy and spin angular momentum considerations would require. Consequently, that theory made it necessary to depend upon some other energy source to make such exchange force possible. Energy in the vacuum, the "physical vacuum" was (is) believed to exist, in the form of fluctuations of fields of various kinds and wavelengths, and ostensibly was tapped as a hypothetical means to create, exchange and annihilate virtual particles (photons, electrons and positrons, etc.), i.e., to embody the electromagnetic force. This in turn called for the development of some process (renormalization theory) to cancel mathematical infinities in calculating the EM force, which resulted from the lack of physical size or internal structure of the electron. (One would think it now long overdue to seriously question the existence of fundamental particles having definite mass-energy and angular momentum and spin properties yet completely lacking in dimension and structure.) Nevertheless, the resulting Quantum Field Theory (QFT) has proved to be extraordinarily successful in reformulating quantum mechanics to conform to the demands of special relativity in accounting for the electromagnetic force.

The GSBP electron enables us to do away with the dependence on the physical vacuum and the "hocus-pocus" of the renormalization process, in the development of a much simpler and intuitively comprehensible model of the electromagnetic force. Having a means of storing (conserving) internal energy-momentum and angular momentum, a free GSBP electron is able to emit real bosons, carrying small quanta of energy-momentum away from the electron in a manner conveying the sign (charge polarity) of the electron's spin angular momentum (helicity of the internal bound photon). In this, the GSBP electron moves in a manner similar to that of a jet engine, by expelling energy-momentum in a direction *opposite* to its travel, with the lost internal energy of the gyrating photon being compensated by the electron's increase in external (kinetic) energy. Thus we refer to the emission of quanta of *thrust* rather than of exchange. This behavior, together with similar action by other electrons in a hypothetical sea of electrons (devoid of other charged particles), would tend, on average, to maximize their mutual separation and minimize the potential energy and EM energy-density of the system. We say "on average" because the process would involve the emission of many bosons from each electron over time, as they are thrust collectively toward a quasi-stationary state of uniform maximal separation and minimal energy-density. Such process would be dynamically similar to the evolution of a Thermodynamic Minimal Gradient Structure (TMGS), as discussed in chapter 1. It is important to note here that the above process makes no mention of the direct exchange of a boson's momentum between two electrons as QFT/QED would have it, since there is little probability of that happening. As explained in chapter 12, a massive charged particle (electron in this case) is responsive to the magnitude and orientation of the local angular momentum gradient imposed by its environment and each photon it emits would travel on only one of a great many possible trajectories, there being no mechanism for precisely directing the trajectory of the emitted photon, though the range of such trajectories would be reduced the greater the local energy-density gradient.

If, in contrast to the above, an electron is near an isolated proton, its kinetic energy would be increased and its negative potential energy decreased by movement, on average, toward the proton, a consequence of its opposite sign (electric charge) of angular momentum, relative to that of the proton. The dynamics of the GSBP electron, assuming it had zero kinetic energy at zero potential energy, would be as follows: The electron, having some small momentum in the general direction of the proton, would emit low energy bosons on some random trajectory away, on average, from the proton, causing the electron to move by recoil-momentum, closer and closer to the proton in random orbits about the proton, thereby gaining kinetic energy and losing potential energy. As the electron approaches the proton, its kinetic energy is manifested increasingly in the form of orbital angular momentum about the proton, leading at some point to deceleration of the approach, i.e., as its velocity increases and orbital radius decreases. In the course of this travel, the electron may reside very briefly in discrete *metastable* states of the conservative system (hydrogen atom), having near-zero orbital angular momentum. Ultimately the electron will reside in the *stable* ground state of zero orbital angular momentum, such that any small fluctuation would lead only to the restoration of the stable state.

This is also a state of zero angular momentum gradient at the surfaces of the GSB photon in the electron and zero potential energy gradient with respect to the nucleus, thus entailing no emission of thrust-bosons, except to counter fluctuations. All of this is in accord with the Dirac theory, with the added advantage of providing a *physical* model of the dynamics of the mysterious electron "jumps" that take it from one allowed state to another, with the emission or absorption of a boson, as discussed in the next section. (We should expect such a model of boson emission would apply straightforwardly in an environment where the local values of c and $\sqrt{G}$ are greater or less than ρ_0, see chapter 11.) In the above account, which assumes the mass of the proton to be effectively infinite, there is no need for the creation and annihilation of "virtual particles" (from energy momentarily "borrowed" from the vacuum) and in particular for the creation of virtual photons to adjust the electron's "self energy" and positron-electron pairs to shield or "dress" "naked electrons." Nor is there any need for renormalization processes to extricate infinities from calculations. All of that became necessary because the effective charge of the electron had to decrease marginally with increasing distance from the proton, in order to account for an apparent anomalous decrease in the Coulomb force, *given that the "point electron" offered no internal means for effecting relativistic corrections to the charge, as well as the mass of the electron, as the conservation of the intrinsic angular momentum of the electron would demand.*[96] (This dependence on the physical vacuum is essentially the same problem we encountered in chapter 6, where it was necessary to resort to the physical vacuum in order to power the superluminal expansion of the Universe, which in turn was made necessary because the speed of light, being absolutely constant, did not allow sufficient time to account for the expansion in a more rational way.)

If quantum electrodynamics, deriving from field representations of the fundamental particles, is a flawed theory, as we now believe, how is it that it applies with such extraordinary precision? We will return to this question later.

GSBP Model of the Hydrogen Atom

The Bohr model of the hydrogen atom, taking into account the reduced mass of the electron (an adjustment relating to the center of orbital motion of the proton and electron), correctly predicts the observed spectrum of the hydrogen atom to within 3 parts in 10^5. (It is curious that the radius and energy of the electron in the ground state and the radiation spectrum of the hydrogen atom can be derived, to this order of accuracy, in a purely classical way, based on the indeterminacy principle.) Part of the remaining error can be removed, in conventional theory, by including a first-order perturbation correction for the relativistic momentum and energy of the electron in the ground state (assuming c absolutely constant). Further corrections take into account (still with c an absolute constant) higher order relativistic terms and the interaction of the electron's spin with its orbital angular momentum, as well as its interaction with the spin dipole moment of the proton. But such corrections became far more difficult for the higher energy states, requiring many years of labor-intensive calculation to achieve thirteen decimal place accuracy in accounting for an *anomalous* increase in the electron gyromagnetic factor $g_e = -2.0023193043737$, relative to the exact value -2 derived from the Dirac equation. According to current QED theory, the coupling strength of the electron must decrease, as the electron orbits farther and farther from the proton, due to "polarization of the vacuum," i.e., the screening of the electron by virtual positron-electron pairs, created by tapping energy in the vacuum. (The electric polarization of positron-electron pairs may be viewed in a similar light to polarized charges in the dielectric of a capacitor.) This reasoning and much of the difficulty encountered in the calculations was necessitated by the assumption that the electron lacks internal structure and dimension (being a "point particle"). In effect renormalization theory emerged as a formal means of correcting for this structural void and attributing some physical and charge dimension to the electron, in order to cancel out infinities in the calculations.

Missing from the above is any reference to the "Lamb shift." According to Dirac theory, the $2S_{1/2}$ and $2P_{1/2}$ states of the hydrogen atom are degenerate, i.e., they have exactly the same energy. Yet, various studies hinted at a displacement of these states, and the Lamb-Rutherford[97] experiments in the mid-twentieth century clearly revealed a separation of the energy levels of about 1000 megahertz, or about nine percent of the spin relativity doublet separation. This provided important empirical support for the field theory development of quantum electrodynamics. Recent experiments with lasers confirm the existence of the "Lamb shift" and with greater precision. The Lamb shift and the "anomalous magnetic moment" of the electron (accounting for the slight departure of the electron g factor from the exact value −2) then appeared to deny legitimacy of the Dirac theory.

In essence, the Lamb-Rutherford experiment created a beam of excited hydrogen atoms in the $2S_{1/2}$ state, which could be selectively deflected and detected, thereby minimizing line-broadening of the $2S_{1/2}$ →$2P_{1/2}$ emission. The atoms in the metastable $2S_{1/2}$ state were subjected to magnetic fields and rf fields of variable strengths resulting in Zeeman and Stark splitting of the $2S_{1/2}$ state, while the atoms in the non-metastable $2P_{1/2}$ state experienced no Stark splitting, because of the extremely short lifetime of the $2P_{1/2}$ state in the beam. In this way the energy *required to populate* the $2S_{1/2}$ state was increased, relative to that of the $2P_{1/2}$ states, when photons of energy $\Delta E = h\nu$ were absorbed by atoms in the metastable state, where h is Planck's Constant and ν is the frequency of the photons. Note, however, the above does not demonstrate that the energy of the $2S_{1/2}$ state is greater than that of the $2P_{1/2}$, only that some additional energy ΔE is necessary to *induce atoms into* the metastable state, from which we might infer that the $2S_{1/2}$ state must owe its metastability to an energy well or barrier, preventing it from instantly decaying to the ground state via the $2P_{1/2}$ state. This barrier is associated with energy stored in the precession of the GSB boson (spin) about the axis of orbital angular momentum of the electron. The energy ΔE is taken here to be the energy required to pass over this barrier or tunnel through it, which *appears* to break the degeneracy of the $2P_{1/2}$ and $2S_{1/2}$ states, as sketched in Fig. 14-1. The evidence for the Lamb shift has been and remains wholly associated with the resonance absorption and decay of energy in the $2S_{1/2}$ state, which we now argue is associated with an energy well characteristic of metastable states. Thus the considerable evidence of resonant absorption of energy ΔE, in hydrogenic and hydrogen-like atoms as well as the hydrogen atom, need not be in conflict with the Dirac theory or the present GSBP model of the electron.

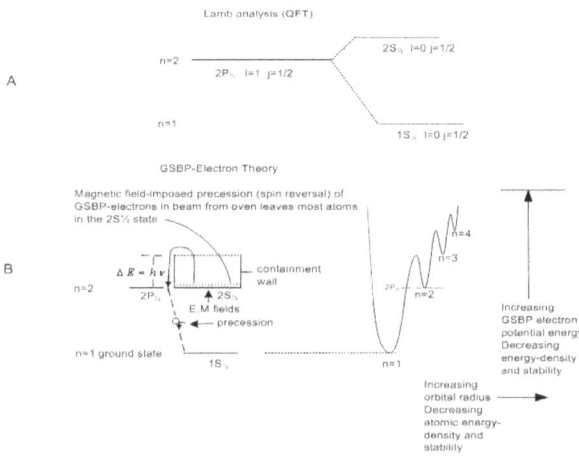

Fig. 14-1 Panel (A) represents the $2S_{1/2}$ state of the hydrogen atom as a metastable marginally-higher energy state, relative to the $2P_{1/2}$ state, yet providing no physical accounting for its metastability while denying existing (Dirac) theory on the matter. These studies were the outgrowth of 2 decades of development of Quantum Field Theory in lieu of any understanding of internal particle structure and dynamics.

At left in (B) we show the $2S_{1/2}$ metastable state of the hydrogen atom occupied by the very few atoms in the oven beam that have been boosted in potential energy $E=h\nu$ (relative to that of the $2P_{1/2}$ state) by a static E-field and an imposed magnetic field precession, where ν

is the frequency of bosons with energy-momentum sufficient to enable the electrons to escape over a containment wall. The imposed magnetic field serves to precess the GSBP-electron in its orbit about the proton in the same way as occurs naturally in the "Thomas precession." It is the precession-selected containment of atoms in the $2S_{1/2}$ state which physically distinguishes the angular momentum of the two states. In the Lamb experiments this energy boost was given only to the atoms coming from the oven in the $2S_{1/2}$ state, since virtually all those in the 2P states traveled immediately to the ground state. The GSBP-electrons in the $2S_{1/2}$ *contained* state could only escape when bosons of sufficient energy-momentum could enable them to surmount the containment wall and travel to the ground state via the $2P_{1/2}$ state. Such microwave excitation and decay was revealed in the Lamb experiments with the squelching of the static field in the sequence of energy level adjustments. But the above could only be seen as an anomaly in any theory in which electrons are said to lack any dimension or internal dynamic structure.

At right in (B) we show, not to scale, the ground state of the hydrogen atom and the n = 2, 3 and 4, l = 0, j = ½ metastable states in which the GSBP-electron must be excited out of an energy well. The atomic energy-density and stability decrease with increasing potential energy and with increasing radius of the electron orbit. Similar reasoning applies to the higher l and j states. In the GSBP-Schröedinger's theory the $nP_{1/2}$ and $nS_{1/2}$ states are not degenerate, since the latter can only be accessed to or decayed from a higher energy level $\Delta E = h\nu_n$, where ν_n, is the frequency corresponding to the depth of the energy well in the n state. And we see that the large negative energy gradients (slope) at the P states strongly attract the electron to lower states, whereas the positive energy gradients at the S states requires kinetic help to escape confinement.

There is more to ponder in the Lamb shift as a major turning point in the way physicists perceived the quantum world. Most importantly, experiment clearly won over theory, partly because Dirac's relativistic equation emerged unheralded, partly because there was direct conflict with the theory, partly because the experiments were so well conceived and undertaken with the help of new technology and perhaps also due to a growing mood for yet another major discovery to advance quantum field theory. But there are deeper concerns. Why should the energy associated with the anomalous magnetic moment of the electron be so close to that relating to the departure of the electron g_e factor from 2 and that necessary to account for the unique *spread* characterizing quantum statistics? In each case, the multiplying offset is ~1.002, suggesting an underlying common causal factor – the requirement in conservative quantum systems that metastable states be marginally shielded, in a small energy well, from destabilizing fluctuations.

Our task now is to model the hydrogen atom in terms of the GSBP structure of the electron. Just as it is necessary to reduce the mass of the Earth in its orbit around the Sun, in order to compensate for the finite mass of the Sun, we need to reduce the mass of the electron to compensate for the finite mass of the proton. But there is a very important difference: by reducing the mass of the Earth (by an amount equal to the ratio of the Earth's mass to that of the Sun) we automatically reduce the gravitational charge $m\sqrt{G}$ of the Earth by the same ratio, thereby preserving the relative kinetic and potential energy balance of the system. There is no way to accomplish a similar compensating adjustment of the kinetic and potential energies of the hydrogen atom, so long as the electron lacks dimension and internal structure and its charge is taken as an absolute constant. *It is on this inherent limitation that all of the difficulties encountered in relativistic quantum mechanics and in particular in quantum field theory have their origin.* The GSBP model of the electron enables us to correct for both the charge and mass of the electron and at the same time to correct for relativity, both special and general.

We approach this task by defining the parameters of the gyrating photon in an electron marginally-bound to a proton, in which the electron is nearly at rest (having near-zero kinetic energy and zero negative potential energy relative to the proton), with both its mass m_r and electric charge e_r reduced by the factor $m_p m_e/(m_p+m_e)$, where m_p is the mass of the proton and m_e the mass of the **free electron). In that state the GSB photon must travel at constant velocity** c in a circular orbit of radius r_0, such that the period of one complete orbit with exact phase closure is:

$$\tau = h/m_r c^2,$$

where h is Planck's Constant and $m_r c^2$ is the reduced energy of the gyrating photon. r_0 may then be derived from the spin angular momentum of the reduced-mass electron

$$r_0 m_r c = \hbar, \text{ with } r_0 \cong 3\times10^{-13}\text{m}.$$

Energy conservation and stability of the ground state of the electron requires the orbital period of the electron, together with exact closure of the phase of the bound photon, be exactly equal to τ times the constant α^{-1}, where α is the fine structure constant given by $e^2/4\pi\varepsilon_0\hbar c$. (We remind here that α is an absolute constant since ε_0 and c vary oppositely with change in ρ or ρ_{local}.) That is, the GSB photon must cycle α^{-1} times, in exact phase closure with one complete orbit of the electron about the proton. Thus the relativistic radius of the electron in the ground state $a_{rel} \equiv \alpha^{-1}\times r_0$. (The value of a_{rel} thus derived must also satisfy the relation

$a_{rel} = h^2/4\pi^2 m_r e^2$ in order to account, together with the Rydberg constant, for the radiation spectra of the hydrogen atom.) The relativistic velocity of the electron in the ground state $v_{erel} = (c^2 - v_{phrel}^2)^{-1/2}$, where v_{phrel} is the relativistic velocity of the GSB photon, which varies inversely with the third power of ρ_{local} in the electron's immediate environment. In the ground state $v_{erel} \equiv \alpha c$. With these relativistic adjustments, the ratio μ_e/μ_B = g_e, where μ_e the angular momentum of the electron = $e v_{erel} a_{rel}/2$, μB the Bohr magneton = $e\hbar/2m_r$, and $g_e \equiv 1$ is the Lande g factor (orbital magnetic moment of the electron), giving $\mu_e/\mu_B \equiv v_{rel}a_{rel}m_r \equiv 1$. This relation is an absolute Universal constant for the hydrogen atom, since v_{rel} and a_{rel} each vary as the inverse third-power of ρ_{local} and m_r as the two-thirds power of ρ_{local}. **In our scenario the gyrating GSB photon is the physical embodiment of a spin dipole moment, which must be defined similarly in relativistic relation to the Bohr magneton.** Substituting μ_{spin} for μ_e, -e for e, v_{phrel} for v_{erel} and r_0 for a_{rel} in the above equation, we obtain:

$$\mu_{spin} = -e v_{phrel} r_0/2, \text{ and the ratio } \mu_{spin}/\mu_B = -e v_{phrel} r_0 m_r/\hbar$$

is likewise an absolute Universal constant. The above follows since the higher velocity of v_{phrel}, in the equation for μ_{spin}, is exactly offset by the smaller radius of r_0. Consequently the total angular momentum in the ground state of the hydrogen atom is on average zero, with the relativistic spin angular momentum of the GSB photon canceling the angular momentum of the electron in its orbit of the proton. If this is so, we see that the electron has a real spin magnetic moment, in accord with our reasoning and the Dirac equation, but contrary to that of QFT/QED theory.

What remains to be said is why an excess of energy is required to reverse the spin angular momentum (charge) of the electron and how that is achieved. Recall (chapter 9) that the same fundamental problem occurred in the requirement to reverse the spin orientation of the d-quarks in a neutron to create a proton, brought on by declining local energy density as a means to relieve stress in the neutron.

In the larger view the transmutation was engendered by the NTI as an ultimate means to minimize entropy production in free neutrons and therefore in the Universe. Those quark spin reversals could only be achieved by increasing the GSB photon's kinetic energy sufficient to climb over and down an angular momentum hill, spinning in the opposite direction. The same applies to the electron's GSB photon similarly to that depicted in Fig. 14-1. Such angular momentum dynamics play key roles in other systems and subsystems, e.g., in the creation and stabilization of molecules, and in particular in accounting for the homochirality of the amino acids in all living cells (see chapter 17).

It might appear that the Lamb shift would depend in a small way on the local energy-density ρ_{local} of the *hydrogen atom*, but as we have seen the value of g_e is independent of the local energy-density of the electron and therefore of that of the hydrogen atom. If that is true, then the spectral emission of the hydrogen atom should be independent of the local energy-density of the atom.

Note the difference in the way relativity is taken into account in our model *vis-à-vis* QFT/QED theory. The latter sought high-order corrections to the electromagnetic coupling strength (the Coulomb force) and the mass of the electron (the "self energy" of the electron) in order to square the Bohr energies and radiation spectrum with empirical data. In this way, the classical charge and mass were adjusted, resulting in a relativistic formulation of quantum mechanics, in keeping with the requirements of special relativity, though making no formal contact with the demands of general relativity. By contrast, general, as well as special, relativity enter, in our schema, via the self-consistent dependence of all parameters in the energy states of the atom *on the local energy-density of the space in which the atom may reside*. If Earth were a satellite of the Sun much closer to the center of our galaxy where the local average energy-density would be greater than in our orbit, we should expect similar self-consistent variations in the values of the above parameters, though clearly the value of the g factors would be unchanged. (In this light, the recently reported evidence that the velocity of light is increasing with the expansion of the universe should have occasioned no surprise.)

Consequently, the energy levels of the hydrogen atom may be calculated, either using the Dirac equation (with the reduced mass of the electron), which represents solutions of a postulated relativistic wave function but having no reference to any internal structure of the electron, or by the GSBP electron formalism taking into account the reduced mass and charge, relativity and spin-orbit corrections for the allowable states. And, as before, we assume that any spectral discrepancies relating to degenerate states are owing to very small energy barriers characterizing the metastable states and deriving from their respective spin magnetic moments. The intensity of these higher energy states decreases as n^3, i.e., with the increase in the ratio of electron radius about the proton to that of r_0, making simultaneous closure of the orbit of the gyrating photon and that of the electron orbit much less probable in the presence of small random fluctuations. The energy barrier ΔE separating the metastable states from their non-metastable partners, and the intensity of the emission lines may be calculated from the indeterminacy relations, as fluctuations would represent an increasing margin of the energy budget.

Why, one may ask, should we go to all this trouble, in theory and experiment, to deepen our understanding of the mechanisms involved in such precision physics? One needs only to note that were this not the case we would have no understanding of how and why the hydrogen atom has been so stable over many giga-years, and will be indefinitely, in the future, defined by the fine structure constant and its self-consistent relations to other fundamental factors.

We note in passing that the GSBP concept of massive particles applies as readily to the analysis of antiparticle dynamics in antihydrogen or antihydrogenic atoms, when the polarization of the bound photon and therefore the orientation of angular momentum (spin, charge) of the gyrating photon, is reversed. There is of course much more to be explored on this matter in a later time. Also, no account has been taken in the above of hyperfine structure of the hydrogen atom, which, in the absence of a magnetic field results in a lowering of the ground state energy level by about 4.4×10^{-6} eV, due to dipole interaction of the proton and electron spins. This small correction has no significant effect on the radiation spectra of low Z atoms, at the present level of experimental verification.

In reflecting on the acclaimed achievements of QFT/QED, we see that the extraordinary decades-long effort was motivated by a perceived need to take relativistic quantum mechanics beyond the Dirac theory, in order to account for the additional energy of the $2S_{1/2}$ state in terms of an "anomalous magnetic moment" of the electron. The initial concern was succinctly stated in the second[98] Lamb-Rutherford paper. "Although the treatment of the hydrogen atom given by the Dirac theory was beautiful and satisfying, the theory carried with it some strange consequences arising from the existence of the negative energy states. It was therefore considered highly desirable to subject the fine structure predictions to careful experimental tests. These were at first, naturally enough, observations on the spectrum of the hydrogen atom. Any departure from theory would be ascribed to one or more of the following causes: (1) error in the Dirac equation, (2) modification of the Coulomb law of attraction between electron and proton, possibly by a short-range non-electromagnetic interaction, or positron theoretic vacuum polarization effects, (3) some finite and physically real remnant of the infinite radiative shift in the frequencies of all spectral lines predicted by the 1930 calculations of Oppenheimer, or (4) unexplained effects." This passage clearly looks to a fundamental change in the representation of quanta and their interaction, so focused were the times on a perceived failure of an "exact" theory and a hope for further theoretical advance. In that time there was relatively little understanding of what physically distinguishes a metastable state from a similar non-metastable state. The closest we find in their writing are the following comments: "The [proposed experimental] program clearly depended on a knowledge of the properties of metastable hydrogen atoms. Such atoms had never been isolated in any clear-cut experiments, although they had been the subject of much speculation. ... In the complete absence of a perturbing electric field, an atom in the $2^2S_{1/2}$ state should have a very long life. (Bethe estimated that the life of an isolated atom would then be several months if relativistic effects were taken into account.)"

What Lamb-Rutherford demonstrated was not the existence of the $2S_{1/2}$ and $2P_{1/2}$ states, as non-degenerates, but the nature and magnitude of the energy barrier separating those degenerate states in the hydrogen atom, making one of them a metastable state; and what the theorists have demonstrated with an extremely counterintuitive formalism, is a way to calculate, with relatively great precision, the energy ΔE of the barrier, making the $2S_{1/2}$ state metastable, and thereby taking theoretical physics down a many decades-long detour. And all of this followed because they could conceive of no rational way to endow massive quanta with internal quantized properties and structure in a manner making them inherently relativistic. As Lamb put it in his 1955 Nobel speech[99]: "the electron does not behave like a point charge as implied in Dirac's equation. ... According to relativity theory, energy and mass are proportional to one another. *In order to obtain a finite mass of the electron on a purely electromagnetic basis, it was necessary to assign an extended structure to the electron* (emphasis added). Attempts to form a satisfactory relativistic theory of an extended charged particle failed."

The thinking at the time is revealed also in Polycarp Kusch's Nobel talk[100] (at the same celebration) On the Magnetic Moment of the Electron in which he states: "one of the great triumphs of the Dirac electron theory was the prediction of these postulated electron properties. The spin and moment of the electron were thus removed from the realm of *ad hoc* assumptions, justified by experimental evidence, to the realm of an integral part of quantum theory." He goes on to say "The electron also has a magnetic moment by virtue of its angular momentum about a spin axis." But both of these statements should be viewed as *ad hoc* arguments so long as we are content to view the electron as a "point particle," which logically can have no spin axis. Kusch's supporting data in his talk (Table 1, observed ratios of atomic g values and the corresponding values of g_s/g_L) "offers overwhelming evidence that the spin moment of the ["free"] electron does indeed differ from its nominal [value 2]". But it is clear that the ratios of atomic g values for the Ga, Na and In atoms all are associated with the S and P transitions in the Lamb experiment, and the only question, as before, is one of interpretation: do the numbers demonstrate significant differences in the energies of those *states* or merely the energy of a barrier or well in which an electron briefly resides and which distinguishes that state from its degenerate partner state. Since this was an earlier experimental program, it is not surprising that the Lamb results made a similar interpretation. But in our scenario it is only necessary to refer to the "magnetic moment" of the internal GSB gyrating photon and to calculate the energy barrier in terms of energy differential ΔE of the photons in the $2S_{1/2}$ and $2P_{1/2}$ states.

However, rarely has this "detour" slowed the pragmatic advance of quantum physics. It is only now, as massive multi-particle quantum systems become more complex and possibly subject to general as well as special relativistic effects, that we are seeing again the need to ask: Are the fundamental questions relating, for example, to the gulf between general relativity and quantum theory (which most physicists have for decades had little or no occasion to be concerned) now causing some to ponder for a second or perhaps a first time?

GSBP Model of Spectral Emission from a Bound Electron

The GSBP model of spectral emission of the hydrogen atom follows exactly that of Dirac's extension of the Schröedinger equation, taking into account the reduced mass of the electron and allowing for the metastability of the $2S_{1/2}$ state residing in a small energy well. The following summarizes the principal features of this spectral emission:

All metastable states, having angular momentum nl are degenerate with the state nl + 1 (Examples: $3S_{1/2}:3P_{1/2}$ and $3P_{3/2}:3D_{3/2}$, etc.)

A metastable state may decay only to its degenerate partner state. The number of metastable states and possible degenerate states at level n = n-1.

All other electron energy transitions may occur only with $\Delta l = \pm 1$ and $\Delta m = \pm 1, 0$ (conventional selection rules).

All emission occurs as the electron spirals down from one allowed discrete energy level to another, such that the time required to make an exact integral number of GSBP orbits in the electron coincides precisely with the closure of one orbit of the electron at relativistic speed about the nucleus and is equal to the inverse frequency v^{-1} of the emitted boson. *That is, the distance traveled by the electron, taking into account its relativistic speed, is equal to the boson wavelength. This ensures exact simultaneous closure of the orbits of the bound photon in the electron and of the bound electron in the atom, thereby minimizing line broadening of the emission, allowing only a maximal error of ±h, where h is Planck's Constant. (Note the implied greater constraint on quantum statistics (see chapter 12). The boson is not emitted as an energy packet. Were that the case, it would not be possible to account for the stability of the hydrogen atom.*

In the radiation of very low energy bosons, e.g., the "Lamb shift" $2S_{1/2} \rightarrow 2P_{1/2}$ transition, the wavelength of the bosons is about 0.3m, equivalent to bosons radiated by a radio transmitter from a half-wave dipole antenna about 15 cm in length. A relativistic electron would then have to travel about $0.3\alpha/4$ m $= \sim 5 \times 10^{-4}$m or about one spiral orbit (from $2S_{1/2}$ to $2P_{1/2}$) over a time period of about 10^{-9}s in the process of emitting a similar boson. (The factor 4 takes account of the increased radius of the electron in the first excited level of the hydrogen atom.) It is interesting to reflect on the similarity between the motion of *many electrons* in a loop antenna (solenoid) at near the velocity of light and the motion of a *single electron* in its orbital travel in the hydrogen atom at velocity $\sim \alpha c/4$, giving rise in both cases to the emission of similar bosons along their respective axes of angular momentum. (As both the antenna and the atom are massive, essentially all of the angular momentum of the electrons is converted in both cases into translational momentum of the boson traveling at the speed of light.) The same dynamics would hold for the emission of a boson in the visible spectrum ($\lambda \sim 5 \times 10^{-7}$m), though the relativistic electron in the hydrogen atom would also need only about one orbit in its state transition, lasting about 10^{-15}s. How is the boson constructed and emitted in the atom? The above would imply that the initial spiral-down of the electron from the higher to the lower energy state initiates the cycle of boson emission, with the orbital closure, contributing to the final freeing of the boson from the atom, the reversal of the process depicted in Fig. 12-3. In so doing it imparts a miniscule recoil momentum to the nucleus. Note also in regions of low charged-particle density, the scattering of an electron would involve boson emission processes similar to the above low energy state.

The reverse of the above process occurs in the absorption of a boson. In all instances the accompanying loss (gain) of orbital angular momentum of the electron results in the gain (loss) of a unit of emitted (absorbed) boson spin (angular momentum of the GSB photon), thereby conserving total angular momentum. The manner in which energy-momentum is "peeled off" of the GSB photon in the electron in the emission or absorption of bosons is discussed in chapter 12.

Hydrogen-Like Atoms: Nuclear Dynamics and General Relativity

The forgoing arguments relating to the dynamics of the GSBP electron in the hydrogen atom take on greater importance in the study of hydrogenic and hydrogen-like atoms, in which the mass, charge, size and dynamics of the nucleus and the energy-density in electron states near the nucleus, open the analysis to general relativistic considerations. Such considerations clearly spurred the development of QED in recent decades, though scarcely with reference to general relativity, there being no fundamental way to incorporate them. These have been fraught with ever more difficult challenges to preserve some consistency in QED theory, as Karshenboim's timely review[101] reveals:

"the bound state QED is not a well-established theory and there are no published common prescriptions for the relativistic quantum bound problem." (p7) "Studies of high-Z ions are related to a strong coupling regime, however, it is unlikely to provide us with more information on bound state QED because of substantial contributions due to the nuclear structure." (p8)

"The Dirac equation itself is an equation ignoring the nuclear motion." p15

"Precision QED tests are realized in atomic systems which can be calculated *ab initio*. Such theory can be in principle developed for any QED effects, but not for the nuclear structure. Importance of the nuclear structure effects increases with Z." (p20)

"The main problem of theoretical calculations for the hydrogen Lamb shift is a lack of an accurate determination of the proton charge radius and an insufficient understanding of uncertainty due to the higher-order QED effects." (p31)

"A pure QED calculation for the hydrogen atom is not enough and before performing any comparison of the QED theory and experiment we need to take into account effects of the internal structure of a proton." (p33)

"A theory of a point-like particle with a non-vanishing value of the anomalous magnetic moment is *inconsistent*. ... The Lamb shift is a result of the electron charge distribution related to the electron structure, however, it is understood only through the QED interpretation. So the internal electron structure effects look rather as a theoretical construction, while the anomalous magnetic moment is a direct experiment fact." (p43)

As stated previously, the proton's spin (the simplest hyperfine contribution of the nucleus) plays a negligible role in the spectrum of the hydrogen atom, inasmuch as the total energy reduction of the atom is an order of magnitude less than that associated with the Lamb shift, when the proton spin opposes that of the electron. That situation is little altered in the case of the deuteron (having the same charge as the proton), even though the nuclear mass is about twice that of the proton.

In hydrogenic, hydrogen-like atoms and ionized high Z atoms, the coupling of the electron(s) to the proton(s) may be significantly increased, resulting in a decrease of the ground state radius, depending on the value of Z. Such systems may involve a significant increase in the local energy-density of the system, thus requiring corrections to the parameters ρ, c, and R_∞ the GSBP radius r_0 and electron orbital radius a and the inclusion of a GSBP charge radius r_p. What "precision physics of simple atoms" is all about, in QED terms, is a patchwork attempts to achieve self-consistency among the various corrections, lacking any understanding of the internal structure and dynamics of massive fundamental particles.

Clearly the challenge for GSBP particle theory is to determine the manner in which the local energy-density of the system varies with the atom's immediate environment, enabling thereby the above corrections to the energy levels of bound electrons in a simple and straightforward way. And in the larger view to determine how such corrections might vary consistently in such atoms in regions of high ρ_{local} (e.g., near the core of galaxies). While this should greatly facilitate the precision physics of simple atoms, the larger benefit initially would accrue from the further realization that Nature's fundamental laws are self-consistent and relatively simple. (See chapter 12)

CHAPTER 15

AN INHERENTLY RATIONAL QUANTUM (PARTICLE) MECHANICS

As the chapter heading implies, our intent is to demonstrate that quantum mechanics, contrary to its reputation, can be inherently intuitive, rational and consistent with the demands of relativity and thermodynamics, once we take into account the internal dynamic structure of the fundamental particles and shed our long-held commitment to the notions of wave-particle duality, the single particle assumption and superposition. Discarding that baggage in no way diminishes the power of Schröedinger's equations to characterize the nonrelativistic dynamics of quantum particles, as those notions were philosophical appendages to the mathematical formalism. Indeed nothing is lost in the more fundamental strictly-particle theory, while still greater precision is gained in its straightforward finite compliance with general as well as special relativity. Here, we seek to demonstrate the inherently rational quantum behavior displayed in a variety of notable quantum experiments, commencing with the simplest double-slit photon interference experiment that still muddies the theoretical waters.

In each case the experiment is characterized by the intentional or accidental creation of some new open, minimal density-specific entropy-production subsystem, as well as by some particular counterintuitive behavior, as we shall note.

As we argued in chapter 10, Feynman's admission to the impossibility of truly understanding quantum mechanics may be traced to the widespread acceptance of what we refer to as the "single particle assumption," on the basis of which superposition theory evolved as an extraneous fundamental proposition. The reasoning goes as follows: *By greatly reducing the illumination from a source of light we can be assured that only a single particle or wave is involved in a double-slit experiment and that one such particle or wave must past through each slit in order to account for the phenomena of diffraction and interference. Consequently the original particle or wave must be latent in two or more physical states or paths from source to detection, consistent with the requirements of energy-momentum conservation.* As cited in chapter 10, two independent experimental studies argue conclusively that such counterintuitive reasoning is invalid, *once it is demonstrated that two or more particles with sufficient phase overlap are required to account for interference phenomena and that even with extremely low illumination from a laser source more than one such boson may be produced to account for the behavior of the system. We note here that the single-particle assumption derives from classical reasoning and hence ill conceived as a tenant in relativistic quantum theory.*

With the above as fundamental arguments, we now review the principal events that led us to accept the inherently strange or counterintuitive behavior of quantum systems, commencing with the famous arguments between Niels Bohr and Albert Einstein nearly a century ago.

In essence Bohr said that the remarkably precise predictions of *non*relativistic quantum mechanics was nevertheless the most complete knowledge we could possibly have of the behavior of quantum systems, and in reply Einstein said, with little or no empirical support, something must nevertheless be missing in the theory, if we are to account for relativity corrections. If this summation is not too simplistic, then Bohr and followers won the debate for lack of a more fundamental theory consistent with the requirements of relativity. That conclusion was supported by numerous experimental tests of Bell's inequality theorem a quarter century later and to the present, with the result being that we simply must accept the irrational contradictions between the postulates of QM and the requirements of relativity. We will have more to say of this later.

We now examine the particularly "weird" results of two widely publicized quantum interference experiments. In an experiment in 1991, Leonard Mandel[102] and students directed photons from a laser to a beam splitter and then to two similar Down-Converters (DC). Signal Photons from DC1 and DC2 were directed, in separate paths via mirrors, to a Signal Detector, while Idler Photons from DC1 were directed to and through DC2, in alignment with Idler Photons from DC2 to an Idler Detector. With proper adjustment of the system, interference was demonstrated at the Signal Detector. However if the Idler Photons from DC1 were blocked before reaching DC2, no interference could be observed. This was interpreted to show that the mere "threat" of obtaining information about which path the photon traveled to the Signal

Detector necessarily forced it to travel only one route, or as Mandel has said: "The quantum state reflects not only what we know about the system but what is in principle knowable." This is perhaps the most counterintuitive statement among many in regard to quantum interference. But spontaneous down-conversion in a nonlinear crystal is a highly nonlinear process, and *it would be surprising if the energy, timing and trajectory of the Signal Photons from DC2 were not altered by the increased local energy-density in DC2, due to the coincident passage of the first Idler Photons through DC2, thereby varying the local energy-density in the crystal and disrupting interference at the Signal Detector.* If so (if the state of DC2 were not too seriously altered) it might only be necessary to readjust the system (keeping the idler path blocked) to reinstate interference behavior. If this were the case, we would have immediately a physical explanation for the weird behavior. If not, the explanation might still apply, since the negative result might only mean that we were not able to readjust parameters sufficiently to reinstate the interference.

An experiment in 1995 by Paul Kwiat[103] and associates was said to demonstrate interaction-free measurement. "Single photons" from a laser were directed to a down-conversion crystal (DC). One of the low energy single photons from the down-converter was directed to Detector (D1) and the other to a Beam Splitter (BS) and thence through BS to a Mirror (M1) and (according to superposition theory) at the same time reflected 90 degrees to another mirror M2. Thus, prior to detection, the photon was reflected from M1 *and* M2 so as to meet back at BS, where they may be made, by proper adjustment of relative path lengths, to annihilate one another in destructive interference. (It has been stressed by T.B. Pittman *et al.* that it is not necessary that the path lengths be made exactly equal, only requiring adjustment of the relative phase of the photons.[104] But if the path from BS to M2 were blocked or diverted by an object, there would be no reflected photon arriving at BS and the photon reflected by M1 would then be free either to go back toward the down-converter or be reflected out of the interferometer to detector D2. The firing of D2 would inform us of the existence of the blocking object even though the photon triggering D2 never encountered the object. As the authors point out, this result is valid for only about one quarter of a large number of tests, since there are several alternative transmissions and/or reflections at the beam splitter. But the statistical details of this system are not our concern here only in that they reveal the unique two photon quantum statistics. As above, should we not expect that two or more photons emerge simultaneously, collinearly and in phase from the laser? If so, shouldn't we also expect that two nearly identical low-energy photons were directed to D1 and at the same time a similar pair could have entered the interferometer?

The answer is yes, as demonstrated by Z.Y.Ou[105] *et al.*, using the stimulated emission process in a pulsed parametric down-converter. As they say "When four photons arrive at a beam splitter, two from each side, a four-photon, six path interference effect occurs to yield a six fold enhancement of the probability for all four photons to exit together from the beam splitter." That is, we should expect an almost perfect knowledge of the existence of the blocking mirror or "pebble.". In different terms we would know that one or the other of the two photons entering the interferometer interacted with the blocking particle (mirror), thus restoring our commonsense understanding that acquiring information always entails some energy or thermodynamic cost.

Recent investigations have extended the range of bizarre quantum interference behavior. Some involve photons in multi-particle entangled quantum states, as for example, in the above-cited investigation of Z.Y.Ou *et al.*, while others seek to demonstrate such strange phenomena involving electrons, atoms, molecules or even macroscopic objects such as a small mirror on a cantilever beam. Four experiments are especially noteworthy, all relating to Schröedinger's cat and the phenomenon of interference.

In one, 1996 C. Monroe[106] *et al.* trapped a beryllium ion and finessed it with laser pulses into a superposition of a spin up state (at one radius) and a spin down state (80 nanometers distant on the radius), as a mesoscopic example of Schröedinger's cat. Clearly, the electron was not in two places at once, a superposition; it was finessed into being alternatively in the ground state by a 1μs laser pulse and then in the Rydberg state by a 10μs pulse in a sequence of 4000 measurements. The two carefully controlled and timed excitations of the atom serve similarly to the two bosons in a double-slit experiment to give rise to interference.

In another experiment, 1996 Michel Brune,[107] Serge Haroche and Jean-Michel Raimond explored the interaction of A Rydberg atom (excited into a superposition of two different energy states) with the electromagnetic field (photons) in a microwave cavity. In the reaction the Rydberg atom transferred its superposed state to the EM field in a cavity, putting it into a superposition of two different phase or vibrational states (another representation of Schröedinger's cat). Again, it must be said that the electron is not in two different states at the same time, but alternately at precisely controlled time and energy; the same holds for the photons and their phase or mode states in the cavity that resonate with the atom. By detuning a second atom sent into the cavity and by varying the time interval between the atoms, they could determine the rate at which the photon states decayed (the "superposition collapsed") and by varying the photon density in the cavity they could see how that rate changed with the density. That is, they were able to monitor the evolution of the system from quantum to classical behavior. This is a beautiful experiment, but it involves no superposition mystery. There occurred no "collapse of superposition states, only the decay of oscillation between excited states of the atom, as the flow of energy necessary to sustain those states declined with the decrease of photon energy-density in the cavity.

A third study, reported by 1996 Michael Noel and C. Stroud[108], in which the electronic state of a potassium atom is put into a coherent superposition of two spatially localized wave packets separated by approximately 0.4 μm at the opposite extremes of a Kepler orbit." When the two wave packets overlap in their oscillation, the phase difference between the de Broglie waves, associated with the two packets, determines the nature of the interference between them. This experiment may be seen as a variation of two-particle interference experiments; in this case it is two coherent nonclassical photons ("wave packets"), which energize the atom, that interfere. The electron cannot interfere with itself. Just as the nearly identical photons or electrons in other experiments were carefully engineered in a source to demonstrate interference, so the photons (wave packets) were carefully tuned here to demonstrate wave packet interference and the apparent existence of the electron in two displaced positions at the same time. As the authors say: "If the two positions are macroscopically distinguishable then this state is similar to a Schröedinger's cat in that it represents a quantum mechanical superposition of two classically distinguishable physical states which are localized in one or the other of the two classical positions only by the act of measurement which destroys the superposition. … Since the superposition is of energy eigenstates all with the same angular momentum, the localization of the electron is in the radial coordinate only. The electron (initially localized to a radial shell near the core) oscillates in and out in a breathing motion at the classical Kepler period."

In all of the above experiments, only two or a few particles were involved, albeit over mesoscopic distances, in the "counterintuitive" demonstration of interference phenomena. In each case the researchers created a very special minimal density-specific open thermodynamic subsystem (atom), at the expense of a very great amount of entropy production in the subsystem's environment.

A fourth experiment is said to demonstrate quantum superposition of macroscopic persistent-current states in a micrometer-sized Superconducting Quantum Interference Device (SQUID), as reported by van der Wal[109] *et al.* They write: "Microwave spectroscopy experiments have been performed on two quantum levels of a macroscopic superconducting loop with three Josephson junctions. Level repulsion of the ground state and first excited state is found where two classical persistent-current states with opposite polarity are degenerate, indicating symmetric and antisymmetric quantum superpositions of macroscopic states. The two classical states have persistent currents of 0.5 microampere and correspond to the center-of-mass motion of millions of Cooper pairs. … At an applied magnetic flux of ½ Φ_0, the system behaves as a particle in a double-well potential, where the classical states in each well correspond to persistent currents of opposite sign. The two classical states are coupled via quantum tunneling through the barrier between the wells, and the loop is a macroscopic quantum two-level system. [Constructive interference occurs within the junction] … The symmetric superposition state is the quantum mechanical ground state with an energy lower than the classical states; the antisymmetric superposition state is the loop's first excited state with an energy higher than the classical states. Thus, the superposition states manifest themselves as the anticrossing of the loop's energy levels near ½ Φ_0." While it is not emphasized by the authors, the symmetric state is energetically lower than the classical states because it minimizes the external flux and energy stored in that flux; no less important, that state is consistent with the requirement of minimum flow-specific entropy production for the system under the imposed constraints. (Similar external flux closure occurs also at the poles of a horseshoe magnet). Such behavior may also be compared to the correlated vortex motion of millions of molecules in a viscous film (Bénard convection), in this case as a means of decreasing the flow-specific entropy production associated with the transfer of thermal energy through the film and minimizing the density-specific entropy in the vortex subsystem(chapter 4). Similarly, a SQUID is a thermodynamic subsystem, involving many millions of correlated electrons, of a larger system, which evolves to a stationary state of minimal density-specific entropy production (maximal energy-density), under constant system constraints.

In the above, the authors are careful, in the title and in the text, to refer, quantum mechanically, to "persistent-current state*s*" and classically to "persistent current*s*" (emphasis added), suggesting, not surprisingly (at the border between micro and macro systems), that the system is equally comprehensible in either analytical framework. We cannot doubt that there are two localized streams of electrons (each of current density ~ 100 Amperes/cm2) traveling in opposite directions in the superconductor, just as two (or more) photons or electrons are necessarily involved in double-slit experiments. What then do we gain or lose with the add-on notion of "superposition?" We gain, if anything, only a reminder in the formalism and culture of QM that what appears to be the behavior of one particle is in fact that of two (or more) physical entities of a kind, locally entangled in energy-momentum and phase; but we risk, by a continued commitment to superposition theory, extending the decades of unproductive theoretical and philosophical arguments that have served progressively to separate fundamental physics from other intellectual enterprise.

Notwithstanding the physical differences of the above experiments, all such are said to have in common the assumption that the particle or particles involved (millions of electrons in the SQUID experiment) may exist in a superposition state prior to its measurement (or even after its measurement in the underdamped SQUID experiment). With each further advance, we are thereby committed more deeply to the counterintuitive demands of the Bohr interpretation of QM and, more to the point, to the denial of the demands of other, equally enshrined, physical theory, in particular, relativity (see chapter 12 relating to quantum statistics). Such experiments tacitly assume that the state of all such particles – photons, electrons, protons, neutrons and similar atoms or molecules – are, among their own kind, intrinsically alike, thereby enabling information about the individual particles to be gained statistically via experiments involving ensembles of particles in the same initial state. However, the amount of information to be gained is very limited, since we cannot work with the same photon or electron, etc. in test after test, while in the study of fabricated macro-"particles" (small mirrors on cantilever beams) we can only work with one and the same such particle, about which we can have relatively little precise information.

Violations of Bell's Inequality Theorem

After decades of general agreement that Niels Bohr won the long-running debate with Einstein (that QM is a complete theory of quantum systems), John Bell set about to cast the argument in rigorous mathematical terms, possibly favoring EPR.[110][111][112] The results of course sealed the verdict in favor of Bohr and QM, but not without continuing empirical, theoretical and philosophical argument. Thompson[113] examined the difficulties encountered in designing and implementing experiments to verify Bell's inequality theorem and concluded: "...*we cannot design loophole-free Bell tests using light – we have been attempting the impossible. Why should this be? If we analyse the experiments carefully, we find that it is because the whole enterprise was undertaken on a false premise: that light could be modeled as particles.*" (emphasis added)

It should be noted at the outset that tests of Bell's theorem are closely related to the various double-slit type experiments discussed above. The existence of two or more photons demonstrating interference in a double-slit or interferometer experiment is evidence of the entanglement of those particles in their creation as assuredly as that associated with the testing of Bell's inequality, though the photons in the one group were created, mass-produced, by a laser, while those in the other were created as unique pairs, one pair at a time in a down-conversion process. Moreover, as Garcia, *et al.* have demonstrated, the extent of entanglement need not be as exact or extended (among many particles) as previously thought.

The evidence clearly favors quantum probabilities in numerous experiments. But the evidence also clearly points to the fact that a measurement of the state of one of the entangled particles specifies the corresponding state of the other, for all time and place (barring perturbation), whether or not it is measured. There is no need, prior to the measurement of particle A, to invoke the notions of superposition or the collapse of a wave function consistent with the dictates of the experimental setup, since the particular setup and detector parameters simply serve to timely select for the record (albeit imperfectly) those particles having the appropriate spin and momentum, among many created over time. Nor is there any need for such notions prior to a coincident measurement of particle B, or indeed whether any such measurement is required, since its state must be correlated with that of particle A, in conformance with the appropriate conservation law. In 1998 Strekalov, Kim and D. Strekalov, Y, Kim and Y. Shih, Experimental study of a Photon as a subsystem of an entangled two-photon state, arXiv:quant-ph/9811060 v1 23 Nov 1998 Shih[114] succinctly argued the central issue: "Quantum theory does allow a complete description of the precise correlation for the spatially separated subsystems, but no complete description for the physical reality of the subsystems defined by EPR. It is in this sense, we say that quantum mechanical description (theory) of the entangled system is nonlocal. ... If the correlations of the pair have been built up in the entangled two-particle state from the beginning, it comes as no surprise that the intensity correlation reflects perfect correlation (EPR, EPR-Bohm or EPR-Bell type correlation) of the pair. The distance between the detectors would not matter. There is no action-at-a-distance involved, if we are willing to give up the classical EPR reality. ... Based on the experimental data, we conclude that the entropy of signal and idler are both greater than zero (mixed state); while the entropy of the signal-idler two-photon system [in our view, subsystem] is zero (pure state). This may mean that negative entropy is present somewhere in the system, perhaps in the form of the conditional entropy. By definition of the conditional entropy, one is tempted to say that given the result of a measurement over one particle, the result of measurement over the other must yield negative information. This paradoxical statement is similar to and in fact closely related to the EPR 'paradox'. We suggest that the paradox comes from the same philosophy." The "classical EPR reality" that we must "give up" in order to preserve causality, locality, etc., is the notion that we have complete control or knowledge of the creation and dynamics of any individual quanta, either in theory or experiment. But this reality also denies legitimacy to the assertion that any experiment involves only one boson, electron, etc. Strekalov *et al.* are prescient in their thermodynamic analysis: The zero entropy of the signal-idler pair is more than compensated by the very large entropy production of the system as a whole, and it is the precise details of the evolution of that entropy production that we can never hope to comprehend.

The employment of a nonlinear crystal or a laser for the creation of entangled particles in an EPR experiment should be seen as the creation of a subsystem, in which precisely, or nearly precisely, correlated particle-pairs may emerge as a means to minimize the density-specific entropy production of the subsystem and the flow-specific entropy production of the system, whether or not the particles remain locally bounded in spacetime.

There have been many variations in the experimental verification of the Bell theorem, including those employing fermions and those demonstrating nonlocality of the entangled particles in delayed choice experiments, but the statistics in each case all make the same argument for QM.

And so the conclusion by almost all in pondering the Bell theorem was and is that QM is complete and there is nothing to be gained by postulating hidden variables. But Bell[115] remained unconvinced after most had their say: *"so long as the wave packet reduction is an essential component, and so long as we do not know exactly when and how it takes over from the Schröedinger equation, we do not have an exact unambiguous formulation of our most fundamental physical theory."* (emphasis added) Elsewhere,[116] Aczel quotes Bell as saying *"the real problem with quantum theory [is] the apparently essential conflict between any sharp formulation and fundamental relativity. It may be that a real synthesis of quantum and relativity theories requires not just technical developments but radical conceptual renewal."* (emphasis added) The former statement implies the need to disavow superposition theory and its requirement for wave-function collapse in measurement and in turn the notion of spreading waves as a complementary facet of quanta. As such it amounts to a plea for a strictly particle theory of the creation, transport and interaction of radiant energy with matter. The latter argues for the need in any new formulation to retain the precision of the linear nonrelativistic Schröedinger equation and enable its extension, in a strictly particle description, to apply to relativistic quantum systems. Taken together, Bell's insight could have inspired our development of GSBP particle theory, though what has been said thus far clearly indicates a different course in our evolution of these ideas.

In GSBP semiclassical theory, entanglement requires the structure of the two entangled photons to store their respective local EM field patterns, polarization and trajectory in arbitrary Euclidean space, as they travel separate paths at the speed of light, as defined in chapter 11. As such the entanglement is completely conserved over the lifetime of the bosons. This holds for each pair created with different polarizations, intensities and trajectories. Thus only one pair, in hundreds of millions or billions produced, would be coincidently detected in a Bell test. But all of this is consistent with the Bell probabilities. The departure of the correlation from a linear relation of angular polarizer settings cannot be due to theory or interpretation but to the physical properties of the crystal and large energy dissipation and entropy production associated with the creation of such extremely ordered pairs of photons, as the very low efficiency of correlated pair production suggests. Does the GSBP theory then imply "hidden variables" in quantum mechanics? No, because quantum mechanics cannot accommodate intrinsic structure in particles. GSBP representation of quanta cannot be seen as an addendum to existing quantum mechanics. Only as a radically reformulated theory, which nonetheless squares with the Schröedinger representation under strictly nonrelativistic assumptions. So, if all of this is sound, then both Bohr and Einstein were deservedly winners in the debate: Bohr was right that the statistics of quanta interference are unique to non-relativistic quantum mechanics, but Einstein was also right in his belief that quantum theory should and could have a more "complete" and fundamental representation in which locality, separability and reality still guide our reasoning, wholly consistent with the demands of relativity. GSBP-Schröedinger relativistic quanta (particle) mechanics qualifies as just such a more fundamental representation.

With this understanding, we shed the probabilistic constraints imposed by nonrelativistic quantum mechanics and gain an abiding commitment to causality, locality, reality, non-separability of entangled quanta, accord with relativity and a common-sense understanding of our world. To be sure, the notion of reality is bounded by Heisenberg's uncertainty principle, but no more so than for a single measurement of a multi-faceted particle about which we have no previous knowledge. But since the terms "uncertainty," "indeterminacy" and "complementarity" associated with this principle suggest an anthropomorphic judgment, it might be more instructive to refer to the principle of minimum flow-specific entropy production (chapter 5). This follows since the principle, whatever its name, represents that no two particles or particle pairs in any new experiment can be prepared exactly in the same discernable state, because the environment in which they are created necessarily changes (evolves) (in part as a result of that event) and that change is accompanied, on average with some measure of energy dissipation. If that were not the case, the world would be static, unchanging at all levels, from quantum to cosmic, and there would be no living thing to contemplate its world. So the challenge for all scientists, and indeed all living matter, is to derive some useful and beneficial model in the reading of successive events or encounters with the ever-changing world, so as to minimize the downside risk, either by learning or via genetic instruction. In what follows we seek to demystify other counterintuitive aspects of the quantum world.

Quantum Tunneling Through a Potential Barrier

The tunneling of quantum particles (photons, electrons, etc.) through potential barriers is a phenomenon unique to quantum mechanics and one that has heretofore been explained, as above, in the context of wave-particle duality and superposition theory, i.e. the finite probability of a particle being in a superposition of states on both sides of a potential barrier blocking the particle's path. In the case of bosons, there appear to be two mysteries: How is it possible for a boson to circumvent a potential barrier? And, if possible, how can it propagate through the barrier faster than the speed of light? Classical mechanics denies the possibility of a particle, with kinetic energy T, passing a barrier of potential energy ϕ, if $T<\phi$. In order to provide a classical account of boson tunneling, it would be necessary to demonstrate the existence of bosons in an ensemble, having a probabilistic distribution of energies greater than T. Is there a "hidden" mechanism to support this proposition? There is. 2001 Milena D'Angelo,[117] *et al.*, "report a proof-of-principle experimental demonstration of quantum lithography. Utilizing the entangled nature of a two-photon state, the experimental results have beaten the classical diffraction limit by a factor of 2." More fundamentally, the experiment demonstrates that when the trajectories of two identical photons (in frequency and phase) converge to occupy the same spacetime position, the local energy-density doubles (the frequency increases and the wavelength decreases by a factor of 2). More generally, the local energy-density increases by a factor N, where N is the number of locally-spacetime-entangled bosons. Thus, to the extent that such local entanglement and resulting local energy-density increase may be engineered or otherwise occur, those entangled bosons in the ensemble will have momenta sufficient to pass over or through any barrier that would just block the passage of non-entangled photons in the ensemble, and the tunneling phenomena would evidently have a semiclassical explanation. Moreover, the reduced wavelength of the entangled bosons should result in a reduced delay in traversing the barrier, relative to that of longer wavelength bosons traversing an equivalent path in air. This would follow, if the detectors respond to a maxima of momentum-transfer, such that the shorter the nonclassical electromagnetic signature ("wave packet") the sooner the detector responds. This gives the appearance of superluminal speed of light, when compared with the time of flight of the non-entangled bosons. However, the above could only hold if the reduction in the velocity of light, due to the higher energy-density of the material barrier, is less than the apparent velocity increase resulting from the reduced EM temporal signature. This is not to say that reshaping the EM signature via the Hartman effect plays no role in the apparent superluminal speed of light. While the D'Angelo experiment utilized spontaneous parametric down-conversion entangled photons, we should expect the same semiclassical results using entangled photons from a laser, i.e., energy and frequency doubling (geometric up-conversion) of bosons (on converging trajectories), enabling some portion of the photons in an ensemble to surmount the barrier.

The tunneling of charged quanta, e.g., electrons, needs similar reexamination. We focus here on two basic physical processes of electron transport through a potential barrier: electron field emission from the tip of an electrochemically machined conductor or scanning probe (Fowler-Nordheim tunneling process) and the tunneling of electrons through a non-superconducting region ("Josephson junction") separating two superconductors or the same superconductor whose ends join at the junction. In field emission the potential barrier is the work function ϕ of a conductor or semiconductor. The rate of electron flow through the barrier (into the high vacuum of an electron microscope or the ambient space of a scanning electron probe) is a nonlinear function of the electric field strength at the tip of the emitter or probe, but little dependent on temperature in the low temperature regime. This is a quantum process and the barrier in question is the work function characteristic of a particular conductor or semiconductor from which the probe is machined. Classically, escaping electrons must have a kinetic energy greater than the work function. In the case of field emission, this can occur only if a boson, of energy $E > \phi$, is emitted by an electron at the tip of the conductor (and absorbed in the body of the conductor), such that the electron gains recoil momentum toward the positively charged screen in the electron microscope or object being scanned in the SEM, in much the same way as the emission of a boson from a free electron propels the electron toward a free proton (chapter 11). The probability of a boson, of energy $E > \phi$, being so emitted by an electron increases as the square of the field strength at the tip of the conductor, in accordance with the semiclassical Fowler-Nordheim equation. The current density would thus be accounted for, not as a mysterious tunneling of electrons through the work function barrier, but in terms of GSBP electrons acquiring recoil energy-momentum at the expense of potential energy. Is there evidence of such boson emission? Yes, Bonard[118] *et al.* report on the observation of luminescence during electron field emission on singlewall and multiwall carbon nanotubes. "The fact that the light emission is observed only in the presence of a finite electron emission current shows … that the luminescence was directly linked to the emitted current, as opposed to the more usual field induced luminescence where ionization occurs through field accelerated electrons. … The narrowness of the luminescence lines and the very small shifts with varying emitted current (<5 meV) show that we are not in the presence of current-induced heating effects. … Intriguingly, the position and width of the narrow Gaussian [luminescence intensity] remained nearly constant from one tube to the next. While the authors "suggest that the luminescence is due to electronic transitions between energy levels at the tip that are participating in the field emission," the superimposed narrow-band and broad-band emission suggest that the latter is characteristic of electrons that emit bosons with energy $E < \phi$ and lack sufficient momentum to escape. The narrow-band

254

emission is an indication that the field strength at the machined tip (subsystem) occasionally rises sufficiently to cause the emission of a boson so as to eject an electron from the tip, by impulse recoil action, leading to a decrease in the density-specific entropy production in the subsystem.

Likewise, the flow of electrons within and between two superconductors separated by a potential barrier, in the form of an island of non-superconducting material (a "Josephson Junction"), can be explained in semiclassical terms, i.e., without recourse to the counterintuitive notions of wave-particle duality and superposition. It is necessary to account first for the spontaneous, resistance-free direct current flow of electrons in a homogeneous superconducting loop and then for its maintenance through a Josephson junction inserted in the loop to enable control of the direction of current flow. For low-temperature, Type 1 superconductors, the BCS theory, involving the interaction of Cooper pairs of electrons with distortion of the crystal lattice (phonon oscillations), argues for the first requirement, if we assume that fluctuations initiate the current in one direction or the other. However, such current creates a magnetic flux through the loop, thereby storing energy in the vacuum (air) and *marginally* increasing the energy of the system, while exposing it to possible dissipative reactions in the environment. (As noted above, the inclusion of a Josephson junction in the loop enables electrons to *tunnel* through the junction, in a *superposition* of both directions in the loop, thereby making possible a lower ground-state energy, free or nearly-free of external flux energy.) In their paper, van der Wal[119] *et al.* argue that *two* persistent macroscopic currents of opposing sign are involved in their experiment, and if each is taken to be a condensate bosonic particle, then we argue that the experiment is defined in terms of two entangled bosons in various phase relation and not as one boson in a superposition of two states. That is, the Cooper-pairs of electrons are not going clockwise and counter-clockwise at the same time in exactly the same place in the loop, as superposition would have it. Indeed, the internal magnetic field forces the clockwise current toward one upper (or inner) surface of the loop and the counter-clockwise current toward the other surface, and this pattern is preserved across the gap, such that for $\frac{1}{2} \Phi_0$ state, no voltage appears across the gap. The resulting elimination or minimization of flux through the loop, and confinement of flux within the loop material minimizes the ground state energy of the system. And with this the flow-specific entropy production in the system and the density-specific entropy production in the subsystem are minimized. As for the tunneling of electrons across the junctions, the tunneling rate, or superconducting current, is dependent on geometry and material properties of the junction and the phase relation of the bosons on each side of the junction. Given the very low binding energy of Cooper-pairs (of the order of $10^{-3}eV$), the energy required to initiate single electron flow through the junction (voltage across the junction or applied magnetic field) is correspondingly low. All of this may be viewed classically either as owing to a reduced work-function at the gap or the extended momentum of the bosons in surmounting the barrier.

At this point it may be helpful to reflect on the "weird" behavior encountered so far. In the case of simple interference experiments (boson or electron), experiment establishes that two or a few entangled quanta, on different trajectories, account for the interference phenomena straightforwardly and without appeal to mystery. In regard to the EPR conundrum, the mystery vanishes once we acknowledge again that two entangled quanta share for all time and place a minimal common state, as required by their common, though incompletely defined, origin. Likewise, localized entanglement of bosons or electrons, effectively increasing their energy-momentum by a factor N, circumvents any need to conjure a notion of quantum "tunneling" through or under a potential barrier. But in the case of superconducting electrons tunneling through a potential barrier (Josephson Junction) at temperatures approaching 0K, cooperative (entangled) behavior proceeds to an extraordinary level; not two or a few but millions of electrons travel coherently, as Cooper-pairs, to minimize the energy and the flow-specific entropy production of a macroscopic system. In the van der Wal experiment two persistent macroscopic currents of opposing sign are involved, each is taken to be a condensate boson whose phase relation may vary relative to that of the other, displaying interference at a macroscopic level. While it is again true that such behavior makes no demand on wave-particle duality or superposition theory, it is the scale of collective dynamic behavior and the manner in which such open, nonequilibrium subsystems emerge spontaneously that warrant special attention here. Though the nature of this collective behavior is unique to particular microscopic laws, it bears comparison in scale and spontaneous origin to behavior found in macroscopically large, inanimate systems of molecules, subject to an imposed external force. Such large-scale behavior is central in what remains for discussion in this chapter. Moreover we should reflect here on the entropy production associated with these experiments. In the case of double-slit experiments, nearly all of the entropy production is associated with the creation of entangled particles (bosons, etc.) in a source and the annihilation of nearly all of those particles at the double-slit screen. Only an extremely small percent of all such particles pass through the slits with their energies dissipated at a sensitive screen or detector. In the van der Wal and similar experiments in which millions of particles correlate their limited motion near 0K, almost all of the entropy production is associated with the processes involved in creating the extremely low-thermal environment. Under these conditions, the total entropy production associated with the external magnetic field, the Josephson Junctions and in the superconductor loop is less than would be the case with unidirectional electron flow in the loop. In both cases we have paid a very large measurable price for the knowledge that has been gained!

The Aharonov-Bohm Effect

Yakir Aharonov and David Bohm[120] argued, about a half-century ago, that a vector potential was itself a fundamental physical entity (the AB effect). In particular it was predicted that a magnetic potential could alter the momentum or spin of a charged particle even if that potential gave rise to zero magnetic field in the space of the particle. In 1982, using a holography electron microscope, Tonomura and colleagues provided a convincing experimental demonstration of the AB effect, following two decades of debate as to the physical reality of the vector potential.[121] A fundamental understanding of the AB effect is important not simply as a QM oddity but because the vector potential in electromagnetism is a simple example of a gauge field, and gauge fields of various sorts are now seen to play fundamental and diverse roles in physical theory. Here we consider a magnetic vector potential as a cylindrical bundle of bosons of constant intensity, bound in constant cyclic external orbits to a high aspect-ratio shielded solenoid. That is, the bound photons, having near infinite wavelength λ_l, are the physical embodiment of the nonradiating electromagnetic flux. (The shielding in the solenoid model simply guarantees against any leakage flux from the photon bundle.) The unchanging magnetic flux (in the boson bundle) flows from one extremity of the solenoid, spreads over all space and returns, unless perturbed, to the other extremity, and the flux density approaches zero in the plane, in the central region and perpendicular to, the solenoid. Thus the trajectory of an electron in that plane in a vacuum would, according to classical theory, remain the same, whether or not the solenoid is energized. Yet experiment demonstrates otherwise, if the trajectory passes very close to the cylindrical bundle of bosons. Proximate to the boson bundle, the electron encounters no magnetic flux, though it may be acted upon by a very large energy-density gradient in the plane and near to the shielded solenoid, not unlike that causing the bending of a photon trajectory in a large gravitational gradient. The gyrating photon in the GSBP electron is torqued by this gradient in toward the bundle or outwardly, depending on the direction of motion of the bosons in the cylindrical bundle. This follows because the velocity of the gyrating photon in the GSBP electron, as it travels nearest the photon bundle, is slightly less, when it is traveling in the same direction as the flux of bosons (the local energy-density is greater) than when it is farthest away, traveling in the direction opposite to that of the flux. The result is an inward torque on the electron's trajectory. Were the electron to travel on the other side of the flux cylinder, other things being the same, the above conditions would be reversed, resulting in a torque outwardly. In an electron double-slit experiment, including a vector potential (created by a shielded solenoid), as discussed above, the effect would be to shift an interference pattern horizontally, either in one direction or the other, from that displayed with the solenoid not energized, depending on the orientation of current flow in the solenoid. There is no need here, as in the conventional QM view, to take account of distant EM fields, endowing remote action on magnetic fields (as in QFT, raising the locality issue), rather to acknowledge the reality of the vector potential A as a localized physical entity, as Aharonov and Bohm proposed. It is the local high-energy-density boson condensate that constitutes that physical entity, but unlike a gravitational scalar potential, associated with a massive body, it is a vector potential, associated with a relativistic momentum, giving rise to a directed energy-density gradient. While the above was restricted to the behavior of electrons, it would apply as well to positrons, protons and ions, etc., taking account of the differences relating to charges and masses. Here again we point to an experimentally created subsystem, the boson flux created by the solenoid, as the embodiment of the vector potential, in perturbing the trajectories of the charged particles. And again a great amount of energy must be dissipated in creating and minimizing the boson bundle in order to minimize the density-specific entropy production in the subsystem. In a recent experiment,

Yau[122] *et al.* have revealed a more complex AB effect involving a fixed electric field and a varying magnetic field, suggesting a spin-dependent modulation of the Aharonov-Bohm oscillation of magnetoresistance The mystery of the AB effect is thus rationalized, but it remains to note (in contrast with the case of superconductivity) that it is the coherent propagation of many millions of coherent bosons (magnetic flux), densely packed in a common trajectory, that is the distinguishing feature of the AB effect, and it is that constant, localized energy-density that acts to alter the momentum and trajectory of a few nearby electrons.

The Quantum Hall Effect

Electrons, trapped at the junction between two crystalline semiconductors, can move only in a plane, revealing quantum variations on the classical Hall effect, in which electrons are given a component of motion normal to both the applied electric field and the magnetic field, at near-absolute zero temperatures. Instead of the Hall voltage rising smoothly with increase in the magnetic field, it increases in steps, with constant values over a small range of field strengths. In addition, the longitudinal voltage – the voltage necessary to maintain the flow of current – nearly vanishes at these Hall voltage plateaus. That is, the electrons become nearly perfectly conducting, though they are not technically superconducting (they do not expel a magnetic field). At each plateau, the Hall resistance RH equals an integer multiple of the quantum of resistance h/e^2, where e is the charge of the electron and h is Planck's Constant. Further investigations have revealed additional plateau levels.

In 2000 Gustavo Gonzalez-Martin[123] Universidad Simon Bolivar, Caracas argues that "It is possible that magnetic flux quanta be associated to the electron orbits corresponding to definite total angular momentum levels in an electron gas in a constant intense magnetic field. … The theory of the IQHE and the FQHE has evolved by the explicit construction of quantum state wave functions that describe many features of these phenomena. Nevertheless it may still be possible to explain certain facts from general principles, as the extraordinary accuracy of the results suggest." He develops his model and shows that "If there are no resistive losses, the voltage induced along the line, by the carried induced flux cutting it, leads to a transverse resistance which is fractionally quantized, $R_t = (f/q)(h/2e^2)$. This is a fundamental general relation which only depends on the quanta q, f carried by the carriers. This resistance value is a high precision number depending on fundamental constants and integers, which is displayed when appropriate microscopic conditions are met. The FQHE is such an experiment, which essentially measures the ratio of charge quanta to flux quanta of the carriers. The generality of the argument may be turned around: The extraordinary accuracy of the measured fractional values indicates that nature is displaying new principles in these experiments and we may expect that current calculations techniques may require corrections. If this is the case, relations like the ones presented here may guide in improving the theory for the FQHE."

We note here that Eisenstein and Stormer[124] expressed some optimism (in 1990) that the even denominator states might be accounted for in terms of the spin states of the electron. As we have seen, the above model directly and necessarily incorporates the electron's spin. Yet it is surprising that there appears to be little awareness in recent years of the work of Gonzalez-Martin. On the other hand it is puzzling that Gonzalez-Martin makes no reference to contemporary efforts to develop the theory of composite bosons. The theory of composite fermions (CF's) is said to correctly predict all the observed fractions, including their relative intensities.[125] However, Shrivastava argues that the CF model cannot be valid, since it is not internally consistent; either it violates Maxwell's equations and creates unreal objects, or it agrees with the data, but the masses, sizes and densities are internally inconsistent. Such efforts were motivated in part to explore possible relevance of superconductivity with FQHE.[126]

The "new principles," that Gonzalez-Martin perceives in FQHE experiments, refer to the *precisely correlated dynamics of countless millions of electrons (a subsystem) adjusting their individual behavior, in response to change of applied force, so as to minimize the energy of the system as a whole, the density-specific entropy production of the subsystem and the flow-specific entropy production of the system, consistent with external and internal constraints.* Each electron is in orbital motion about a quantum of magnetic flux (normal to the electron plane), which is itself moving in that plane. The orbital motion results, not because it is acted upon by the applied magnetic field, but because the quantum of magnetic flux constitutes a magnetic vector potential (as in the AB effect), that results in a radial energy-density gradient, causing the gyrating photon in the GSBP electron to follow a circular path. In the plateau regions the longitudinal conductivity is very large, though not infinite (zero resistance), as is the case for superconductors, requiring most of the inevitable dissipation to occur in the leads. Similar behavior is seen in the conductivity of single-walled carbon nanotubes.[127] Similar behavior is seen also in the vortex mechanics, characterizing the fractional quantum Hall effect, on the one hand, and the Bénard convection vorticies, discussed in chapters 4 and 5, in both cases driven by the laws on nonequilibrium thermodynamics.

Thus we see in regard to FQHE there is no need to speak of wave-particle duality or superposition or counterintuitive particle behavior - only of extremely complex, many-body behavior that yields to relatively simple semiclassical analysis.

Dilute-Gas Bose-Einstein Condensates (BEC)

What, if anything, is "weird" or counterintuitive about Bose-Einstein condensates? Must we refer to the composite bosons (atoms) that make up a BEC, as particles sometimes but waves otherwise, or as being collectively in a "superposition" of a single state? If several million such bosons comprise a BEC, can they be said to occupy a single quantum state (all physically in the same spatial position) and if so how could that be reconciled with the implied extreme energy-density? Clearly, the result is not one massive "superatom" atom, rather a collection of atoms occupying a minimum allowable volume in a common lowest energy state, consistent with external constraints ("macroscopic occupation of a single quantum state"). While the GSBP fermions that make up the bosonic atom have (at all times) associated wavelengths, characteristic of their individual kinetic energies, there is no need to speak of waves or wave-packets, though the latter may be algorithmically useful. Nor does the notion of superposition add to our understanding of the behavior of the

BEC. The above applies as well to boson lasers. The bosons exiting the laser are all entangled in the same energy state, but certainly not as one boson of enormous energy on a single trajectory. While lasers may have had an aura of mystery at earlier times, we would speak today only of their extraordinary, but well understood, behavior.

Nevertheless, as E.A. Cornell[128] *et al.* have said, dilute-gas BEC are of great interest because the interaction of particles in such is much greater than that possible in a laser beam and the realm of behavior is much richer. But this brings us to the heart of the distinction. What is there physically to distinguish composite bosons from bosons or from composite fermions? "Why," the authors ask "should the presence or absence of distinguishing tags on the atoms so profoundly change their statistical behavior? For the particular case of bosons, the problem may be more palatably expressed in terms of fields, rather than particles." But they note "as we try to apply the notion of fields back to atoms, we are left with a new set of troubling notions – can the probability amplitude for a large collection of atoms ever really be like an electric field – coherent, continuous, macroscopically occupied, macroscopically observable? The answer is 'absolutely yes!' Troubling or no, this is exactly what a Bose condensate is about." The "distinguishing tags," referring to Bose statistics and Fermi statistics, are merely names to label the results of statistical experiments; they provide no physical insight, which any physical model of quantum behavior should provide. Why, we should ask, are photons capable of attracting one another, under the right circumstances, as evidence of "self-focusing" in laser beams indicates?[129] We conjecture along the following lines an answer to the above question.

Two phase-coherent bosons (particles characterized by *local* EM fields), on sufficiently close parallel trajectories, appear, each to the other, as electromagnetic vector potentials tending to inwardly torque each boson, eventually resulting in the merging of the two bosons into one of total energy-momentum equal to twice that of one boson (see Fig. 12-2). In the formative stage, where many such bosons are involved, we would refer to this behavior as "boson bunching." If, on the other hand, two or more such bosons, were to have opposite relative phase, we should expect the EM vector potentials to engender outward torque on the bosons, tending to disperse the boson trajectories. Moreover, the above behavior should be greater or less, depending on the energy of the bosons and the initial separation of the trajectories. Without some such model, we would be at a loss to account for the fact that no boson (physical particle) reaches a detector or a sensitive film in the regions between the fringes, consistent with the conservation of energy-momentum. Similarly, electrons must be physically directed to and from the fringe regions of interference.

Could a similar model be helpful in understanding the self-attraction of composite bosons? In the GSBP model, the kinetic energy of a composite boson (atom) would be associated with the "external" velocity (momentum) of an "internal" gyrating photon, whose energy is the rest mass of the atom, in the same way as for a moving electron (chapter 11?). The moving atom is thus represented as a boson moving on a helical path of a particular pitch (wavelength), and hence appears to other nearby similar atoms on similar trajectories as constituting an EM vector potential, causing their trajectories to merge or diverge to the extent enabled by phase coherence and other factors, including nuclear repulsive forces. When many if not all of the composite bosons in a BEC have the same phase and momentum, in a trap, the density will be maximal under the system constraints, i.e., the attractive and repulsive forces will offset one another. The notions of quantum phase and coherence in a dilute-gas BEC system are thus embodied in the GSBP model.

There are three areas of research on dilute-gas BEC that merit some discussion here, in that they reveal some sense of mystery (though certainly not on a par with the violation of Bell's inequality) and for which the GSBP model of composite bosons may be relevant. The first concerns the process by which a non-condensate dilute gas transforms over time and temperature change into a BEC. Miesner[130]*et al.* evaporatively cooled sodium atoms close to the onset of Bose-Einstein condensation and then "suddenly quenched to below the transition temperature. The subsequent equilibration and condensate formation showed a slow onset distinctly different from simple relaxation. This behavior provided evidence for the process of bosonic stimulation, or coherent matter-wave amplification crucial to the concept of an atom laser." *This process is characterized by bosonic stimulation, in which the presence of bosons in one state enhances the probability that other identical bosons may be induced into the same state in a subsystem as a means of minimizing the density-specific entropy production (maximizing the energy-density) of the subsystem.* As the authors note, the theoretical description of BEC in weakly interacting dilute gases has a long history and has accounted for most of the experimental results, though a full description of the dynamics of condensate formation has not yet been developed. Can the GSBP model of composite bosons facilitate this development?

A second area of study relates to the formation of large vortex aggregates in a rapidly rotating dilute-gas BEC. As Engels[131] *et al.* state "The phenomenon of quantized vortex formation is a unifying feature present in many quantum mechanical systems. Vorticity is intimately connected to superfluidity and thus is a premier means for fundamental and comparative studies of different 'super' systems. ... In our case ... the giant vortex formation arises as a dynamic effect. Nevertheless the lifetime of our giant vortices can extend over many seconds, which we attribute to a stabilization of density features in a rapidly rotating condensate due to strong Coriolis forces. The influence of Coriolis forces can also induce oscillations of the giant vortex core size in the early stages of its evolution." In other related vortex studies, the atom-atom interaction can be changed over a wide range, from repulsive to attractive using Feshbach resonance, leading to controlled collapse of a BEC. Other efforts aim at precision measurements of vortex lattice spacing and vortex core size[132], or structural properties.[133] Solitons, localized disturbances in a continuous nonlinear medium that preserve their spatial profile due to a balance between the effects of dispersion and nonlinearity, are observed to decay into vortex rings in a BEC[134]. *Such vortex formations are clearly macroscopic subsystems created consistent with the demands of 2nd Law.*

Lastly, we refer to the phenomena of "slow light," in which the speed of light in a dilute-gas BEC is inversely related to the cube of the local energy-density in a dilute-gas BEC. Experimentally, this involves sending a laser probe beam through the condensate, utilizing the phenomenon of electromagnetic induced transparency (EIT) to enable zero absorption of the probe beam, and precisely measuring the transit time delay of an optical pulse through a BEC of known diameter. From this and the calculated energy-density of the condensate, we can determine the speed of light in the condensate in the trap, as the value c times the energy-density in space divided by that in the BEC. For a sodium vapor BEC, we thus obtain a value of 20.8 m/s, in good agreement (considering the uncertainty in the estimated average energy-density in our local region of the universe) with that derived from the pulse time delay measurement of 17m/s and that resulting from classical calculations based on the group index of refraction as a function of the wavelength of light[135]. This is a very important result, not simply because theory and experiment agree, but because it provides an initial quantitative verification of our assumption in chapter 7 that the speed of light varies inversely with the cube of the local average energy-density. It is important also for what it does not say: There can be no "dark energy" in the universe, or the present estimated energy-density would be low by an order of magnitude. Indeed our value for the speed of light in the condensate suggests that the present estimated energy-density is too large by twenty percent. Were there such a large amount of dark energy, making the present energy-density less than 10^{-29}kg/cm^3, the speed of light in the BEC would be less than 8m/s, a factor of two or more off from the measured value. Finally the above experimental result prompts us to look upon a dilute-gas Bose-Einstein condensate as a universe in microcosm, which *if left unto itself, must be inevitably subject to dissipation and entropy increase.* Again we are left to contemplate the enormous energy cost of turning the clock back in order to create the nearly quiescent, highly-ordered (low entropy) micro-universe in its near-lowest energy state.

In their discussion of the early expansion of a rotating condensate, Coddington[136] *et al.* clearly refer to such microcosm: In the limit when the radial Thomas-Fermi radius expands very rapidly, "…every point in the cloud expands radially outward at a rate proportional to its distance from the cloud center. In this limit, the 'fabric of the universe' is simply stretched outward, and all density features, including vortex core size expand at the same fractional rate." Boyd[137] *et al.* have studied the limits of maximum time delay (minimum velocity of light) induced by slow-light propagation and concluded that there appear to be no fundamental limits. However, we should expect that if there are any such limits they would likely be associated with an increase in the repulsive force or decrease in the attractive force (or both) between atoms, as the temperature declines further toward absolute zero.

If a dilute-gas BEC is in some sense a model of the very early universe, we should expect the attractive force binding the atoms together in maximal order to be simply that due to gravity, but curiously there is little or nothing in the BEC literature about the attractive force and in particular about the force of gravity. According to our assumption in chapter 7, G must vary inversely as the sixth power of the average energy-density in order make the gravitational charge m$\sqrt{G}$ an absolute constant (similar to the value of the electromagnetic charge *me* taken as an absolute constant) in an expanding universe. If we assume the total mass (~11kg) of a sodium vapor BEC to be concentrated at the center of a sphere of radius 50μm , a single atom (~1×10^{-25}kg) one micron separated from the BEC boundary would experience a force of about 10^{-38}N (the binding force at the boundary's surface would be several orders of magnitude greater). This force would cause the atom to be accelerated at the rate ~10^{-13}m/s^2, thus requiring about 30 minutes for the atom to be absorbed in the BEC, where the atom spacing is of the order of 10^{-8}m. Numerous other atoms, similarly spaced from the boundary, would be attracted to the BEC much more rapidly due to the growing mass of the BEC. This of course neglects any repulsive force, thermal or other, acting on the atoms. These estimates are based on the presently accepted values of G and the rest mass of the sodium atom, since we have taken the gravitational charge m$\sqrt{G}$ to be an absolute constant, independent of variations of the local energy-density. *Here the boundary of the BEC could be seen as a "membrane" enclosing the BEC subsystem.*

There is in the above no deeply mysterious behavior associated with Bose-Einstein condensates. We see, on the contrary, the long-range highly-ordered behavior of millions of atoms in a very low entropy state.

The importance of the arguments in this chapter, involving the creation and evolution of entangled particles or long-range correlated (organized) particle systems, warrant here the strongest possible statement: **All such systems are created as open thermodynamic subsystems, characterized by minimal density-specific entropy production, within a larger system, characterized by minimal energy flow-specific entropy production in the stationary state.**

Transition from Quantum to Classical Systems

As we have seen above, there is no deep mystery or counterintuitive truth about the behavior of *unperturbed* quantum systems, and we should expect therefore no such unexplainable phenomena to arise in the transition to *macroscopic* classical systems, so long as we understand that all of the latter evolve and persist, for a time, subject to the laws of Nonequilibrium Thermodynamics. The implication is that quantum systems, described by Schröedinger's equation, are essentially *isolated* from the considerations of nonequilibrium thermodynamics (so long as no dissipative measurements are made) and there is therefore a fundamental distinction between the two kinds of systems; the one closed to the flow of energy, from or to, its environment (relying solely on the 1st Law and ignoring the *2nd Law*), and the other open and dependent on the flow of energy, which may facilitate the growth and evolution of organized, structured subsystems that serve to minimize the flow-specific entropy production of the system.

This distinction is clearly revealed in the emergence and growth of a dilute-gas condensate of millions of atoms in a single, lowest-energy state, the consequence of a very large and controlled flow of energy required to cool the gas. And it is later revealed in the measured loss of those atoms at the boundary separating the subsystem and the system, once that sustaining flow of energy is removed. Such experiments are thus defined by three chronological epochs, in the first of which energy flow through a strongly-nonlinear thermodynamic system (dilute gas) nucleates and evolves a long-range ordered subsystem (condensate), while decreasing the flow-specific entropy production in the system and the mass-specific entropy production in the subsystem. The second period is characterized by the system being in a stationary state of large entropy production but minimum flow-specific entropy production, far from equilibrium, and the subsystem in a stationary state of minimum mass-specific entropy production and relatively low entropy. And the third period reverses the action in the first period, with the destruction of the subsystem and the system evolving toward a state of thermal equilibrium with the system's environment. The boundary we seek between the quantum and classical domains is the moving "membrane boundary" between the subsystem and the system, as discussed in chapter 4.

The above is couched strictly in the language of thermodynamics and as such may serve to characterize the fundamental features of the transition from quantum to classical systems, independent of the specific kinds of quantum systems, whether they involve only a few particles in interference phenomena or many-particle systems, e.g., squids, FQHE systems, etc. Not surprisingly, the literature on the transition from quantum to classical systems reads largely in QM terms and specifically in terms of wave-particle duality, coherence/decoherence and the superposition of particle states, since the current equilibrium or linear thermodynamic theory offers relatively little insight. Nonetheless, a great deal has been learned from recent experiments designed to elucidate the nature of the transition for particular systems.

We should expect also that the transition be seen differently from the perspective of a strictly-particle-ontology quantum mechanics, as presented here, vis-à-vis conventional QM. In our perspective, particles and particle dynamics are the focus of analysis in both regimes; fields and waves are strictly local properties of particles and a single particle is never in a superposition of two or more states (never here and there simultaneously) and the system is always accountable to a logic of reality, locality and causality. Yet the two regimes are distinguished by certain fundamental, as well as emergent, properties (relating to the number, variety, density of particles and relatively weak forces of interaction among them). The requirements of symmetry, indeterminacy relations, underlying the concepts of non-commutivity and intrinsic spin are fundamental and unique to the quantum regime, while the local energy-density (in our corner of the universe or in the nucleus of atoms), that determines the local values of the "constants" c and G, governs in both regimes. A system of particles, will evolve quantum mechanically to the extent that it can be isolated (shielded from all energy-momentum exchange) from its environment and all of its internal interactions are dissipation-free. In practice the probability of realizing these two requirements decreases rapidly with the number and kinds of particles per unit volume (local energy- density) in such systems because emergent, relatively weak dissipative interactions (e.g., van der Waals) play an increasing role, contributing to decoherence in the system. Thus it becomes increasingly difficult to isolate even simple inanimate systems from their environment and impossible, by definition, in the case of living systems and, with their dependence on many-particle weak interactions, such systems can only be comprehended in terms of classical mechanics, subject to thermodynamic requirements.

CHAPTER 16

GSBP QUANTUM MOLECULAR THEORY

Hydrogenic Atom and Small Molecule Theory

The task of creating a first-principles theory to account for the structure and dynamics of all the atoms in the periodic table, beyond the hydrogen atom, is all but impossible in the present "point particle" representation of the electron. And the difficulty of creating such theory increases exponentially as atoms combine to form molecules. Indeed many of those who helped to create QED would likely acknowledge that QED does not derive from first principles. The problem stems from the lack of any direct means in QFT/QED for including a reduced charge, along with the reduced mass, of the electron, as discussed in chapters 11, 12 and 14. As the number of nucleons in the nucleus of high-Z atoms increases, the agreement between theory and experiment, achieved via the elaborate calculations of QED, rapidly declines, requiring still greater departure from first-principles reasoning. Thus the problem faced by theorists in modeling molecular systems turned on the question: How and in what form can we include the corrections, needed to square theory with experiment in the study of increasingly complex molecules, so as to pragmatically minimize computational costs? Efforts toward this end have been focused on the development of Molecular Orbital Theory (MOT) in dealing with small molecules and on Density Functional Theory (DFT) to apply to more complex systems, including electron dynamics in solids.

In principle, the shape of a polyatomic molecule, the specification of all of its symmetry, it's bond lengths and angles, can be predicted in MOT by calculating the total energy of the molecule for all of the various possible conformations and identifying that conformation which results in the lowest energy. But molecular orbital theory cannot adequately model even the higher-Z atoms in small molecules or excited states thereof; consequently the errors and calculation costs quickly escalate. With these limitations, the real costs in software design and computer running-time has become a major factor in guiding these studies.

In an effort to circumvent these problems, density functional theory evolved in recent decades toward the development of a generalized Schröedinger-like equation to model the ground-state density of all the electrons in a molecule or solid. Considerable progress has been made in recent decades in this effort, though we see, by definition ("ground state"), that excited states of such systems cannot be modeled in this way. The principal problem is the lack of a suitable exchange-correlation functional in any equation of the above form, applicable to a wide variety of molecules and solid-state systems. This is a *many*-body problem in which the motion of one electron affects, and is affected by, that of all the others and so we should expect that the fundamental problem traces to the lack of physical dimension and internal dynamic structure of the electron. We might expect the GSBP model of the electron to facilitate the development of MOT and a general exchange-correlation functional in DFT, but it is unlikely that a first-principle theory of molecular evolution may emerge from either a MOT or DFT strategy, because they could not take account of the *growing local energy-density of the molecule as other atoms are enabled to bind to it*. On the other hand that is exactly what a growing molecule, taken as a open thermodynamic subsystem, might do, given suitable empirical input and software tools to make the necessary calculations.

GSBP-Electron Structure and Dynamics of High-Z Atoms

Modeling the electron dynamics and spectra of hydrogen-like atoms and that of molecules must rest fundamentally on our capability for modeling high-Z atoms with suitable precision and acceptable economy. The GSBP model of the electron should greatly facilitate the modeling of high-Z atoms in that it directly, precisely and economically (calculation costs) adjusts the electron's mass-energy and charge (in accordance with the accompanying mass increase of the nucleus), as the electron is added in the *Aufbau* process. And it should do so, in keeping with the requirements of general relativity in which the local energy-density of the molecule might vary. Thus the electron's mass-energy and charge in the ground state of the deuteron atom would both be "reduced" somewhat less, due to the larger mass of the nucleus, and the excited states should be corrected accordingly. As for the hydrogen atom, these corrections, derived from the GSBP model of the electron, take into account spin-orbit, relativity, etc., factors, while ignoring the structure and dynamics of the nucleus.

The electronic structure of the helium atom may be modeled in the same way, with corrections to the ground state of the paired-electrons being determined principally by the reduced-mass factor. The GSBP structure of the two electrons is identical, save for their spin orientation. But here, in what Karshenboim refers to as the precision physics of simple atoms, we cannot ignore the structure, dynamics and spin of the protons and neutrons in the nucleus because the ground-state electrons are closer to the nucleus and more sensitive to its structure and dynamics. However the question in all of this ought not to be whether we *can* ignore these details but whether we *can* accurately model the nucleus economically, together with the electronic structure, and thereby obtain a more complete theory of possible relevance to future studies. Such modeling is all but out of the question, even for the helium atom, in terms of the "point particle" representation of neutrons, protons and electrons. As seen by the GSBP electron, the nuclear charge derives from a GSBP proton, whose structure and dynamics differ from the electron only in its spin orientation, much smaller radius of photon gyration and much larger mass-energy. And like the electron the orbiting photon in the proton must exactly close in n orbits as the proton completes one orbit around the center of mass of the atom. Moreover, as we have seen, the proton's greater mass relative to that of the electron) and its charge derive from the mass-energy and combined spins of three quarks which make up a proton in nuclear Euclidean space. And finally in that space the protons and neutrons in the helium atom combine structurally so as to maximize the separation between the parallel oriented spins of the two protons, within constraints, and minimize the energy of the nucleus. In so doing it also maximizes the energy-density of the atom by minimizing the volume of the nucleus and that of the electron space. We may conclude from this that the adjustments, which decrease the energy of the nucleus and the electrons in the ground state, are more than matched by the accompanying decrease in the volume of space occupied by the atom. In this we see that stable high-Z atoms come into being, consistent with the principle of minimum density-specific entropy production and the requirements of the extended *2nd Law*, as a means of maximizing the energy-density of massive particles (subsystems) and thereby minimizing the entropy production per unit volume of space.

The above arguments apply broadly to all high-Z atoms, within the constraints imposed by the quantum rules and provide a theoretical foundation for constructing models of hydrogenic and hydrogen-like atoms and exploring their radiation spectra, and in consequence the same applies in the context of small molecules.

Molecular Evolution: The Thermodynamic Imperative

We have seen in previous chapters the inception and evolution of all quanta (accompanying the expansion of the Universe), the later creation of charged particles and with that the creation of the simplest atoms and their eventual gravitational collapse into dense cosmic bodies, the creation of high-Z atoms in stars and their scattering throughout the Universe (as the Universe cooled), and now, in a low energy-density environment, the orderly agglomeration of those atoms in a wide variety of molecular structures. Each such evolutionary event was and is orchestrated by the extended *2nd Law* of thermodynamics, together with the demands of symmetry, quantum indeterminacy and relativity. Specifically, they are the open-subsystem embodiments of the principle of minimum density-specific entropy production, alternatively viewed as conditionally evolving to a state of maximal energy-density, within the constraints on energy flow in the Universe. We seek here to sketch a thermodynamic foundation, leading hopefully to the development of a first-principles theory of molecular evolution.

A molecule may be viewed as an open subsystem in a closed strongly-nonlinear thermodynamic system and one that grows, via the transport of one or more atoms through a virtual membrane (boundary of the molecule), in accordance with the principle of minimum density-specific entropy production, as discussed in chapters 4 and 5. The evolution of such subsystems may also involve the coordinated transport (exchange) of one kind of atom, which is more limited in its capability for building a molecule, for another less constrained, under given conditions. In the process the energy-density of the subsystem is maximized and some energy is dissipated in (radiated from) the molecule's environment, the larger system. At the same time the total space occupied by the molecule and the atom to be added may be minimized, such that the energy-density of the evolved molecule is maximized in a quasi-stationary state. Here we see the fundamental interrelationship between the minimization of density-specific entropy production in an open subsystem and the minimization of entropy production per unit energy flow in the evolving molecule: *Any action that minimizes the space occupied by an evolving (growing) molecule (subsystem), while typically increasing its stability (lifetime), tends to decrease the entropy production per unit density in the subsystem.* The same requirements characterized the repeated down-conversion evolution of all the massive particles, in the adiabatic quantum-number expansion of the Universe.

The adjustment of bond lengths and angles that accompanies the evolution of a molecule, specific to a given symmetry-based conformation, may be seen in the same light as the GSBP fine structure adjustments to the dynamics of the electron in the hydrogen atom. Such adjustments should apply to excited states of a molecule because they take account of the varying energy-density of the molecule, and so we should expect the model to agree closely with the radiation spectrum of the molecule. And, as before, the GSBP model of the electrons (and the protons in the nucleus) take account of variations in the local energy-density of the system and its environment. In the evolution of a single molecule of water, for example, two hydrogen atoms, in a local suitably high energy-density environment, may form σ bonds, one at a time or simultaneously, with the unpaired $O2p_x$ and $O2p_y$ electrons in the oxygen atom, with the $2p_x$, $2p_y$ and $2p_z$ bonds all mutually orthogonal. But this configuration does not result in the optimal bond lengths and angles. Symmetry is broken enabling fluctuation-induced variations in bond lengths and angles which together decrease the volume of the molecule, thereby maximizing its energy-density and minimizing the density-specific entropy production in the evolution of a more stable ground-state conformation. This prompts four important observations: (1) Many-body problems, those that may be characterized in terms of a potential energy function, as in MOT and DFT, do not admit of *exact predictable* solutions, only relatively coarse, probabilistic characterization. They are only coarsely sensitive to structural change as it contributes to variation in the energy-density of the molecule. (2) "Exact" retrodictive solutions may be obtained *only* via physical evolutionary processes; Nature formulates and solves her own problems in rigorous detail and in a timely fashion. (3) Such physical evolutionary processes may be monitored and perhaps controlled in computer models as well as in the laboratory setting. (4) Though necessarily limited in predictive power, such evolutionary models enable the testing of *first-principle theory of many-body systems vis-à-vis empirical results in the laboratory*. Thus linear molecules, e.g., a hydrogen molecule, may be modeled exactly, consistent with special and general relativity, in terms of the GSBP electron, since only bond lengths are subject to fine-structure corrections. In contrast, angular molecules, nominally classed in one or another symmetry group, may be subject to significant bond angle variations, making such classification in some cases almost meaningless. For all such, the "exact" retrodictive solutions, referred to above, may be found only via precise empirical studies of the end product, i.e., employing X-ray or other technological tools for characterizing molecular structure. But perturbation of the angles and bond lengths in a computer model would enable a direct test of GSBP first-principle molecular evolution. Technological advances in modeling the surfaces and volumes of molecules offer a very useful, though limited strategy, for correlating the spatial properties of molecules with variations in their energy-density, as a molecule evolves with the addition, or exchange, of one or more atoms.

With recent advances in femtosecond optical excitation technology, it might be possible to characterize the step-by-step evolution of molecules, as open subsystems in a closed thermodynamic system, and perhaps verify the corresponding time evolution (decrease of density-specific entropy production in the subsystem and flow-specific entropy production in the closed system), as for example, in the evolution of H_2O, CH_4 or molecules including higher-Z atoms.

The foregoing should apply with even greater potential for advancing our understanding of the evolution of the macromolecules in the living cell and indeed of the origin of such cells.

PART IV
GRAVITY <u>IN</u>
THE RELATIVISTIC MOLECULES
AND SYSTEMS OF LIFE

It may be well to reflect here on what we have been proposing: In Part I we adopted (adapted) the Darwinian concept of *natural selection* (loosely referred to as *evolution*) as a strategy for the development of a rigorous theory of nonequilibrium thermodynamics, applicable to systems far from equilibrium. The resulting principles of minimum flow-specific and density-specific entropy production and the extended *2nd Law* of thermodynamics were then employed, in Part II, to assess the validity of the current scenario of the origin and evolution of the Universe and to structure the development of an alternative one, taken as a quasi-adiabatically expanding nonequilibrium thermodynamic system. This led in turn to the requirement that the parameters c, and $\sqrt{G}$ remain constant throughout the quantum era, while the mass m_N of the particles and h (Planck's Constant taken as a parameter in the quantum regime) halve with the doubling number N. The extended quantum era functioned principally in the creation of the 10^{80} matter particles and free photons in the Universe today, the residue from the annihilation of 10^{79} each matter and antimatter particles created in down-conversion epochs in the quasi-infinite past. What is important here is that all such particles emerged as open subsystems, step by step, at the behest of the *Nonequilibrium Thermodynamic Imperative*, as is the case in the creation of the stable macromolecules of living systems and of the living systems themselves at various hierarchical levels. The scalar era of cosmic physical expansion commenced with the cessation of the down-conversion process, elimination of antiparticles, the restriction of h to its present value and beyond and the seeding of all the large-scale structure in the Universe, principally brought about via the force of gravity and the NTI.

This intuitive, simpler model for the origin and evolution of the Universe, makes no demands today for "dark energy" or "dark matter," and provides a straightforward accounting for the creation of the massive matter particles, with no need for a "Higgs field." We then sought in Part

III to discern the implications of our developing scenario for resolving the many problems in joining the quantum world with the scalar expanding Universe, given the dependence of c *and* $\sqrt{G}$ on the local energy-density of particles. That led to simpler and more intuitive explanations for various strange behaviors of quantum systems, including that of charged particles (quarks, electrons), said to have mass-energy and spin but no spatial dimension or internal structure. Throughout, open thermodynamic subsystems played a central role in the initial evolution of the massive quanta, galaxy clusters, galaxies, stars and atoms, planetary systems, geological systems, quantum systems and small molecules, in accord with the Nonequilibrium Thermodynamic Imperative to minimize the rate of entropy increase in the Universe. As stable open subsystems, the massive quanta have well-defined radii in any epic, as the Universe expands; they are characterized by well-defined maximal energy-densities, in their lowest-energy stationary states, bounded by virtual membranes, thereby minimizing density-specific entropy production in the open system, tending to decrease the rate of entropy increase in the expanding Universe. The heavy atoms formed in massive stars. Similarly all gravitationally created open systems, from giant walls, galaxy clusters, galaxies, stars, planets and moons, etc. may be seen as virtual membrane bounded bodies, with relatively definite dimensions in any epoch, which minimize their density-specific entropy production as open systems, as the Universe expands. And, as we have seen in the previous chapter, all molecules evolve as virtual membrane-bounded open subsystems subject to the *Nonequilibrium Thermodynamic Imperative*. This brings us now to ask: How can we better understand the extraordinary complexities of animate matter in light of these new insights, and in particular, how can the principles of nonequilibrium thermodynamics and the extended *2nd Law* inform our understanding of the origin and early evolution of living systems?

That is our concern in chapter 17, i.e., with the emergence of what we *assume* to be *proto-life*, under *assumed prebiotic conditions*. The emphasis is appropriate, as we have seen how fundamental assumptions may seductively lead us astray in our reasoning and in the development of fundamental theory. As all attempts to construct a rational account of the emergence of life have thus far been stymied in one form or another of the "chicken-egg" problem, our aim is not to offer yet another chemical model of life's beginning, but to explore the enabling potential of nonequilibrium thermodynamics, given biochemical and geological constraints, for driving inanimate matter uphill toward proto-life. In particular we seek to understand how the principle of thermodynamic selection may function in delimiting the possibilities of random *statistical selection* in biochemical reactions, as for example in the non-template ordering of amino acids in proteinoids. Indeed *our concern here and in all of the remaining chapters* would be in part *to understand how the principle of thermodynamic selection acts, not only in the formation of organic monomers and polymers at the most elementary levels, but how it undergirds Darwinian natural selection in the evolution of biological systems at various hierarchical levels.* In follow on chapters we would hope to articulate a self-consistent path from proto-life to the emergence of the earliest-known biotic systems, in which the basic apparatus of cellular growth and replication are readily discerned.

We hope that this effort may lead others toward the development of a rigorous general theory of evolution that informs us about systems of various kinds (involving the exchange and flow of matter-energy of various forms), at various levels in the Universe and dynamically interdependent at various levels. This would seem to be a worthy goal. The postulates and extended *2nd Law*, as presented in chapter 5, would appear to provide a base for the development of such theory, but the peculiar role of gravity as a means for the stable, long-range storage of mass-energy (slowing the growth of entropy) and its indiscriminant coupling to all forms of matter and energy suggests that the ubiquitous play of evolution hides much insight yet to be gleaned.

CHAPTER 17

THE ORIGIN OF LIFE –
A THERMODYNAMIC PERSPECTIVE

The uniqueness of living matter: its dynamic complexity, so dependent on the steady constrained flow of energy in a state far from equilibrium and its commonality of minimal atomic constituents, yet uncommon dependence on multiple weak forces which delicately fashion and stabilize such complex systems in myriad shapes and forms, all set it apart from inanimate material systems as objects of scientific study. Compounding this, the religious, spiritual, moral and ethical issues and questions that are uniquely relevant to living systems and in particular human life have led many to place such systems off limits to scientific study and to thwart the teaching of any such reasoning. But they have a point: Such reasoning has yet to account for the *origin* of life at the most fundamental level.

Given the relative immaturity of the biological sciences and the even shorter history of paleontology and laboratory investigations of the origin of life, it is not surprising that no generally accepted theory has emerged, despite important advances over the past century. These efforts have given rise to several phases of optimism followed in each instance by awakening to yet another unforeseen challenge. There is no need here to review the history of these efforts, as it has been extensively surveyed by Iris Fry[138] and others. Instead we focus on what has been largely *missing* or *lacking* in these efforts, relating to the conceptualization or theoretical underpinning of the problem and the consequent design of experimental programs.

Progress and Limitations in Fundamental Theory

A century of scientific speculation and research on the origin of life has been defined as much by what has been lacking in fundamental theory as what has been learned in the laboratory. All would no doubt agree that nonequilibrium thermodynamic theory should play a fundamental role in these studies, yet few authors even refer to the work of Prigogine and others, relating to the evolution of *dissipative structures* in systems far from equilibrium. Others make occasional reference to the notions of *open systems*, *self-ordering* (sequencing), *self-organizing* or *self-assembling systems*, all of which, in our scenario, are logically mislabeled and add little to our understanding of how such ordering is orchestrated; it is the Nonequilibrium Thermodynamic Imperative that drives and selects the ordering, etc. Of course, this void in the literature is not surprising because such fundamental theory of the driving forces, constraints and selection processes was and remains nonexistent in the literature, as J.L. Fox[139] has said: "The formation of macromolecules is interpretable as the production of a metastable state. The energetics of this process must be handled by non-equilibrium thermodynamics. …Unfortunately the formalism to fully handle this does not exist in a satisfactory state. The forces which determine amino acid sequences are largely unknown. These macromolecules can be considered as energy storage states. Since they are obviously stable they must represent an entropically and energetically favored state. Insofar as the interpretation of elevated energy states as informational levels is correct, these macromolecules may, thus, be considered as information storage molecules." Matsuno[140] has articulated the underlying problem: *"Physics equipped with the adiabatic approximation conspires to make matter inanimate. If we free ourselves from adiabatic approximation, matter can be seen as an active agent simply by virtue of the fact that interaction changes propagate at a finite velocity. If it is to be applied to material evolution, non-equilibrium thermodynamics must be extended so that it can rid itself of the fetters imposed by the adiabatic approximation."* (emphasis added) This was our central objective three decades ago, in the publication of A Thermodynamic Theory of the Origin and Hierarchical Evolution of Living Systems[141] and in its further development a decade ago, as set forth in Part I. But, as explained in chapter 1, developing a formal thermodynamic theory, applicable to systems far from equilibrium, requires freeing ourselves of the formalism of the variational calculus, which has restricted us to the linear regime, i.e., to states very near to equilibrium (the "addiabatic approximation").

Jeffrey Wicken[142] has offered important theoretical insight regarding the nature of living matter. "Hierarchy," he says, "implies ontological 'layers,' which communicate with each other somehow. Biological nature is not like this. It is a web of relationships that fit with, and are defined by, each other. Nature is *organismically* constituted… 'Organization' itself is not even part of the lexicon of physics and chemistry, which deals with systems structured by physical forces rather than those structured for function. ... Organisms might be described as *informed* kinetic structures, in which the information coded in such static structural elements as DNA provides organized kinetic pathways for degrading matter and energy." [pp 7-17] Here, Wicken relates the emergence of organized systems (subsystems in our view) to the *degradation of matter and energy,* where, again in our view, it is the *minimization* of such degradation in the total system (under given constraints) that correlates with the emergence of such organized subsystems. And on page 64 he makes the connection between self-organization and entropy production even more explicit: "Self-organization is thus one of nature's means for serving the universal directive to entropy production." Unfortunately Prigogine's concept of "dissipative structures" has left its mark on others who have sought insight into the nature of thermodynamic systems far from equilibrium. As noted above, that attempt to free us from the "linear regime" has led to little insight of analytical or predictive utility. For Wicken and others the notion of dissipative structures in states far from equilibrium implies a direct relation between dissipation and the evolution of organized systems. Indeed, as Margulis and Sagan observed (chapter 5): " 'dissipative structures' [is] a rather awkward term because it focuses on what the structures -- actually, systems, not structures -- throw away rather than what they retain and build..." In contrast, in our scenario such open *sub*systems evolve toward a state of *minimal density-specific entropy production* (*maximal energy-density*) *under fixed system constraints,* as required by the extended *2nd Law.* It is that light which should guide our thinking on the origin of life. Thus the emergence of proto-life should be seen as a recent evolutionary product of the Nonequilibrium Thermodynamic Imperative that has in the past driven the open-subsystems creation of: all the matter quanta of the Universe and their gravitational aggregation in galaxy clusters, galaxies, stars and planetary systems, along with all the atoms, small inorganic molecules and, most recently, the macromolecular monomers and polymers essential to living systems. *In all of this the same underlying fundamental thermodynamic principles form the basis of a self-consistent, general theory to account for the emergence and evolution of open subsystems that serve to store (lockup) some portion of energy-flow in the larger systems in which they evolve, thereby slowing the rate of entropy increase in the Universe.*

The void in thermodynamic theory that might have guided us in origin-of-life research is evidenced in the terms that have been used to characterize the reactants and products of reactions in experiments. Sidney Fox and associates and followers of the work in his and other laboratories have made much of the "self-ordering," "self-sequencing," "self-organizing" and even "self-sequencing-predilection" of amino acids in the partial ordering of proteinoids. Matsuno goes further in trying to link such terms to some sort of "organizing principle" which might point toward a more fundamental and general mechanism to account for the partially-ordered sequencing of monomers in laboratory experiments. The "self" connotation in these terms conveys a sense of mystery or "judgment" unsuited to a serious scientific study, but the real loss in insight is that such terms appear to be unique to living things and therefore make no connection with similar selective or organizing processes in the inanimate world, as for example selection of one or another additional atom in the creation of an inorganic molecule. The forces involved in different cases may vary but we might hope that the underlying fundamental principles would apply in the process of such selection, organization, etc. That is the work of the *Nonequilibrium Thermodynamic Imperative* in the evolution of the Universe.

Also lacking in the literature, is *focused* concern for the development of a general theory of hierarchical organization and the roles of symmetry, stress buildup leading to symmetry-breaking and hereditary information transfer, as they may inform our understanding of the origin and evolution of living systems. Since these considerations would appear to be fundamentally important in the evolution of the Universe and all the open thermodynamic subsystems referred to above, they should likely play a similarly critical role in the origin and evolution of living subsystems, albeit in a different context. Here we draw attention to the contributions of a number of authors, whose theoretical insights are consistent with the Nonequilibrium Thermodynamic Imperative driving the evolution of animate matter.

The early contributions of the physicist, Howard Pattee[143], toward the development of organization theory are especially noteworthy. He was well aware of the challenge facing theory development in biology when he said "The significant facts of life are indeed more numerous than the facts of inanimate matter. But physicists still hope that they can understand the nature of life without having to learn *all* the facts." Pattee was concerned with "what is perhaps the most fundamental problem in biology — the question of how large a system one must consider before biological *function* has meaning. Classical biology generally considers the cell to be the minimum unit of life. But if we consider life as distinguished from nonliving matter by its evolutionary behavior in the course of time, then it is clear that the isolated cell is too small a system, since it is only through the communication of cells with the outside environment that natural selection can take place. … How large a system must we consider in order to give meaning to the idea of life?" … In my picture, it is the constraints of the primeval ecosystem which, in effect, generate the language in which the first specific messages can make evolutionary sense. The course of evolution by natural selection will now produce better, more precise, messages as measured in this ecological language; and in this case signals from the outside world would have preceded the autonomous genetic controls which now originate inside the cell." *As these observations relate to the origin-of-life problem, the constraints of the primeval ecosystem are in the form of physical, chemical and thermodynamic laws, there being as yet no biological demands. At this stage all of our models should be guided by these demands, as if we were unaware of living cells.*

Elsewhere Pattee[144] reasons: "If there is to be any theory of general biology, it must explain the origin and operation (including the reliability and persistence) of the hierarchical constraints which harness matter to perform coherent functions. This is not just the problem of why certain amino acids are strung together to catalyze a specific reaction. The problem is universal and characteristic of all living matter. It occurs at every level of biological organization, from the molecule to the brain. It is the central problem of the origin of life, when aggregations of matter obeying only elementary physical laws first began to constrain individual molecules to a functional, collective behavior. [Fox's microspheres?] It is the central problem of development where collections of cells control the growth or genetic expression of individual cells. It is the central problem of development where collections of cells control the growth or genetic expression of individual cells. It is the central problem of biological evolution in which groups of cells form larger and larger organizations by generating hierarchical constraints on subgroups. *It is the central problem of the brain where there appears to be an unlimited possibility for new hierarchical levels of description.* These are all problems of hierarchical organization. *Theoretical biology must face this problem as fundamental, since hierarchical control is the essential and distinguishing characteristic of life."* (emphasis added)

In an earlier publication on hereditary organization in primitive chemical systems, Pattee[145] said "**I believe that the concept of 'hereditary process' is fundamental at all levels and for all types of organization. ...by** *hereditary process* **I shall mean any process by which information is transmitted from one structure (the parent) to another structure (the heir) so as to result in a net increase in the physical order of the total system (comprising parent, heir, and environment).**" (emphasis added) He goes on to define "replication as only a special case of hereditary process in which the parent and heir are (1) similar, and (2) separated. If in addition the parent and heir are (3) isolated from environmental sources of information I shall speak of *self-replication*. These three conditions are fulfilled, for the most part, in higher organisms, but they are by no means necessary for hereditary, transmission, for acquisition of information, or for evolution. ... Finally, anticipating the biological interpretation, we may consider the linear sequence as a hereditary memory, while the sequence-dependent folding represents the functional expression of the memory which couples with the environment. But unlike higher organisms, this coupling works in both directions. It is clear that, in polymers propagating under these restrictions, the information contained at any time in the linear sequence may he acquired from the environment just as well as from a parental sequence. Both types of information are equally heritable, and in fact they are indistinguishable without knowledge of the full history of a particular sequence." Summarizing, Pattee states "I am proposing that primitive hereditary molecular evolution may have actually started with a kind of training by direct, selective feedback interaction with the local environment, and developed only gradually by a range of less and less direct and increasingly *delayed* selective processes to the final stage of completely indirect natural selection. From this point of view, it is not productive to ask when hereditary propagation of order became replication or self-replication, or when direct feedback interactions with the environment became natural selection. The difference is only in degree, and both evolved continuously."

The above may be viewed as a measure of a particular parameter in the evolution of a living system, in this case the degree or extent to which it manifests hereditary propagation. However, if we grant that even primitive life forms can exhibit very complex internal and external (with the environment) behavior, then classification of the evolution of a life form should require an appropriately large parameter space. Lacking that, we have not been able even to argue rigorously whether a system is *alive* or when and to what *degrees* it exhibited animate behavior. Several authors have skirted around this issue, without addressing it as a serious void in our theoretical framework. There is much in Pattee's arguments that resonates with the implications of the NTI and with our focus (later) on the role of geological mechanisms in the emergence of living systems.

In keeping with Pattee's fundamental perspective, Fox and Dose[146] [molecular evolution] offer this counsel: "The researcher who wishes to comprehend evolution must move from simple to complex rather than from complex to simple, and he must use his structures to 'seek' functions (as evolution did) rather than employing functions to identify structures and components of structures. ... Prebotic molecular evolution would have employed structures to 'search' for functions. Natural (Darwinian) selection occurs at the level of function. Evolution must have been constructionistic, putting together components to yield functional systems, upon which selection then operated. ... nucleic acids can hardly be conceptualized as coding for proteins that did not already exist. ... The proteins are chemically adapted to an evolving set of enzymes, and derivatively to metabolism. This follows since protein molecules possess the flexibility necessary for enzymic interaction with substrates; they permit sufficient variegation within a molecule and sufficient variation between molecules to allow for the evolution of an array of specificities." [pp 238-40] Here we might add that the earliest function would be to follow the bidding of the NTI to maximize the energy-density of an emerging molecule in the interest of minimizing entropy production in the immediate environment.

Harold Morowitz has contributed much to origin-of-life studies in terms of fundamental theory, as well as empirical propositions. In *Energy Flow in Biology*[147] he says "Molecular organization and material cycles need not be viewed as uniquely biological characteristics; they are general features of all energy flow systems. Rather than being properties of biological systems, they are properties of the environmental matrix in which biological systems can arise and flourish." This directly pertains to cyclic mechanical or fluid energy flow, as for example Bénard convection, referred to in Part I, in which we speculated on its possible classification as a very primitive member of a larger class of strongly nonlinear systems displaying one or more attributes of living things. (Some have sought the development of a suitable parameter- or phase-space for the characterization of living systems.) One such feature, Morowitz notes[148] is that "Sustained life under present-day conditions is a property of an ecological system rather than a single organism or species. A one-species ecological system is never found. ... Traditional biology has tended to concentrate attention on individual organisms rather than the biological continuum. The origin of life is thus looked for as a unique event in which an organism arises from the surrounding milieu. *A more ecologically balanced point of view would examine the ecological cycles and subsequent chemical systems which must have developed and flourished for a considerable period before anything resembling organisms appeared. ... the minimum functioning biological unit is membrane-bounded and distinct.* (emphasis added) ... The cellular mode of organization provides a method of preventing the loss of intermediates by diffusion. It also provides a method of exerting control on the internal environment. ... we might note that the mode of cellular organization provides a method of partially and temporarily isolating a portion of the ecosystem." And on page 136 he says "a nonequilibrium system implies material cycles in the system. The features usually involved in functional complexity are connected with the cycling, while a related process in terms of covalent bonding leads to structural complexity. The functional and structural complexity are ... closely related aspects of the flow of energy."

Elsewhere Morowitz[149] offers the following: "Systems which are being ordered by the energy flux principle can be described by a parameter t_{bar}, the mean residence time or, by analogy with matter flow, the average time that electronic energy spends in the system. ($t_{bar} = T/F$, the electromagnetic energy stored in the system divided by the energy flux through the system.) We now postulate that t_{bar} is a measure of how far the system has evolved or how chemically complex the system is. Evolutionary and ecological considerations suggest a second postulate: Systems linking sources and sinks tend to an extremum value of t_{bar} consistent with their constraints. The value of t_{bar} is influenced by the rate of loss of energy to heat. In a complex system, because a long chain of chemical reactions is involved in the final energy degradation, t_{bar} can become large. Physical barriers such as membranes also reduce the rate of energy dissipation. We can present a number of qualitative arguments to suggest that t_{bar} is a good measure of complexity and that many present-day biological systems are adapted to maintain large values of this quantity. … Physics and biology can be drawn much closer together if we study t_{bar} from the point of view of physics. The immediate goal is to attempt to deduce the extremum principle from either kinetic theory or irreversible statistical mechanics." Unmistakably, this is the reasoning and voice of a physicist seeking to comprehend hierarchical organization in biotic systems.

But in his more recent publication, Morowitz[150] unintentionally reveals two disturbing inclinations, shared by many others, in his approach to the origin-of-life problem. The first has to do with the need to account for the homochirality of amino acids and sugars; one finds no reference to the issue in 195 pages of his book, as much as to say this is not *the*, or even *a* key issue, conditional to all the others. The second issue relates to the importance of the very early evolution of a proto-cell enclosure in which he says "To be an entity, distinguished from the environment, requires a barrier to free diffusion. … The necessity of thermodynamically isolating a subsystem is an irreducible condition of life. … The isolation is most generally achieved by membranes, amphiphilic bilayers of a thickness of around 10^{-8} meters. … It is the closure of an amphiphilic bilayer membranne into a vesicle that represents a discrete transition from nonlife to life." (p8-9) But on page 105, in his only reference to Fox's work, he states "…we exclude coacervate-like proteinoid microspheres as proto-cells (Fox, 1965) in that they lack the universal barrier of a liquid crystalline, nonpolar phase, present in all contemporary cells. The simplest proto-cell that fulfills the principle of continuity is a bilayer vesicle made from a single type or mixture of small amphiphiles." Morowitz makes no mention of the bilayer-like properties of the enclosure, enumerated and acknowledged by others, or of Fox's electron micrographs, revealing a bilayer structure whose thickness is of the order of 10^{-7} meters in a microsphere of about 3 micron diameter. Why should he expect that the very first barrier have the same complex structure and properties present in all cells today, when he does not make the same requirement on life's initial metabolic apparatus? This is especially puzzling since on page 12 he says "a system of *primitive* vesicles could have developed molecular complexity." (emphasis added) How can one account for such summary dismissal of decades of work by many well-respected researchers, especially from one who has contributed so much to the development of physical biological theory in this and other publications?

Jacob Bronowski[151] gives a different voice to the notion of hierarchical organization in saying: "The stratification of stability is fundamental in living systems, and it explains why evolution has a consistent direction in time. Single mutations are errors at random, and have no fixed direction in time, as we know from experiments. And natural selection does not carry or impose a direction in time either. But the building up of stable configurations does have a direction, the more complex stratum built on the next lower, which cannot be reversed in general (though there can be particular lines of regression, such as the viruses and other parasites which exploit the more complex biological machinery or their hosts). Here is the barb which evolution gives to time: it does not make it go forward, but it prevents it from running backwards. The back mutations which occur cannot reverse it in general because they do not fit into the level of stability which the system has reached; even though they might offer an individual advantage to natural selection, they damage the organization of the system as a whole and make it unstable. Because stability is stratified, evolution is open, and necessarily creates more and more complex forms. … The stable units that compose one layer are the raw material for random encounters which will produce higher configurations, some of which will chance to be stable. So long as there remains a potential of stability which has not become actual, there is no other way for chance to go. … It is evolution, *physical* and biological, that gives time its direction: and no mystical explanation is required where there is nothing to explain." (emphasis added)

(We note, as an aside, that the *hierarchy problem* that disturbed Weinberg (see chapter 13), relating to the very large range of massive quanta energy densities, has a counterpart in the enormous range of energy-densities of living systems, from that of simple prebiotic forms to that of the largest dinosaurs, not to mention that of social systems of animals and in particular of humans. However, it is interesting that the energy-densities of the quanta evolved downward in the early quantum-number expansion of the Universe, while that of living systems evolved upward in the most recent several billion years. In neither case do these data point to a problem; rather they testify to the wondrous creativity wrought by the Nonequilibrium Thermodynamic Imperative, via the orderly flow of energy in the expanding Universe.)

The physicist-biochemist Koichiro Matsuno[152] has written extensively on evolution's dependence on non-random variations: "The generation of non-random variations is always under the influence of the ones generated previously because of the internal selective capability of the generation. *An earlier non-random variation becomes a cause for a later one, and consequently non-random variation acts as an evolutionary agent. In fact, the internal generation of non-random variations underlies the regulative principle for material self-organization. Evolutionary mechanism comes into existence through evolution itself.*" (emphasis added) Matsuno seems close to saying we could discard "self" from all such discussion if "the regulative principle" were rigorously defined at the most fundamental level. In the present context that insight would come only with the discovery of a rational answer to the question "How does a molecule, at one stage of evolution, *act* to select this or that element from a set of available candidates, which best satisfies some fundamental requirement or law governing the growth of such molecules?" Since we cannot invest the molecule with knowledge of the requirement, we have to assume that each of the different elements is distinguished by some specific property (form) and that noise (energy fluctuations) in the environment *acts* to *sample* and thereby *select* that element which, on average, best *fits* with the existing state of the molecule to satisfy some potential *functionality* (emphases added). The most fundamental requirement is, in our scenario, the demand to minimize, on average, the density-specific entropy production of the evolving molecular subsystem.

Szoke[153] *et al.* have asserted that "Essentially all biochemistry is catalyzed; therefore, altered biochemistry implies altered or new catalysts. In that sense catalysis is the medium of evolution. We propose that a basic property of enzymes, at least as fundamental as reaction rate enhancement, is to *adjust the reaction path by altering and eventually optimizing the reversible interchange of chemical, electrical and mechanical energy among themselves and their reactants. ... Accordingly, we will use the term catalysis in the broadest possible sense for any alteration of any detail of a chemical reaction by its environment, provided that the latter stays unchanged at the end of the reaction.* (emphasis added).

… Evolution selects enzymes based on catalytic rate enhancement as well as the ability to preserve energy in high-grade, usable form. Balancing these two criteria results in the selection of an optimized reaction path in which a significant fraction of energy can be maintained in a high-grade form for subsequent use. … *A fundamental property of living systems as well as evolving pre-biotic replicating molecules is that they exploit chemical reactions that already possess both a kinetically preferred pathway and a thermodynamically preferred pathway, and they do this in such a way as to optimize the balance between the two. The optimization process is an evolutionary one, and optimization of the balance can be defined loosely but operationally as changing the propensities of the two pathways relative to one another in successive rounds of replication and natural selection in a way that maximizes the ability of the system to grow and to replicate itself with reasonable fidelity."* (emphasis added) At a time in which origin-of-life research is confronted with several imponderable chicken-or-egg problems, the foregoing should give high priority to solving the homochirality problem and greatly extending thereby the catalytic potential of long chain proteinoids.

Also we should expect that symmetry and symmetry-breaking theory would play a fundamental role in the life sciences – from the level of biochemistry to that of our emerging global human intercourse, as it has in cosmology, relativity and quantum physics. But, according to Peter Wolynes, little progress has been made toward that goal, noting that "there has been no general unifying theory of symmetry in biology. Exact symmetry in biology would even seem to be antithetical to the notions of complexity, variety, and metamorphosis that are central to the idea of life as we know it. … X-ray diffraction and nuclear magnetic resonance have led to a view of proteins and nucleic acids as complex three-dimensional structures with some precision. At the same time, dynamical studies show that the situation is much more complex than the static models alone suggest and that the proteins fluctuate through a variety of conformational substrates whose average is represented by the beautiful pictures now in biology textbooks. This complexity is reflected in the notion of the energy landscape. … We will argue that, while the relationship between symmetry, stability, and dynamics is different for biological physics than for the subatomic world, there are deep similarities. … Symmetry of its biomolecules may provide some essential advantage to an organism and thus may reflect some important constraint on function. Alternatively, biomolecular symmetry may be a 'frozen accident' of natural history. ... Fast-folding proteins can access their thermodynamically stable organized (folded) structure at temperatures where glassy dynamics and traps can be avoided. To do this, the interactions in the folded structure must act in concert more effectively than expected in the most random case. This idea is called the 'principle of minimal frustration' ... it can be shown to be equivalent to the assertion that the ground state energy of the protein is significantly (i.e., many standard deviation of the energies in the rugged part of the landscape) lower than the bulk of the collapsed states of the protein. Such an energy landscape allows folding into the correct state at temperatures where glassy traps are unimportant. When the ground state energy is very low, structures similar to the ground state will also be low in energy, thus leading to a funnel structure of the energy landscape ... the most rapidly foldable proteins have the most pronounced funnel structure to their landscape and are quite stable. ... It is natural to suppose that symmetry, minimal frustration, and the funneled nature of the landscape are somehow connected. ... We must unfortunately admit the current lack of a deep understanding or mathematical proof of why symmetric arrangements are always so stable for such systems with symmetric (and therefore minimally frustrated) interactions. But the observation of *extra stability through symmetry* is nevertheless not only very plausible for systems of equal particles but also a fact." *To this we add that it is also plausible that the principle of minimal frustration can be derived from the principle of minimal density-specific entropy production (maximal energy-density of an open subsystem)*, as we will pursue later.

Hoang[154] *et al.* "present a simple physical model that demonstrates that the native-state folds of proteins can emerge on the basis of considerations of geometry and symmetry. ... We show that a simple model that encapsulates a few general attributes common to all polypeptide chains, such as steric constraints, hydrogen bonding, and hydrophobicity, gives rise to the emergent free-energy landscape of globular proteins. The relatively few minima in the resulting landscape correspond to putative marginally compact native-state structures of proteins, which are assemblies of helices, hairpins, and planar sheets. A superior fit of a given protein or sequence of amino acids to one of these predetermined folds dictates the choice of the topology of its native-state structure. Instead of each sequence shaping its own free energy landscape, we find that the overarching principles of geometry and symmetry determine the menu of possible folds that the sequence can choose from. … Our work here underscores the importance of hydrogen bonds in stabilizing both helices and sheets simultaneously…allowing the formation of tertiary arrangements of secondary motifs. Indeed, the fine tuning of the hydrogen bond and the hydrophobic interaction is of *paramount importance* in the selection of the *marginally* compact region of the phase diagram in which protein native folds are found. It is also important to note that proteins are relatively short chain molecules compared with conventional polymers. These are special features of proteins, which distinguishes them from generic compact polymers. A free-energy landscape with 1,000 or so minima with correspondingly large basins of attraction leads to stability and diversity, the dual characteristics needed for evolution to be successful. Proteins are those sequences that fit well into one of these minima and are relatively stable. Yet, the fact that the *marginally compact phase* lies in the vicinity of a phase transition to the swollen phase allows for an exquisite sensitivity of protein structures to the right types of perturbations." This work enables Wolynes' observation (above), regarding *extra stability through symmetry*, to be explored in relatively simple models and likewise to suggest that such models might yield greater insight when evaluated in terms of density-specific entropy production rather than the *energy penalty* and *solvent mediated interactions energy* parameters.

Avetisov and Goldanskii[155] make the following argument in regard to the phenomenon of mirror symmetry breaking at the molecular level: The main characteristic of the chemistry of chiral compounds is associated with the invariance of mirror reflection or space inversion of electromagnetic interaction. "This type of interaction as a rule dominates coupling of intermolecular electrons and nuclei, and therefore, the states of a chiral molecule are described by a symmetric double-well potential with minima corresponding to the L and D configurations. In compounds with an asymmetry center, e.g., with a carbon atom bonded to four different substituents, the characteristic time of tunneling between L and D is much greater than that of elementary chemical reactions initiated, e.g., by thermal excitation at room temperature. Therefore, the L and D configurations can be regarded as quasistationary states of a chiral molecule on ordinary chemical time scales. The enantiomers have the same ground-state energies and, hence, the same reactivities. This is the most representative feature of the chemistry of chiral compounds, which stems from the symmetry properties of electromagnetic interaction." However, in living nature, the primary structures of DNA, RNA, and enzymes, the informational and functional biocarriers, are homochiral without exception. "Control exerted over enantiomers in the course of biosynthesis of RNA, DNA, and enzymes is so exact that for every 10^6 to 10^8 DNA links, there is less than one failure." Studies involving the oligomerization of nucleotides from a chirally pure solution have revealed that "in the vicinity of a chiral defect, an informational carrier loses its main property, namely, its ability to be the matrix for a complementary replica. That is where the most outstanding distinction between chiral defects and mutations lies. Mutations, when they appear, also upset the complementary correspondence between the links of the matrix and its replica; however, the mutant replica retains its matrix properties in that it can serve as a matrix for producing its duplicate sequences. ... Truly sophisticated constraints arise on the prebiotic level of complexity, i.e., at $N = 50 - 150$ links. On the one hand, the discrimination energy of enantioselective functions must be well above kT, and hence, macromolecular carriers of these functions, as well as carriers of biochemical functions, should be 'dense' and rigid structures. On the other hand, these structures should be able to enhance their enantioselectivity drastically even with an insignificant increase in complexity of information carriers, because the threshold discrimination energy (ΔE_{min}) increases dramatically with the rise of N." The authors scarcely deal with the possibility that geophysical or geochemical conditions in the early earth may have broken symmetry in the sense that Bredig argued (see below) and made no mention or reference to his work.

In his article on Asymmetric Autocatalysis and its Implications for the Origin of Homochirality, Blackmond[156] notes: "It is interesting…that the definition of absolute asymmetric synthesis originally proposed by Bredig in 1923 placed an important emphasis on the need for the intervention of physical forces to create an imbalance in ee (*enantiomer excess*, a point that has been echoed in numerous authoritative texts since." If Bredig's arguments received significant attention in his day, there is scarcely any evidence of that at this time.

But we have seen that open sub-systems, of various kinds and at various hierarchical levels, emerged as a consequence of the buildup of stress, accompanying the growth of a larger system, and with the relief of that stress in the breaking of some symmetry property. This happened in the down-conversion-creation of all the matter and antimatter particles in the Universe and again in the creation of the charged particles, as presented in chapter 9. The same holds in the creation of cosmic bodies, though the notions of "symmetry" and "stress" may be less easily perceived in the mind's eye. It holds also in the creation and evolution of molecules, macromolecules and the polymers and cellular systems of life, as the stress buildup with each added atom, in accommodating the needs of the various atoms, requires all to adjust their relative bonding positions and angles so as to maximize the energy-density (minimize the density-specific entropy production) of the subsystem as a whole. Where would we find the stress and dissipation associated with the transformation of racemic proteinoids into pure long-chain proteins and enzymes or attending their creation from homochiral amino acids? While racemic molecules have the same ground state energy in the L or D form, it requires some energy dissipation to change the one form into the other or ensure that all will be homochiral in their forming. This is reminiscent of metastable electron states in atoms.

We have much to learn in creating a general theory of stress buildup in systems and its role in symmetry-breaking. In regard to symmetry and symmetry-breaking in the origin of life, there are several *which-came-first* type of problems that have effectively blocked the advance of research. First and foremost of these is the need to demonstrate a straightforward path, consistent with geophysical and chemical assumptions, to the creation of stereospecific L amino acids and their polymerization in the making of proteins and enzymes, on the one hand, and D ribose sugars, in the creation of nucleic acids and their polymerization into RNA and DNA, on the other hand. The others make increasing demands on the resolution of this problem, which we may refer to as (1) the homochirality problem. They vie for evolutionary urgency, relating to the need: (2) for the early creation of a *bounded, selectively open protein enclosure*, in which there may then be demonstrated (3) a straightforward path leading to the creation of *template apparatus enabling self-replication and evolution of the cellular system via Darwinian natural selection* and coevolving with (4) a *simple, limited metabolic pathway*. A later requirement would be for the transformation of the protein enclosure into a *lipid-protein membrane* capable of more selectively constraining the flow of matter into and out of the proto-cell. Such seemingly-circular reasoning, suggesting the need to resolve multiple demands nearly simultaneously, as seen from our understanding of present-day cellular dynamics, can be resolved only via some *external* symmetry-breaking force or mechanism, as we shall argue later.

T.O. Fox[157] observes that "Levels of evolution involve systems that are distinguished by their primary constituents, the interactions of these constituents, and the focus of selection by which each system changes in time. ... The focus of selection of a system is the major property, or set of properties, that selectively changes in increments. These properties are both the major sources of variability in the system and the portion of the system upon which selective pressures are 'focused.' The selective pressures arise from the interaction of the changing environment with expressed properties of the system. Thus the term 'focus of selection' applies to the overall process of selective pressures upon the source of variability. ...When levels of evolution develop to particular stages, emergent systems can assemble from the components present. These systems will, if selected, yield new levels of evolution. The synthesis of proteinoids and the assembly of microspheres are examples of new systems. When the organic level of evolution contains sufficient amounts of an appropriate mixture of amino acids, an energy-requiring synthesis of proteinoids can occur. ... Likewise, the appropriate proteinoids assemble in water to yield microspheres. ... It is, in fact, this feature of assembly with the resultant 'whole...greater than the sum of the parts' that provides us with the notion of discrete levels of evolution. ... A new system can be selected and can evolve only if it provides a new focus of selection that allows a faster rate of selection than does the selective focus of the older system at that time. More rapid selection is essential to enable the new system to adapt to a changing environment more quickly than do the components of the older system. If the older level evolves too rapidly, those of its components that are needed as inputs to, or as entities of, the new system may be depleted. Since other assembled systems from the older level could cause these depletions, we see that there is competitive selection of levels of evolution. ... In addition to potential evolution, some of the properties that were manifest for a given component in the older system are incompatible with the newer system. These properties are not observed within the newer level. Thus properties of older components are restricted within a newer level of evolution to the extent that they can limit expression of new properties and components. ... It must be emphasized that the competition among levels of evolution differs from Darwinian competition within a single level. Darwinian evolution can be conceptualized as occurring within a single plane, and the selection of one individual may lead to the extinction of competitors. Evolution of levels involves an added vertical dimension. Preceding levels necessarily do <u>not</u> disappear, although they continue to evolve with the restrictions described above." Fox concludes, arguing that the above reasoning should apply at still higher levels of the evolution of living systems. He is concerned more with how hierarchical levels emerge than with their fundamental or analytical character or classification.

Braham[158] offers similar insight in a general systems context. Patterns and pathways of relation (or communication) that are maintained over time are what we understand, by 'structure.' Holism is implicit in the program of an organization, whether we are dealing with the electro-chemical encoding at the organic and inorganic levels of nature or with the ideational encoding of intentional human organizations. Balance, harmony, proportion, unity, are inherent to atoms, leaves, scientific theories and works of art. ... the loss of independence of the parts and the achievement of inter-dependence, a generally stable and thus routinized structure is built up as the framework of organizational stability: the parts lose freedom as the whole gains it." Where Braham refers to, or implies, the development of organizations at various levels, the same could be expressed in terms of the evolution of hierarchical organizations of open thermodynamic systems, which depend progressively on increasing numbers of weaker and weaker binding forces, that shield the lower level systems from potentially harmful events in the environment. Braham argues further that "As every organization at every stage of its development is subject to the homeostatic principle of self-maintenance, every stage that has been achieved tends toward its own perpetuation (i.e., the maintenance of its own equilibrium). From this standpoint, every new development is a threat to the equilibrium thus far achieved." Here again, in our view, the appropriate terms should rather be *evolution* and *stationary state* in place of development and equilibrium, the important point being that the hierarchical evolution always carries with it the potential for loss or destruction of some preexisting functional capability. Also we should italicize "self" in the above term in which maintenance of the subsystem is defined in part by the larger system. Lastly, he notes that "An organization that can assimilate a wide variety of free energies has greater survival potential over those that cannot. Or, put differently, an organization with a narrow adaptive range has a less survival potential than one with a wide adaptive range." This is particularly relevant to the *emergence* of living systems, whose metabolic options were likely very limited in the form of available energy, though perhaps distributed in geographic location and energy level.

Kenneth Boulding's interests ranged far beyond his principle field of economics. In an article seeking a general theory of growth, Boulding[159] offers the following: "Problems of structural growth seem to merge almost imperceptibly into the problems of structural change or development, so that frequently 'what grows' is not the over-all size of the structure but the complexity or systematic nature of its parts." He was especially concerned with how "innovations" (mutations) come about. "In society as in physics … we find that very small amounts of "impurities" (say one Edison per hundred million) produce effects fantastically disproportionate to their quantity, because of the nucleation principle." And, he notes, "the nucleus does not have to be homogeneous with the structure that grows around it; thus the speck of dust at the heart of the raindrop!" Another "general principle of structural development might be called the principle of "non-proportional change. As any structure grows, the proportions of its parts and of its significant variables cannot remain constant." He cites two important corollaries of this principle. "The first is that growth of a structure always involves a compensatory change in the relative sizes of its various parts to compensate for the fact that those functions and properties which depend on volume tend increasingly to dominate those dependent on area, and those dependent on area tend increasingly to dominate those depending on length. Large structures therefore tend to be "longer" and more convoluted than small structures, in the attempt to increase the proportion of linear and areal dimensions to volumes. It follows therefore that as a structure grows it will also tend to become longer and more convoluted. The second corollary follows immediately from the first: if the process of compensation for structural disproportion has limits, as in fact seems to be the case, the size of the structure itself is limited by its ultimate inability to compensate for the non-proportional changes. This is the basic principle which underlies the 'law of eventually diminishing returns to scale' familiar to economists." He goes on to say "Growth creates form, but form limits growth. This mutuality of relationship between growth and form is perhaps the most essential key to the understanding of structural growth. … It is often the 'loose ends' of systems that are their effective growing points—too tight or too tidy an organization may make for stability, but it does not make for growth." Lastly, we refer to another of Boulding's insights, the 'carpenter principle.' "In building any large structure out of small parts one of two things must be true if the structure is not to be hopelessly mis-shapen. Either the dimensions of the parts must be extremely accurate, or there must be something like a carpenter or a bricklayer following a 'blueprint' who can adjust the dimensions of the structure as it goes along." It was difficult, he says, to find any detailed evidence in the biological sciences in support of this principle, but, in fact, we find it at the most fundamental level in the origin of life, in which all of the atoms in the intermediate and macromolecules adjust their bonding positions and angles in the interest of minimizing density-specific entropy production of an open thermodynamic subsystem. There is much more in this article, relevant to our interests here.

Progress and Limitations in the Laboratory

Contemporary investigations commenced in the latter-half of the twentieth century, with early interests in the autosynthesis of amino acids under various assumed prebiotic conditions and soon came to focus on one or the other of two principal strategic approaches. (Note, in this view we pass over much earlier well-documented studies by Oparin, Haldane, Calvin and others.) One strategy, which we will call the "Miller-Urey approach," was guided in part by the belief that the molecules of life could be destroyed as easily as they might be made even in relatively low-heat reactions and led to experiments principally involving radiation energy sources. Those pursuing this path were also motivated by the presumed need for an early demonstration of template-directed molecular replication, as Darwinian natural selection would require in order to evolve around the many chicken-egg problems that confronted them. The other strategy (the "Fox approach") simply asked: What might result if we cooked the various constituents of amino acids in an aqueous solution, under various conditions of temperature and time? This assessment, of course, greatly simplifies the reasoning that led to each set of investigations.

Several observations may be noteworthy now in characterizing these early investigations and their results: While the "Miller-Urey approach" did indeed reflect an early concern in the need to cope with the problem of molecular self-replication, the "Fox approach" essentially ignored that problem, only to be surprised when a possible answer to it emerged spontaneously in the course of their work. Lacking any rigorous non-equilibrium thermodynamic or quantum molecular theory, there was little beyond classical thermodynamics and radiation physics to guide the design of experiments in either strategy. Curiously the literature lacks much concern or focus on the homochirality problem in either approach; this is most puzzling, especially in regard to the Fox experiments, as a solution to that problem would likely have opened the door widely to the creation of real proteins and enzymes. The "hand of the chemist" is seen in the *design* of both experimental programs in the control and cycling of materials and energy conditions. (These are very important considerations for, as we shall see, the homochirality, materials cycling and some other problems may trace to a common solution.) The results of these quite different kinds of experiments show that roughly the same kinds and quantities of amino acids were obtained while yielding no saccharides, lipids, or nucleosides in the early efforts. No peptides were formed in the "Miller approach", while, in the "Fox approach," short chain-length "proteinoids," with statistically non-random sequencing, were readily formed in the heating and condensation of amino acids. The proteinoids displayed weak catalytic behavior, no doubt limited by the fact that the amino acids making up the proteinoids were racemic. But the defining feature of this approach emerged in the spontaneous formation of large numbers of cell-like "microspheres," capable of growing and budding to form new, similar entities, upon heating and cooling of concentrated solutions of proteinoids. Numerous other lifelike features were claimed.

A Path Ahead: A Deeper and Wider Perspective

Since the above might appear to favor Fox's research program, we need to ask: Why did many, including later researchers, largely dismiss that effort, opting for a path toward an "RNA World?" (In 1998 Leslie Orgel[160] reviewed "facts and speculations" on the origin of life without mention of the investigations of thermal synthesis of amino acids in Sidney Fox's laboratory and all but ignoring the fundamental importance of the homochirality problem.)

One might speculate that nonequilibrium thermodynamics could play an important role in this regard, but in its absence there was little reason for the biochemist to focus attention on it. This leaves four possible explanations, or combinations thereof: (a) proteinoids and microspheres lacked sufficient properties to be seen as precursors to real life; (b) Fox endowed the microspheres with far too much life, in his anthropomorphic comments on their structure and behavior; (c) the program simply ran out of steam, seeing no way to resolve various problems, in particular the homochirality problem; or (d) with new discoveries on the properties of RNA molecules and new investigators to explore the issues, the "RNA World" beckoned. But item (c) can now be seen also to characterize the RNA path. Looking back, it is especially puzzling that Fox did not make a major issue of the need to search for a solution to the chirality problem, or that others did not pursue that critical need, as Bretig urged in 1924. But Fox was in good company, as there was and remains relatively little emphasis on that problem in the literature.

In a more fundamental and broader perspective, the chirality problem may be seen not simply as the need for some physical mechanism that may autosynthesize a select enantiomer of amino acids or sugars but one that operates similarly in all proto-cells or microspheres as if to define thereby a common Euclidean space for the evolution of all the functioning parts in each microsphere. And of course such mechanism should provide sufficient fidelity in the replication of the proto-cell as well as potential for variance in that replication. Moreover, as we will see later, such mechanism should facilitate a straightforward transition from the proto-cell's dependence on heterotrophic growth and evolution toward that of autotrophic systems.

Much has changed in the past several decades, which might reinvigorate origin-of-life research. The further development of nonequilibrium thermodynamics, as a general imperative driving the creation of various kinds and hierarchies of open thermodynamic subsystems, could be key to the development of a deeper and wider perspective. Adding to this in the same period was the discovery of deep hydrothermal vents and the strange life forms they support. However, it is the physical and chemical features of these vents that hold great promise for circumventing a number of problems that have stymied research in the laboratory. In fact the existence of unique present-life in these regions greatly complicates empirical research on the origin of life in these vent areas.

All laboratory studies of the origin of life reveal the inability to autosynthesize, in an abiotic environment, homochiral L-amino acids, identical to the 19 found in proteins in all living cells. Also the productivity in the prebiotic synthesis of even mixed-stereospecific amino acids is very low, due in part to the relatively low fraction of amino acids produced by energetic reactions that are not subsequently destroyed by destructive reactions under the same conditions, whether involving radiation, thermal or other energy sources. This implies that there must have existed some physical mechanism or property, characteristic of the assumed prebiotic environment, which gave rise to higher concentrations of precursor molecules and to fractionation of *generation over degeneration* of amino acid product. The same holds for the synthesis of sugars, nucleotides and other macromolecules.

Moreover, since there is no reason to believe that sugars, fatty acids, lipids, nucleotides and nucleic acids were created in different or separate reaction local environments, from that which gave rise to amino acids, proteinoids and microsphere structures (indeed there must be some physical network for their interdependence), should we not expect their emergence in the same or similar reaction conditions? While there is occasional mention of the lack of nucleic acid intermediates in the Fox efforts, the literature reveals no significant progress in this direction. However, there is in these studies a concern with, if not a focus on, hierarchical evolution, the simultaneous interdependent evolution of macromolecular evolution of the essential components of living cells, and of the continuity and stability of the environment in which they evolve, as many theorists have counseled.

The point in all of this is not to downplay the early contributions of Oparin, Miller, Fox, etc., and many others of more recent times, to origin-of-life studies – any one of the above considerations might well have represented an insurmountable task – but to prioritize the heretofore *missing* links in our efforts to discover a self-consistent evolutionary path resulting in the emergence of template-governed cell replication, yet open to marginal template variation.

HYDROTHERMAL VENT AS A MESOSCOPIC, QUASI-BÉNARD CELL

If we focus our attention on a hydrothermal vent, as a localized stable heat source and sink, we see that the vortex-flow of saline water appears in part as a mesoscopic, localized, single Bénard cell. It is a kinetic, open thermodynamic subsystem of the larger energy-flow, geophysical system, which serves to minimize entropy production and maximize the efficiency of heat transfer in the system. It does this through the very long-range organization of molecular motion. The open subsystem (vortex) stores some small portion of the energy (that would otherwise be dissipated) characterized by maximal average energy-density under given *system* constraints. As we speculated in chapter 5 about interpreting Bénard convection as a very primitive quasi-living system, so we might perceive the vent vortex subsystem in which billions upon billions of water molecules have been constrained to move in a strongly ordered or organized manner (quasi-laminar flow). But much more is going on in the "system" and the "subsystem;" the water cycles with it ions, other atoms, molecules and solutes, some small portions of which are newly added from within the earth in each cycle. The vortex is driven and constrained by mechanical (hydraulic) forces as well as thermal and gravitational gradients, and the existence and flow of ions results in electromagnetic forces and flows, giving rise to a very complex open subsystem involving the interplay of four forces and their respective flows. For the present, we shall ignore these other atoms or molecules, with the exception of the cycling of the negative chlorine and positive sodium ions. In such a system chlorine and sodium ions would cycle in quasi-laminar flow and in vertical, bounded semi-planes, angularly displaced about the orifice axis, in which gravity plays an important role, as in the case of Bénard convection (we neglect here the relatively small contributions of sulfate and magnesium ions). However, since the mass of Cl⁻ is about 1.5 times that of Na⁺, we might expect that chlorine ions would be driven (by thermophoretic forces) at a slower velocity and on smaller vortex orbits than for the sodium ions, in the stationary state. But the difference in masses and kinetics would be significantly reduced since each ion is enclosed in a "cage" of six water molecules, making the effective mass of the sodium ion ~6(2+16)+23=~131amu and that of the chlorine ion ~143amu. The mass of the sodium ion complex being about 0.9 that of the chlorine ion complex should then result in the average velocity of the former being about 1.09 that of the latter, in the stationary state, still a substantial difference in computing their relative electromagnetic properties. But the net positive ion current and therefore the strength of any resultant magnetic field should be expected to increase as the square of their mass ratio, since the relative ion densities would vary also inversely with their mass ratio (due to the greater constraint on potential energy of the Cl⁻ complex). Such vortex kinetics would thus tend to break the local charge-null symmetry that would otherwise characterize a uniformly-heated reaction pot. The effect of this ion-flow asymmetry might be to create a moderate-to-strong magnetic field in the regions of relatively large net positive current flow. The asymmetry of reaction zones should play an important role in the simultaneous synthesis of reactants at one level and of reaction products at a higher level.

It should be noted that no magnetic field would exist (ideally) outside the vortex system, though the total energy stored in the subsystem (vortex) is increased by the energy associated with the magnetic field and the orbital angular momentum of the neutral molecules. (As seen from the outside, the system reveals no angular momentum.) Thus, the entropy production associated with the vortex subsystem might be relatively large in a stationary state far from equilibrium, while the entropy production per unit of energy flow through the open subsystem necessarily would approach a minimum under the imposed conditions, the difference being the negative entropy production associated with the creation and maintenance of the open subsystem, including that associated with the magnetic field. Any stable, endothermic reaction products occurring in the subsystem would necessarily further increase entropy production in the system while decreasing the density-specific entropy production in the subsystem, but only if they tend to decrease the total angular momentum of the orbiting product molecules. That is to say, the spin angular momentum of the reaction products would act to oppose the orbital angular momentum of the ions, as in the case of the spin of the electron opposing the orbital angular momentum of the electron in the ground state of the hydrogen atom. *The orientation of the magnetic field thus might provide a common sterospecific constraint on the spin of all carbon atoms involved in a reaction, including those associated with the creation of L-amino acids and their polymerization in peptides, depending on the strength of the magnetic field.*

Such relatively simple, ubiquitous and asymmetric physico-chemical systems in the prebiotic ocean could account for a significant enantiomer excess (ee) of amino acid monomers if the magnetic field strength, produced by the cycling net charge, were sufficient, *relative to thermal noise*, to suitably align the spins in the C—N bond in the vast majority of amino acid reactions, whether via a single-pass through the reaction zone or multi-cycle amplification of stereoisomers. The same applies to the carboxyl and side-chain groups and the peptide bond.

The issue of homochirality would be so very important in an early period of transformation from inorganic to organic chemistry because that option would enable the most efficient utilization of energy and precursor molecules, as compared with other later options in which the unwanted monomers would be selected out. No less important in this consideration is the realization that the greater the number and kind of L-amino acids available in the early phase of the polymerization process, the sooner would be the yield of stereospecific enzymes to catalyze other reactions leading to the creation of D-ribose sugars, bases, nucleosides, nucleotides and ATP-like energy carriers.

In the following we seek coarse estimates of the magnetic field intensity H (normal to the planes of ion cycling), the torque acting to align the spin of a carbon α-atom in the amino acids with the field, and the angular acceleration of spin rotation. From the latter we can estimate the minimum time required to ensure the creation of each amino acid of a specified enantiomer excess (ee).

We assume that the early ocean salinity was similar to the present value (about 30 each Cl⁻ and Na⁺ for every thousand water molecules), which is relatively insensitive to that ratio, and we argue that the Na^+ ions would travel, on average, ~1.1 times the velocity of the Cl^- ions in response to electrophoretic forces. Also the charge density of Na^+ would be about 1.1 that of Cl^- (see above). And since the ion current would be the product of the ion velocity times the ion density, the net ion current density would be approximately $[(1.1)^2 \times 30-30] \times 10^{-3} \times 10^{24} \times 1.6 \times 10^{-19} \times 0.2 =$ ~ 200 Amperes/cm², assuming an average velocity for the Na^+ complex of 0.2 meter per second, in which the factor 10^{-3} is the thickness in meters of the semi-planes of water and ion vortex flow. Finally, we find that the maximum magnetic field intensity H_{max} produced by this current density would be about $20 \times 60 \approx 1200$ Ampere-turns per meter, with the field oriented according to the right-hand rule, relative to the direction of travel of the Na^+ to that of the magnetic field (Fig. 17-1). Here we assume about 60 channels of 0.1cm² average cross section (1mm×1cm²) displaced around the vent at a radius of 10cm. We note also that the right-hand rule in ion flow translates into the left-hand rule in electromagnetic interaction with the spin of charged particles and it is the latter that defines the orientation of atoms in molecular systems. The torque $\mathbf{T}$ acting on the magnetic dipole moment $\mathbf{m_c}$ of a carbon atom would then be given by $\mathbf{T} = \mathbf{m_c} \times \mathbf{H}$ and the maximum torque $T_{max} = m_c \times H_{max} = \mu_B \times H_{max}/(12 \times 1840)$, where μ_B is the Bohr magneton $= 9.3 \times 10^{-24}$ J/tesla × 1200AT/(12×1840) and the denominator is the factor representing the increase of the mass of the carbon atom over that of the electron. Thus, $T_{max} \cong 5 \times 10^{-25}$Nm. The maximum angular acceleration of the dipole moment would then be $a_{max} = T_{max}/C_{mass} \cong 5 \times 10^{-25}/(2.2 \times 10^{-26})$ $\cong 23$ radian/s². In the first 0.1s (a fraction of one vortex cycle) $[\theta = 1/at^2]$, a carbon atom could be maximally rotated about four radians in the hot thermal reaction region of the vortex, as a result of the differential velocities and densities of the sodium and chlorine ions. A more accurate estimate could be achieved with a finer grid partition, given the various nonlinear relationships in three dimensions, but the result should not differ substantially from the above. The foregoing takes no account of dispersion and dissipation losses, which could decrease the magnetic field strength and acceleration rate by a factor of 2.

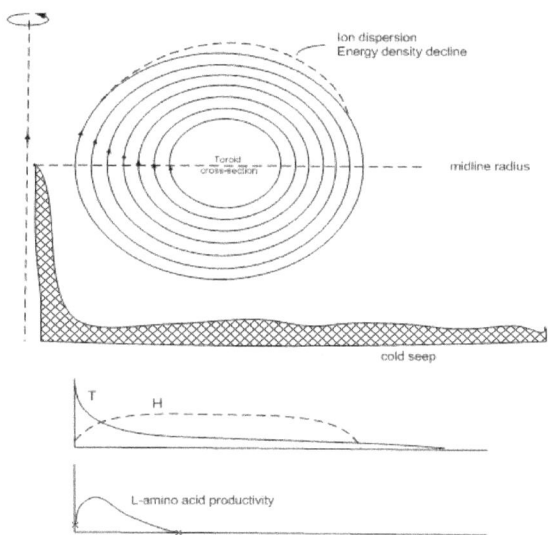

Fig. 17-1 Cross-sectional view of a hydrothermal vent-vortex subsystem in a stationary state. Thermal energy, flowing from the vent, drives the maintenance of a cyclic laminar-flow of water, ions (principally Na^+ and Cl^-) and other solutes, in a toroidal form around the vent axis (note the resultant flow dispersion). The directed water flow drives a similar orderly flow of the ions at reduced velocities. The difference in the masses of Na^+ and Cl^- would result in a net average positive current flow in all regions of the vortex (arrows), which would in turn result in a steady-state magnetic field acting to orient the spin of charged particles, in accordance with the electromagnetic left-hand rule (see text). Were the vent temperature T, magnetic field strength H and reactant concentrations, at the midline radius, sufficient, we should expect a large percentage of amino acids created in the vent-vortex subsystem to be L-amino acids and to polymerize in the creation of proteins similar to those in all living cells.

Thus, we might expect that many if not all amino acids would have left-handed asymmetry in one pass through a vortex cycle. Moreover, since the amino acids remain in the same high field environment before and during a polymerization process, there is every reason to expect the products to conserve or extend their enantiomer excess (ee), enabling the creation of natural proteins and enzymes. If this reasoning is sound, the doubts over where and how absolute homochirality could have emerged in the origin of life would be resolved.

The submarine hydrothermal system sketched above would enable the abiotic polymerization of L-amino acids in long chains in a confined aqueous reaction zone at elevated temperatures, while yet enhancing the stability of the product polymers in the much colder water in which they reside most of the cycle time. Such dehydration-condensation polymerization process, in which the product in one flow cycle becomes a reactant in a subsequent cycle, has been demonstrated in the laboratory, displaying strong catalytic properties. In an attempt to simulate in part a hydrothermal vent system, Imal[161] et al. demonstrated in a 1999 report elongation of oligopeptides of glycine in which the yields of triglycine, synthesized from glycine, were more than a 1000 times greater than the yields expected from equilibrium reactions at the high temperature available from the simulated hot vents. Also, Braun and Libchaber[162] showed that a 96-pair aqueous laminar-flow vortex, in a relatively small radial temperature gradient (95°C axial temperature and 60°C at 2.5mm radius). The same conditions enable thermophoretic trapping (associated with the Soret effect), which enable a great increase in the local concentration of DNA. (See also[163] [164] [165]) [It is appropriate here to call attention to the fact that the dynamics of hydrothermal vent systems are undoubtedly more chaotic than the carefully created laminar-flow system referred to above. Nevertheless, this was a very important experiment and the authors are clearly aware of the larger implications of their work: "It is interesting to note that, even in one convection chamber, the laminarity of convection produces trajectories which are isolated from each other by diffusion. For many geometries and molecules the diffusion is much slower than the temperature cycling. A single chamber can thus host a number of thermal cycling conditions which interact at time scales slower than the convection itself. Therefore different species of molecules could replicate in parallel under different conditions and allow for coevolution and evolution of subpopulations." About the same time as these experiments, Amend and Shock[166] showed that "The energy yield from the autotrophic synthesis of amino acids in a 100°C mixed hydrothermal fluid may ultimately be used to overcome the energy requirements of protein synthesis."

The above reveals four additional benefits deriving from the hydrothermal vent subsystems: increased concentration of reaction products due to fast cold water quenching of reaction degradation; further concentration of reaction products due to thermophoretic trapping; the potential for "one pot" autosynthesis and interaction of multiple molecular products; and a demonstrated capability for the polymerization of nucleic acid monomers, with catalytic support, under relatively low-temperature conditions. Two additional benefits are immediately obvious: the existence of multiple vents in the early ocean floor (with similar, though independent and marginally varied features) and their relative isolation from local or global cataclysmic events would provide great redundancy with variation in the early experiments of life.

Proposed "Fox" Strategy:
Research Under Simulated, Deep, Hydrothermal-Vent Conditions

As we have noted, Fox's origin-of-life research strategy was simply to ask with little presumption: What might result if we cooked various constituents of amino acids in an aqueous solution, under various assumed prebiotic conditions? The same strategy should apply here, except that the cooking would proceed under simulated hydrothermal-vent conditions and constraints, as characterized above, in which the reactor (pot) is repeatedly cycled, between upper and lower temperature limits, in a kinetic open-thermodynamic-subsystem, driven by the Nonequilibrium Thermodynamic Imperative. In this, we see that Nature's laboratories at the bottom of the oceans would have had little need for the "hand of the chemist" to cycle the pots from one temperature limit to another. Indeed, these laboratories would have come equipped with a unique property, a magnetic field capable of biasing the spin orientation of some, many, or perhaps all, α-carbon atoms involved in the linking of amino acids into short- or long-chain life-like proteins or enzymes, depending on the strength of the magnetic field in the vortex. The same applies to D-sugars.

This would have had profound consequences, even in the case of a relatively weak magnetic field, as it would have enabled the amplification of populations of the L-amino acids and D-sugars in the vent environment. In the weak-field case, we might expect the initial result to be a replay of Fox's experiments in which proteinoids evolved quickly to form relatively porous microsphere enclosures, leading to the evolution of slightly biased (elongated) peptides within the enclosure. But such bias might give rise to amplification and elongation, over many vortex cycles, of stereospecific proteins and enzymes. Alternatively, relatively strong magnetic fields might result directly in the formation of long-chain proteins, capable of folding in the cold-water cycle into highly active enzymes, in particular to assist perhaps in the creation of phospholipids. A third possibility is that higher magnetic fields would result initially in the creation of protein enclosures, similar to the proteinoid microspheres, though providing a more rugged and stable enclosure. And a variation on this might lead to the subsequent rapid evolution of one or more enzymes, leading to the formation of phospholipids and the creation of a phospholipid bilayer enclosure within the protein enclosure.

We argue that the creation of such proteins critically distinguish the molecules of life from their abiotic, racemic counterparts, though, as we have suggested, a less critical analysis would perhaps include Fox's proteinoids and microspheres and even the open-thermodynamic vortex subsystem itself, as evidencing some degree or form of living subsystems. Whereas the proteinoids were capable of combining to form microspheres, they were clearly limited in chain length and in their capacity to fold into highly specific and active catalysts. With those limitations decreased or removed, the protein vesicles would be free to continue to grow (mainly in density) in response to the thermodynamic imperative and the relatively rapid and extensive cycle of thermal energy in the vortex. Such growth in the form of long-chain proteins and their folding in the internal milieu to form enzymes could assist in the creation of phospholipid structures and signal a major structural advance, giving rise to a real membrane and the legitimacy of the enclosure as a real cell, as Morowitz emphasized in distinguishing living systems from Fox's microspheres. *It is important to note that neither these or the previous "advances" occurred in anticipation of living cells; they emerged, driven by the Nonequilibrium Thermodynamic Imperative and the physical and chemical forces associated with the hydrothermal vent. Had we known nothing of cells and life, but observed Nature's hydrothermal experiment, we might have discovered cells and the beginning of life.*

Thermodynamic Selection of Reaction Products

We saw in chapter 1 that Prigogine's theorem of minimum entropy production did not apply to systems in states driven far from equilibrium, rather that the rate of entropy production in such systems must increase the farther is the state from equilibrium. No less important, Prigogine and others showed that such systems could become unstable as constraints are varied, leading often to unpredictable bifurcations, i.e., transformations of state of one kind or another. The reason that Prigogine's theorem of minimum entropy production does not apply to systems driven far from equilibrium is that it derives from assumptions and a formalism, which necessarily limits its province to the linear regime, to systems very near a state of equilibrium or a stationary state. It is thus ineffective as a predictive tool in the study of systems far from equilibrium. And we argued, in order to significantly extend the province of nonequilibrium thermodynamic theory, we need to circumvent these limiting formal constraints. The notion that *all* systems *evolve in space and time* toward equilibrium or a stationary state is in line with Matsuno's argument "Physics equipped with the adiabatic approximation conspires to make matter inanimate. *If we free ourselves from adiabatic approximation, matter can be seen as an active agent simply by virtue of the fact that interaction changes propagate at a finite velocity. If it is to be applied to material evolution, non-equilibrium thermodynamics must be extended so that it can rid itself of the fetters imposed by the adiabatic approximation.*" One of our fundamental assumptions in chapter 1 states: "Matter-energy exchange is conservative in all irreversible processes and the flow of matter-energy in thermodynamic systems occurs at finite velocity." Thus the nonequilibrium thermodynamics developed in Part I derives explicitly, from distinctly different fundamental assumptions, to provide a rigorous basis for the analysis of systems far from equilibrium, as exemplified in living systems and their molecular prebiotic ancestors.

In chapter 2 we saw that evolution of a closed thermodynamic system led in space and time to a relatively stable unique physical structure in a stationary state, characteristic of the system's internal and external constraints, by processes of thermodynamic selection in every microscopic region of the system, as the system evolves. As we saw in later chapters such selection processes must apply generally to all nonequilibrium systems, whether simply kinetic in nature (e.g., ideal gas systems) or involving chemical reactions. In the latter case the process of thermodynamic selection may become much more dynamically complex and the stability of the stationary state may vary greatly depending on numerous system parameters and external constraints. Fortunately, almost by definition, the stationary states of the fundamental chemical systems of concern to us are, within certain external constraints, relatively stable, otherwise biological systems could not exist. This holds at various hierarchical levels in the evolution of living systems, including social systems of biota, ecosystems and the biosphere itself.

How may nonequilibrium thermodynamics account for the delimitation of statistical probability in the successive polymerization of amino acid monomers? If we assume that each of the 20 kinds of monomers were available in the same concentration and activation energy sufficient to effect a peptide bond between any two selected monomers the probability of any specific reaction occurring would be extremely small, assuming a uniform bond energy. But the assumption is already simplistic at the dipeptide stage in the consideration of the probability of forming an additional peptide bond. It is simplistic in that it ignores the fact that the energy level of the product may vary greatly or slightly, depending on the nature of the added peptide, as Matsuno[167] argues "The generation of non-random variations is always under the influence of the ones generated previously because of the internal selective capability of the generation. An earlier non-random variation becomes a cause for a later one, and consequently non-random variation acts as an evolutionary agent." We might expect that the polymerization product would favor that added monomer, relative to others, which resulted in the lowest energy of the system. However the situation is more complicated in that the polymer constitutes a nascent open thermodynamic system in an early stage of evolution toward some stationary state of minimum density-specific entropy production (maximal energy-density), in a larger system evolving toward a state of minimum flow-specific entropy production. Thus the energy constraints on the larger system play a role in the evolution of the total energy of the polymer and most importantly of the flow-specific entropy production of the system and of the density-specific entropy production of the polymer itself, taken as an open subsystem. Already we see that thermodynamic selection must favor numerous very weak secondary or tertiary bonds, unrecognized by statistical theory. In effect the system is beginning to define, in broad brush, a specific, preferred, nonrandom, energetic and thermodynamic evolutionary path.

Using language appropriate to a higher level of discussion, Szoke[168] *et al.*, make a similar point: "Evolution selects enzymes based on catalytic rate enhancement as well as the ability to preserve energy in high-grade, usable form. Balancing these two criteria results in the selection of an optimized reaction path in which a significant fraction of energy can be maintained in a high-grade for subsequent use. ... A fundamental property of living systems as well as evolving pre-biotic replicating molecules is that they exploit chemical reactions that already possess both a kinetically preferred pathway and a thermodynamically preferred pathway, and they do this in such a way as to optimize the balance between the two." The notion of thermodynamic selection also may be seen to underlie the relatively new development of *energy landscape* theory. As Wolynes[169] states "Fast folding proteins can access their thermodynamically stable organized (folded) structure at temperatures where glassy dynamics and traps can be avoided. To do this, the interactions in the folded structure must act in concert more effectively than expected in the most random case. This idea is called the 'principle of minimal frustration'..." The state of "minimal frustration" in energy landscape theory would thus appear to be closely related to the more general and fundamental condition in which the stationary state of an open thermodynamic subsystem is characterized by minimal density-specific entropy production. There will be much more to say about the contributions of nonequilibrium theory later.

Does nonequilibrium thermodynamic theory offer predictive insight? No and Yes. In keeping with the classical role of thermodynamics, it argues only to deny in the long term certain possibilities and therefore it lacks a capability for *precise quantitative* predictions from first principles, yet it makes comprehensible the evolution of extremely complex systems that might otherwise appear in limbo, unattached to any underlying physical rationale.

Homochirality: Further Thoughts and Implications

However important the above arguments, they would scarcely matter if origin-of-life studies were inescapably stalled for lack of resolution of the homochirality problem. As Blackmond remarked "It is interesting ... that the definition of absolute asymmetric synthesis originally proposed by Bredig[170] in 1923 placed an important emphasis on the need for the intervention of asymmetric physical forces to create an imbalance in ee (enantiomer excess), a point that has been echoed in numerous authoritative texts since." On the face of it, classical thermodynamics favors the synthesis of neither enantiomer of amino acids.

How might biomolecules evolve, were it possible to conduct experiments in a closely-simulated hydrothermal vent? Note, this question presumes the creation of open systems at five or more hierarchical levels and possibly requiring only a very short span of time in the same "one pot" geophysical system: At the first level, thermal energy flow from the vent would create a kinetic open thermodynamic subsystem (vortex), in which a relatively constant net ion current would be established (due to unbalanced thermophoretic forces). This, in turn, would give rise to a unidirectional magnetic field which would favor the creation of relatively stable L amino acids, constituting a second level of open subsystems. The NTI-selection and dehydration condensation-linking of these amino acids would yield simple structural proteinoids or *proteins* and multiple weak-forces would bond them in sheet-like structures, capable of growing and closing to form relatively porous membrane-like microspheres, i.e., stable open subsystems at the third hierarchical level. Many similar proteinoids or proteins might continue to form selectively as open systems outside or within the microspheres, enabling the microspheres to grow and multiply, constituting a fourth hierarchical level of organizational activity. Alternatively, at the fifth level, long-chain proteins might assemble within the microspheres to fold into globular open subsystems (enzymes) capable of catalyzing still higher-level reactions within or between microspheres. At each higher level, the potential for increasing the energy-density in the subsystem increases, most notably at the aggregation-of-microspheres stage in which folded protein chains and other dense macromolecules may accumulate with little or no increase in the size of the microsphere subsystem.

But this raises many questions: Where are such high molar-mass, dense macromolecules most likely to be created, within the microspheres or in the larger vent subsystem? What are the arguments favoring one place or the other, i.e., tending to characterize the microsphere as either an autotrough or heterotrough subsystem? Other than *greatly increasing the energy-density of the microsphere (decreasing its density-specific entropy production)*, as an open subsystem, what function(s) might be served by these internal macromolecules? If Fox's proteinoids were replaced by comparable small proteins, composed only of simple L-amino acids, would they bond together in aqueous solution to form microspheres, and if so, how would that change the microstructure and membrane-like behavior of the microsphere? And at what stage might such a protein microsphere enclosure be augmented or replaced by a lipoprotein plasma membrane, a first step toward enabling controllable import and export of molecules of various structure, size and function? How are the monomers in the long chains of macromolecules selected, in lieu of any template-ordering mechanism and how may that process evolve into a template-directed process? How can such evolving molecular hierarchy give rise to metabolism, eventually freeing the proto-cells from dependency on the vortex subsystem? And assuming the Nonequilibrium Imperative, together with quantum chemistry, holds answers to the above, can we see a straightforward path leading to nearly-precise self-replication of the proto-cell and its further evolution via Darwinian natural selection?

In the larger view of the vortex subsystem, the microsphere would appear as a vesicle subsystem, requiring constrained two-way communication (molecular transport) with its organized vortex environment, in order for it to grow and divide or bud off. Lacking lipoprotein structures in a prebiotic vesicle, such transport would depend on the porosity of the structural protein enclosure. A structure with relatively large pores would function as a heterotrough, importing small monomers, e.g., amino acids, along with precursors, yet limiting the outflow of larger macromolecules. Conversely, small pores would limit the inflow to small precursor molecules, requiring the microsphere to grow and function more like an autotrough. There are advantages and disadvantages, for example, relating to reactant densities and confinement of reaction products, in both options. Here, we assume that the size and nature of the pores would evolve as an inverse function of the size (diameter) of the microsphere, with the number of relatively large pores decreasing as the microsphere grows, resulting ultimately in the blocking of further growth or requiring the continuation of structural growth via the creation of a new microsphere. Alternatively, the microsphere might form as a constrained enclosure, with a single relatively large opening enabling the transport of monomers and precursors, which might later serve in the budding of new microspheres. In either case the growth of the microsphere would be accompanied by the buildup of stress, culminating in breaking the symmetry of the microsphere and relieving the stress, thereby decreasing its density-specific entropy production and enabling a further decrease in the flow-specific entropy production of the system. These considerations might be subject to empirical investigation and/or computer simulation. Such empirical support of the above-conjectured autotroughic creation of *protein* microspheres in a simulated hydrothermal vent might reveal paths to the creation of molecular open subsystems within the microsphere at still higher hierarchical levels.

It is sufficient initially that some large macromolecules might be induced to accumulate or to be created *internally* in the microsphere, simply as a means of greatly decreasing the density-specific entropy production of that subsystem, as implied above. But it is easy to imagine that some could be long-chain *proteins* free to fold into very dense macromolecules and serve as steriospecific *enzymes* to catalyze and accelerate the up-hill synthesis of other stable high-order macromolecules, in particular ATP-like molecules and nucleotides, within the microsphere. We see in this muse of speculations the potential for very rapid increase in complexity, stability and decline in density-specific entropy production.

As to the selection of L-amino acid monomers in the making of *proteins*, the arguments and evidence presented by Fox and his associates regarding "self selection" should still apply, though the credit for such selection should go at root, as we have said, to the Nonequilibrium Imperative. Lacking such revelation, Fox[171] reasoned along the following lines: "…experiments have shown that sharply definitive instructions are obtained from the reacting amino acids themselves; this emphasis fits with the rudimentary evolutionary requirement that new phenomena emerge from matrices." "The newer view," he says, "is that the proteinoid served as an initial informational macromolecule, having arisen as a product of favored rather than random, polymerization. … The history of biochemical evolution appears to include a history of modulation of reversible bidirectional reactions (of a kind that constitute organic chemistry) to unidirectional pathways of reaction sequence (of a kind that constitute biological chemistry.) This emergent unidirectionalism may apply as well to recognition between amino acids and nucleotides. ... the forces operative in the thermal poly-condensation of amino acids are similar to the forces that control the expression of the genetic code. This view has molecular logic, in that individual assemblies of nucleotides correspond to individual amino acids, the whole relationship being a code. ... The significance of the self-ordering of amino acids extends beyond the chicken-egg question. For example, this principle demonstrates how information latent in the environment (mixed amino acids) could have been the informational matrix bridging to the first organism. Informationally, the missing link between environment and proto-organism was the macromolecule ordered by its monomers." Clearly, Fox was grasping for some deeper understanding of the forces leading from inanimate to animate matter.

Can we see a pathway for the coevolution of a long chain protein, as both a structural and informational molecule and its informational, nucleic acid counterpart, within the microsphere? At some point in the expansion of the microsphere, its thermodynamically-mandated growth, toward decreasing density-specific entropy production, proceeds more and more via the *selective* addition of amino acid monomers to the internal informational protein, capable of folding to maximize its mass-energy density and perhaps beginning to function as an enzyme. In so saying, we see that a new, higher level of open subsystems emerges within the confining microsphere subsystem. But to speak of the protein functioning as an enzyme is to imply the catalysis of a still higher-order reaction product and yet another open subsystem within the microsphere. Moreover, the information stored in the protein could not be read (to initiate and instruct some new course of action) without destroying its higher-level, enzyme-like structure to expose the message in its lower-level ordering. This would effectively restrict the protein to function only as an enzyme and would therefore require a separate message molecule (set of instructions) to evolve, along a parallel molecular path, together with the selective polymerization of amino acids in the protein and its folding into an enzyme. This in turn implies that the selection of a particular amino acid and its polymerization, in the evolution of the protein, is accompanied by a corresponding selection of monomers in the parallel evolution of a write-able, read-able, transport-able and reliably store-able Information Macromolecule (IM). All of this could be realized within the microsphere subsystem (within the hydrothermal vent subsystem) in the stationary state flow of energy and materials from the vent, but only if: (1) the flow of energy and materials into and out of the microsphere, together with the evolving enzymic activity of the folding protein, suffices to activate and extend the polymerization of the monomers in the IM; (2) those monomers, as in the case if amino acids, are characterized by complementary homochiral specificity (3) the binding constants, characterizing the interaction of the amino acids with the monomers of the IM, enable reliable transcription of the evolving protein message in a long-chain informational polymer that is read-able, copy-able and tranport-able; and (4) the total set of instructions result in the relatively precise replication of the microsphere subsystem in structure and function, within the environment of the hydrothermal vent subsystem. (Item (2) would clearly signal the emergence and evolution of yet another high-level subsystem in the microsphere.) The foregoing might then characterize the autosynthesis of L-amino acids and their polymerization and coalition into evolving *protein* microspheres in a simulated hydrothermal vent subsystem, which might then in turn give rise internally to the autosynthesis and evolution of an enzyme enabling the autosynthesis of D-ribose nucleotides and their polymerization into an RNA information macromolecule. (We assume here that the D-ribose homochiral sugars would be selectively formed in the same magnetic field that gave rise to homocchiral L-amino acids.) Needless to say there are many facets of this scenario, which would require empirical support for it to receive serious attention. However, in its defense it is distinguished by the fact that it derives from and rests upon a self-consistent nonequilibrium theory, applicable to hierarchical systems and subsystems far from equilibrium, which we have referred to as the Nonequilibrium Thermodynamic Imperative.

Lacey and Mullins[172] have explored the origin of the genetic code and in particular the possible role of thermal proteinoids and their interaction with polynucleotides in that quest. They say "Since the amino-nucleotide relationships of the genetic code are not known to have significance outside the context of protein synthesis, it seems logical that whatever studies are carried out should simultaneously attempt to understand not only the origin of the coding mechanism, but also how it allowed the origin of a protein synthesizing system. ... As our thinking on the problem has matured through the years, we have also become convinced that the code and the process of protein synthesis must necessarily have coevolved. ... Perhaps the most succinct statement of the problem can be asked as follows: *'Is there a set of organizing principles which operated between amino acids and nucleotides, and which encourages the synthesis of selected peptides?'* [emphasis added] Based on our present information, *we have generated the following working hypothesis as an answer to this question: genetic coding arose due to selective affinities between amino acids and their anticodonic nucleotides; this, in turn, allowed selected enhancement of reactions leading to peptide synthesis.''* [emphasis added] In our scenario, the "set of organizing principles" is, of course, the Nonequilibrium Imperative, which would drive the coevolution of a long-chain protein message and its anticodanic long-chain nucleic acid message, capable of promoting "the synthesis of selected peptides."

Plotting the data on the relative hydrophobicities of all the amino acids and nucleotides in a three dimensional representation, they "could see that when viewed from a particular angle, most of the points fell in a curved plane. ... This rather remarkable outcome shows again a direct correlation of the properties between all twenty amino acids and their anticodonic nucleotides. ... Again testing the anticodonic idea, we plotted the data for the homocodonic amino acids verses that for their anticodonic nucleotides, again obtaining a direct correlation, but naturally inverted from the hydrophobicity data." The above lends support to their stated working hypothesis and their argument that the coding origin and the peptide synthesis are two facets of a common problem. These biochemical discoveries would lend empirical support also to our above-conjectured coevolution of a folded, long-chain, internal protein, to function principally as an enzyme, and a long-chain macromolecule, to serve largely as a means of storing and transporting information required in the replication of the microsphere, if the polymerization of the amino acids enabled the emerging enzyme to catalyze the nucleic acid chain without metabolic support, that is, energized only by the hydrothermal vent subsystem. If that were the case, then we might expect the polymerization of the nucleotides to proceed very rapidly, as the experimental studies of Braun and his colleagues demonstrates. This would leave us with the need to sketch a path by which the information stored in the nucleotide chain might result in the directed reproduction of peptides and their roles in the structural and functional replication of the microsphere subsystem.

The earliest polymerization products should serve to extend the mass-energy, complexity, stability and flexibility of the open thermodynamic system of autosynthesized L-amino acids, thereby minimizing the density-specific entropy production of that hierarchically evolving system and minimizing the flow-specific entropy production in its energy and materials environment. This implies thermodynamic selection among various possible orderings of amino acids – a *natural experiment, carried out in time* – to achieve in some degree the above open-potentialities appropriate to this early stage of molecular evolution. Homochirality should yield proteins with much greater molecular mass and stability and more complex dynamic enzyme structures with increased specificity. The constraints imposed by available external energy and materials and internal kinetic limitations determine whether such experiments are more or less extended in evolutionary time and far-reaching in scope; the greater the number of options, as for example in the selection among 20 amino acids in each addition of a peptide, the more likely will such experiments go on and on, at one or more hierarchical levels, and conclude with relatively unique and critically important results. Nature's thermodynamicist-biochemist conducts these *experiments* with great precision and patience, *not because she knows the outcome is critically important to the emergence of living systems in some distant future, but because it contributes step-by-step to the conversion of some portion of the kinetic energy in the vortex into high-energy stable chemical bonds and in so doing serves to minimize entropy production in the Universe. That is to say, she is supremely curious but has no systemic design in mind.*

Parallel Synthesis of D-Saccharides

What was said above concerning the spin alignments of reactants, vis-à-vis the orientation of the magnetic field, in the synthesis of L-amino acids, holds also for the synthesis of D-monosaccharides, except that the products, and in particular the ring formations of glucose and ribose, require counter-alignment of the spins of the aldehyde and ketone groups with the orientation of the magnetic field. While the magnetic field would be common to the two factories, the products would be quite different and we might expect the synthesis of D-sugars and their polymerization to be compartmentalized in channels of the vortex distinct from (possibly adjacent to) those of amino acid synthesis. The polymerization process would be similar to that of amino acids, with the glycosidic bond taking the place of the peptide bond and the process benefiting from the physical separation of high temperature reaction zones and low temperature-, magnetic field-supported condensation regions, significantly enhancing the stability of sugars in the aqueous magnetic field environment.

In the above scenario, we see that the magnetic field created by the vent-vortex subsystem was by far the most important feature of that subsystem in that it gave rise to all that followed in the evolution of living matter to the present time. This evolving process can be succinctly defined as follows:

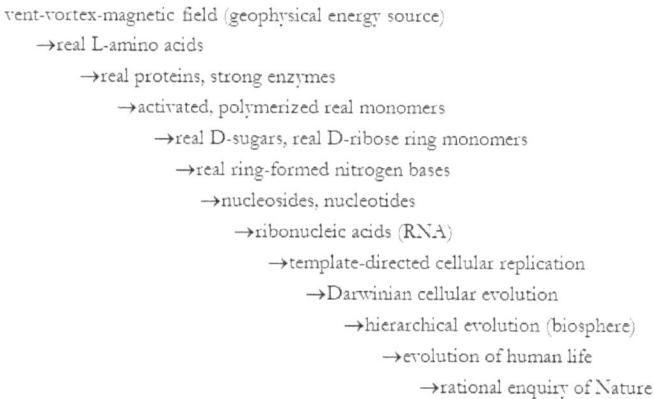

vent-vortex-magnetic field (geophysical energy source)
  →real L-amino acids
    →real proteins, strong enzymes
      →activated, polymerized real monomers
        →real D-sugars, real D-ribose ring monomers
          →real ring-formed nitrogen bases
            →nucleosides, nucleotides
              →ribonucleic acids (RNA)
                →template-directed cellular replication
                  →Darwinian cellular evolution
                    →hierarchical evolution (biosphere)
                      →evolution of human life
                        →rational enquiry of Nature

Energy-density, time increase with each line down

We recall here Howard Pattee's counsel "it is the constraints of the primeval ecosystem [the vent-vortex subsystem] which, in effect, generate the language in which the first specific messages [Nature's systemic energy flow generates multiple open subsystems characterized by a symmetry-breaking magnetic field] can make evolutionary sense."

Proposed Experimental Program

The above broadly defines a proposed program of experimental research, commencing with robotic magnetometer measurements of the intensity and handedness of DC toroidal magnetic fields (H) surrounding active vents on the ocean floor. We should expect to find the intensity of such fields, an order of magnitude or more, greater than the earth's magnetic field. If this proves not to be the case, much of the above and what follows would be called into question.

Assuming our expectations are fundamentally sound, we should then undertake the design and construction of laboratory apparatus capable of simulating and quantitatively characterizing the cyclic and magnetic behavior of vent-vortex subsystems in various kinds of experiments. One such fixture is sketched in Fig. 17-2. The publications of Braun *et al.*, previously cited, are relevant in such effort. We assume that the velocity of the circulating ions would vary linearly with the physical dimensions of the vent-vortex subsystem (for $T > T_{critical}$ and for T_{hot} and T_{cold} constant) and that the number-density of ions involved scales as r^{-2} so that the magnetic field intensity would be unvarying. Initial tests should serve to verify or refine and alter this assumption.

The first simulation experiments should seek to repeat Fox's quest to see what results from cooking the prebiotic constituents of amino acids and in particular to determine the extent to which the left-handedness of resulting amino acids correlate with the orientation and strength of the magnetic field and the efficiency of amino acid synthesis.

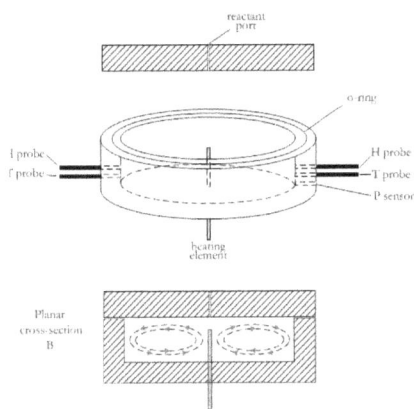

Fig. 17-2 .(A) Sketch of a laboratory reactor for the simulation of hydrothermal vent-vortex subsystems, showing a heating element (hot vent) in a pressure vessel, probes for measuring the temperature and magnetic field gradients and the pressure in the vessel. Also shown is a microport in the vessel lid to enable the entry of reactants and removal of samples of reaction products. (B) Sectional pattern of the flow of water and ions.

Next, we need to know whether or not the thermo-cyclic (dehydration-condensation) polymerization of L-amino acids results in short polymers capable of aggregating to form microspheres, as is the case with largely racemic polymers. If so, is the identifying membrane, of most or nearly all microspheres, double-layered? How do the physical and behavioral properties of such membranes differ from those in previous experiments and in particular how does this change affect their stability and their growth and proliferation?

A third major area of investigation would involve the potential for synthesis of large macromolecules *within the protective* confines of the microspheres and with particular focus on the synthesis of large globular proteins and enzymes, as they depend on the strength of the magnetic field. The weak catalytic capability of the "proteinoids" was severely limited by their short range of elongation which in turn traces to the racemic structure of the synthetic amino acid polymers.

Similarly the inability to homoselect isomers in monosaccharides severely constrained early efforts toward the synthesis of polysaccharides. (The asymmetric number one carbon atom in ribos monomers exists in one or the other of two isomer states, just as is the case for the α-carbon atom in amino acids. Thus there was no means of achieving homochirality, L-amino acids in the one and D-monosaccharides in the other. The former is key to the evolution of proteins and enzymes in all living cells and the latter is key to evolution of polynucleotides and therefore of the further evolution of life itself.) The task here is to explore quantitatively the relations between the temperature gradient, pressure, magnetic field strength, time and physical parameters of our simulated vent-vortex subsystem as they pertain to the synthesis of D-ribose sugars and large polysaccharides.

Could the vent-vortex-magnetic field subsystem give rise to the synthesis of adenosine monophosphate (AMP) macromolecules, given the supply of requisite intermediate molecules, and aided by the NTI-ordered stereospecific enzymes created in earlier experiments? This would open a path toward the heterotraughic synthesis of the nucleotides and, no less important, the creation of an initial metabolic process. De Duve and others have sought what amounts to interim, less complex, lower energy but still effective free-energy carriers to initiate the evolution of metabolic systems, as for example thioesters in de Duve's approach. Ronald Fox[173] summarized the principal early metabolic possibilities as follows: "Several kinds of chemistry were occurring on the prebiotic earth, including acid-base reactions, in which protons transfer from one type of molecule to another, and oxidation-reduction reactions, in which electrons transfer from one molecule to another. During some oxidation-reduction reactions proton exchanges also take place. The proton-transfer and electron-transfer reactions began early on the primitive earth, since they derived from geophysical energy flows, and no doubt played prominent roles in prebiotic energy transfer." And he continues, almost as an afterthought: "Another type of chemistry, based on phosphorus, just as assuredly played a special role. Ultimately, phosphorus in the form of phosphate bond energy enabled geophysical oxidation-reduction energy to flow through the organic matter that existed on the primitive earth. *Through conversion into phosphate bond energy, oxidation-reduction energy is capable of driving the polymerization of monomers.* Contemporary organisms use the phosphate bond energy in adenosine triphosphate (ATP), a nucleoside derivative, to drive polymerization. The useable energy in ATP is associated with the pyrophosphate moiety, the anhydride linkage between two phosphates in ATP or—in the simpler case—in pyrophosphate." Later (p 48) he favors iron-sulfur oxidation-reduction reactions as a possible early path but adds *"Another possible mechanism involves the compound carbamyl phosphate, which forms spontaneously in aqueous mixtures of phosphate and cyanate, CNO. Cyanate has a positive free energy of formation relative to the elements carbon and nitrogen, but it forms easily when energy (electrical discharges and UV light) flows through mixtures of gases containing carbon- and nitrogen-rich molecules (such as CO_2 and NH_3). Carbamyl phosphate is energy-rich and could be a source of phosphate bond energy for the synthesis of the much more versatile phosphate carrier, pyrophosphate."* We emphasize this possibility since (1) CO_2 and NH_3 would be readily available in the early atmosphere, (2) UV light would be intense in the absence of O_2 and Ozone and (3) phosphate concentration in the ocean would perhaps be significantly greater then than now, giving rise to relatively high concentrations of carbamyl phosphate and its spontaneous conversion to high-energy pyrophosphate for activation of monomers and polymerization of amino acids and nucleic acids. *And we suggest that this simple 3-stage process might have served as an initial efficient photosynthetic metabolic path toward the heterotrophic evolution of microsphere systems in vent-vortex environments perhaps two billion years before the evolution of green plants.* If so, these investigations would be taking us higher and higher up the energy (energy-density-complexity) ladder in the evolution of early living subsystems.

If and when the above studies result in the synthesis of adenosine monophosphate, GMP, etc., then and only then can we begin to explore, with our simulated vent-vortex laboratory facility, the possibility of polymerizing nucleotide monomers to form short or long chains of RNA macromolecules.

Notwithstanding the above, it would be a mistake to think that life may have suddenly emerged with the onset of hydrothermal vents. Thermodynamic selection occurs at the microscopic and macroscopic levels via numerous processes of trying-this and trying-that, subject to kinetic and thermodynamic constraints, with more or less stationary states resulting only after considerable passage of time: some reaction products may not be amino acids and some may not be L-amino acids and some of these products may interfere, as reactants, in the polymerization process and limit or inhibit the synthesis of proteins and enzymes, and so on in reactions at higher levels, all at best slowing the pace of evolution and muddying the waters. Nevertheless, the path sketched above, if not too far-fetched, could well have required as little as several thousand years or as much as a few million years, as the abiotic earth became more hospitable.

Chapter 17 Summary

The foregoing may be seen as a departure from other efforts to discern the beginnings of life on Earth in two principal respects: (1) the belief that nonequilibrium thermodynamics holds the key to comprehending the dynamic complexities of abiotic molecular evolution and (2) a growing awareness that the "chicken or egg problem" could be resolved only, or best, if nature first provided some dynamic, mesoscopic or macroscopic mechanism in which thermodynamics, chemistry and energy flow might conspire in the evolution of high energy macromolecules. This belief was first evidenced in the author's 1977 publication of "A Thermodynamic Theory Of The Origin And Hierarchical Evolution Of Living Systems" and the awareness emerged with mounting evidence that equilibrium chemistry is inadequate to the task and that it was necessary first to elaborate nonequilibrium thermodynamic theory.

EPILOG: A PATH BEYOND

How might the fundamental theory and geophysical systems sketched in chapter 17 lead to a next most critically important advance toward understanding the evolution of living systems, that is, to articulate a self-consistent path from proto-life to the emergence of the earliest-known biotic systems in which the basic processes of cellular growth and replication in interaction with the environment are explainable in Darwinian terms? It is in this interaction at various hierarchical levels that we should expect the NTI to play critical roles in the evolution and development of living systems from the earliest time to the present, with Thermodynamic Selection providing an underlying physical basis for Natural Selection of mutant individuals of a population, species, etc. And it is during this very long period that heterotrophic life may make the transition to autotrophic forms. Later studies would focus on the emergence of Eukaryotic cells and still later efforts to explore the evolution of multicellular organisms, as the organizational complexity of systems and subsystems increases hierarchically.

Lastly, it is our hope that this effort may help to restore our faith in critical common sense reasoning, as the bedrock of past, present and future scientific research. Hopefully also such reasoning may find its way in the education of the "common man," enabling her or him to gain and enjoy a deeper understanding of our evolving wonderful world. The future of our species and of all life on Earth is increasingly and urgently dependent on such general understanding.

REFERENCES

1 H. J. Hamilton, A Thermodynamic Theory of the Origin and Hierarchical Evolution of Living Systems, Zygon, 12, no. 4(December 1977).

2 R.Reiss, H.Hamilton, L.Harmon, G.Hoyle, D.Kennedy, O.Schmitt, C.Wiersma, Neural theory and Modeling, Stanford University Press (1964)

3 L. Onsanger, L, "Reciprocal Relations in Irreversible Processes," Physical Review 37 (1931): 405(pt. 1), 38 (1931): 2265 (pt 2).

4 I. Prigogine, Introduction to Thermodynamics of Irreversible Processes, Interscience Publishers (1961)

5 I. Prigogine, From being to becoming: (1980), W.H. Freeman and Company, p xiii

6 H. Matsuno, Beyond Neo-Darwinism, ed: Mae-Wan Ho and Peter T. Saunders (1984) p72

7 I. Prigogine, From being to becoming: (1980), W.H. Freeman and Company, p xiii

8 Human Biology, ed. By L. Reed, A. Lotka, The law of evolution as a maximal principle, Vol. 17 No. 3 1945

9 A. Zotin, Thermodynamic Bases of Biological Processes: Physiological Reactions and Adaptations, Walter de Gruyter (1990), p55

10 G. M. Edelman, Neural Darwinism: The Theory of Neuronal Selection, Basic Books (1987)

11 R. Potts, Humanity's Descent: the consequences of ecological instability, William Morrow (1996)

12 L. Margulis, D. Sagan, What is Life, University of California press (1995)

13 A. Zotin, Thermodynamic Bases of Biological Processes: Physiological Reactions and Adaptations, Walter de Gruyter (1990), p55

14 H. Kragh, Cosmology and Controversy, Princeton University Press (1996) p10

15 P. Coles, G. Ellis, Is the universe open or closed, Cambridge University Press (1997) p53

16 H. Kragh, Cosmology and Controversy, Princeton University Press (1996) p22

17 H. Kragh, Cosmology and Controversy, Princeton University Press (1996) p46

18 H. Kragh, Cosmology and Controversy, Princeton University Press (1996) p47

19 Kragh, Cosmology and Controversy, Princeton University Press (1996) p50

20 Kragh C Cosmology and Controversy, Princeton University Press (1996) p74-5

21 H. Kragh, Cosmology and Controversy, Princeton University Press (1996) p75

22 Kragh Cosmology and Controversy, Princeton University Press (1996) p100

23 Kragh Cosmology and Controversy, Princeton University Press (1996) p108

24 H. Kragh, Cosmology and Controversy, Princeton University Press (1996) p142

25 H. Kragh, Cosmology and Controversy, Princeton University Press (1996) p343

26 A. Guth, Physical Review vol. 23 No 2 15 p 347 Jan. 1981

27 J. Barrow, Theories of Everything, Oxford University Press (1991) p87

28 J. Barrow, Theories of Everything, Oxford University Press (1991) p104

29 S. Hawking, A Brief History of Time, Bantam Books (1988) p156

30 S. Hawking, A Brief History of Time, Bantam Books (1988) p148

31 P. Peebles, Principles of Physical Cosmology, Princeton University Press (1993) p80

32 H. Kragh, Cosmology and Controversy, Princeton University Press (1996) p75

33 H. Kragh, Cosmology and Controversy, Princeton University Press (1996) p47

34 J. Barrow, F. Tipler, The Anthropic Cosmological Principle, Oxford University Press (1988)

35 M. Livio, M. Rees, Anthropic Reasoning, Science, 12 August 2005 Vol 309

36 T. Siegried, A 'Landscape' Too Far Science, 11 August 2006, Vol 313

37 M. Tegmark, Cosmological Constraints from the SDSS Luminous Red Galaxies. ArXiv:astro-ph/0608632 v1 30 Aug 2006

38 K. Freeman, The Hunt for Dark Matter in Galaxies, Science, 12 December 2003, Vol 303

39 R. Irion, Galaxies, Black Holes Shared Their Youths, Science, 16 June 2000, Vol 288

40 R. Irion, Galaxies, Disks of Destruction Science, 16 June 2000, Vol 288

41 C. Alcock, The Dark Halo of the Milky Way, Science, 7 January 2000, Vol 287

42 R. Irion, As Galaxies Turn, Science, 7 January 2005, Vol 307

43 J. Ostriker, P. Steinhardt, New Light on Dark Matter, Science, 20 June 2003, Vol 300

44 M. Livio, M. Rees, Anthropic Reasoning, Science, 12 August 2005 Vol 309

45 T. Siegrried, A 'Landscape' Too Far Science, 11 August 2006, Vol 313

46 W. Scheider, http.cavendishscience.org/phys/tyoung/tyoung.htm

47 D. Bohm, Quantum Theory, Prentice-Hall, Inc. (1951) p60

48 D. Bohm, Quantum Theory, Prentice-Hall, Inc. (1951) p24

49 R. Mills, Space, time and quanta: an introduction to contemporary physics, W. H. Freeman (1994) p382

50 D. Wick, The infamous boundary: seven decades of controversy in quantum physics (1950) p39

51 D. Wick, The infamous boundary: seven decades of controversy in quantum physics (1950) p31

52 R. Hughes, The structure and interpretation of quantum mechanics, Harvard University Press (1992) p108

53 D. Bohm, Wholeness and the implicate order, Routledge (1981) p73

54 G. Milburn, The Feynman Processor, Allen & Unwin Pty Ltd (1998) p1

55 G. Milburn, The Feynman Processor, Allen & Unwin Pty Ltd (1998) p20

56 R. Omnès, Understanding Quantum Mechanics, Princeton University Press (1999) p47

57 R. Omnès Understanding Quantum Mechanics, Princeton University Press (1999) p54

58 S. Auyang, How is quantum field theory possible? Oxford University Press (1995) p62

59 S. Auyang, How is quantum field theory possible? Oxford University Press (1995) Auyang p140

60 D. Wick, The infamous boundary: seven decades of controversy in quantum physics (1950) p128

61 Ed: M. Bell, K. Gottfried, M. Veltman, John W. Bell on the foundations of quantum mechanics, World Scientific Publishing (1995) p49

62 D. Wick, The infamous boundary: seven decades of controversy in quantum physics (1950) p123

63 A. Whitaker, Einstein, Bohr and the quantum dilemma, Cambridge University Press (1996) p179

64 R. Omnès, Understanding Quantum Mechanics, Princeton University Press (1999) p63

65 S. Weinberg, What is quantum field theory and what did we think it is?, arxiv:hep-th/7902027v1 4 Feb 1997

66 S. Schweber, QED and the men who made it: Dyson, Feynman, and Tomonaga, Princeton University Press (1976) p87

67 R. Omnès Understanding Quantum Mechanics, Princeton University Press (1999) p91

68 R. Garcia, I. Saveliev and M. Sharonov, Time-resolved diffraction and interference: Young's interference with photons of different energy as revealed by time resolution, Philosophical Transactions of the Royal Society of London (2002) May 15

69 R. Austin, L. Page, Interference of a single photon, http://www.ophelia.princeton.edu/user/page/single_photon.html

70 W. Jackson Classical Electrodynamics, John Wiley and Sons (1967) p178

71 R. Feynman, QED, Princeton University Press (1988) p128

72 S. Schweber QED and the men who made it: Dyson, Feynman, and Tomonaga, Princeton University Press (1976) p70-1

73 S. Schweber, QED and the men who made it: Dyson, Feynman, and Tomonaga, Princeton University Press (1976) p xxv

74 S. Schweber, QED and the men who made it: Dyson, Feynman, and Tomonaga, Princeton University Press (1976) p567

75 D. Wallace, In defense of naivete: The conceptual status of Lagrangian quantum field theory, December 23 2001 http::/phitsci.pitt.edu/view-quantum-field-theory.html

76 D. Bohm, Wholeness and the implicate order, Routledge (1981) p134

77 S. Weinberg, Dreams of a Final Theory, Pantheon Books (1992) p85

78 J. Barrow, Theories of Everything, Oxford University Press (1991) p77

79 S. Schweber, QED and the men who made it: Dyson, Feynman, and Tomonaga, Princeton University Press (1976) p353

80 B. Greene, The Elegant Universe, Vintage Books (1999) p129-30

81 B. Greene, The Elegant Universe (1999) p87

82 S. Schweber, QED and the men who made it: Dyson, Feynman, and Tomonaga, Princeton University Press (1976) p58

83 P. Dirac lecture http://www.nobelprizes.org 1933 Physics

84 W. Scheider, Do the "double slit" experiment the way it was originally done, originally published in The Physics Teacher 24 217-219, 1986

85 N. Garcia, I. Saveliev, M. Sharonov, Time-resolved diffraction and interference: Young's interference with photons of different energy as revealed by time resolution http://www.physic.ut.ee/instituudid/efti/loengumaterjalid/opt/optika/young%20timeresolved_Garcia.pdf

86 R. Austin, L. Page http://ophelia.princeton.edu/~page/single_photon.html

87 http://www.hqrd.hitachi.co.jp/em/doubleslit.cfm

88 E.Gouliemakis, et al, Direct Measurement of Light Waves, Science, 27 August 2004 vol 305

89 G. Foster, L. Orozco, H. Castro-Beltran, H. Carmichael, Quantum State Reduction and Conditional Time Evolution of Wave-Particle Correlations in Cavity QED, Physical Review Letters, 9 October 2000 vol 85, number 15

90 Carleton DeTar and Steven Gottlieb http://.physicstoday.org/vol-57/iss-2/p45.html

91 A. Aczel, Entanglement, the greatest mystery in physics, Four walls Eight Windows (2002 p137

92 K. Moriyasu, An Elementary Primer for Gauge Theory, World Scientific Publishing Co (1983) p34

93 S. Weinberg, A Unified Physics by 2050?, Scientific American, December 1999,

94 G. Kane, The Standard Model, Scientific American, June 2003 p70

95 I. Ciufolini and J. Wheeler, Gravitation and Inertia, Princeton University Press 1996 p109

96 I. Vincent, The Force of Symmetry, Cambridge University Press (1995) p259

97 W. Lamb and R. Rutherford, Fine structure of the Hydrogen atom by a Microwave Method, Physical Review volume 72, number 3, August 1, 1947

98 W. Lamb and R. Rutherford, Fine structure of the Hydrogen atom, Part I, Physical Review, Vol. 79, No. 4, August 15. 1950

99 W. Lamb lecture http://www.nobelprize.org 1955 Physics

100 P. Kusch lecture http://www.nobelprize.org 1955 Physics

101 S. Karshenboim, Precision physics of simple atoms: QED tests, nuclear structure and fundamental constants, arXiv:ph/0509010 v1 1Sep 1 2005

102 J. Hogan, Quantum philosophy, Scientific American, July, 1992 p93

103 P. Kwiat, H. Weinfurter, A. Zeilinger, Quantum seeing in the dark, Scientific American, November 1996 p72

104 T. Pittman et al, Can Two-photon intererence be considered the interference of two photons?, Physical Review Letters 2 September 1996 p1917

105 Z. Ou, J. Rhee, L. Wang, Observations of four-photon interference with a beam splitter by pulsed down-conversion, Physical Review Letters, volume 83, number 5, 2 August 1999

106 C. Monroe, A "Schrodinger's cat" superposition state of an atom, Science, 24 May 1996 Vol 272 p1131

107 P. Yam, Bringing Schrodinger's cat to life, Scientific American, June 1997 p124

108 M. Noel and C. Stroud, Jr., Excitation of an atomic electron to a coherent superposition of macroscopically distinct states, Physical Review Letters, volume 77, number 10, 2 September 1996

109 C. van der Wal, et al, Quantum superposition of macroscopic persistent-current states, Science, vol 290 27 October 2000 p773

110 M. Bell, K. Gottfried and M. Veltman editors, John S. Bell, On the foundations of quantum mechanics, World Scientific Publishing, 2001 p11

111 A. Whitaker, Einstein, Bohr and the Quantum Dilemma, Cambridge University Press, 1996 p179

112 Aczel, Entanglement, the greatest mystery in physics, Four walls Eight Windows (2002 p137

113 C. Thompson, Timing, "Accidentals" and other artifacts, EPR experiments, arXiv:quant-ph/9711044 v2 25 Nov 1997

114 D. Strekalov, Y, Kim and Y. Shih, Experimental study of a Photon as a subsystem of an entangled two-photon state, arXiv:quant-ph/9811060 v1 23 Nov 1998

115 M. Bell, K. Gottfried and M. Veltman editors, John S. Bell, On the foundations of quantum mechanics, World Scientific Publishing, 2001 p49

116 A. Aczel, Entanglement, the greatest mystery in physics, Four walls Eight Windows (2002) p137

117 M. D'Angelo, M. Chekhova and Y. Shih, Two-photon diffraction and quantum lithography, Physical Review Letters, volume 87, number 1, 2 July 2001

118 J. Bonard et al, Field emission-induced luminescence from carbon nanotubes, Physical Review Letters volume 81, number 7, 17 August 1998

119 C. van der Wal, et al, Quantum superposition of macroscopic persistent-current states, Science, vol 290 27 October 2000 p773

120 Y. Aharonov and C. Gohm, Significance of Electromagnetic Potentials in the quantum theory, Physical Review, second series, vol. 115, No 3 August 1, 1959

121 http://.hqrd.hitachi.co.jp/em/doubleslit.cfm

122 J. Anandan, Putting s spin on the Aharonov-Bohm oscillations, Science, 6 September Vol 297 p1656

123 G. Gonzalez-Martin, Charge to magnetic flux ratios, http://prof.usb.ve/ggonzalm/

124 J. Eisenstein and H. Stormer, The fractional quantum Hall effect, Science,I new series, vol. 248, No. 4962 June 23 1990 p1510

125 David Leadly 10-02-1997 http://www.warwick.ac.uk/

126 K. Shrivastava, Comments on "Composite fermion (CF) model of quantum Hall effect, arXiv:cond-maat/0304014 v1 1 Apr 2003

127 S. Frank, et al, Carbon nanotube quantum resistors, Science, vol 280, 12 June 1998 p1744

128 E. Cornell et al, et al, Experiments in dilute atomic Bose-Einstein condensation, arXiv:cond-mat/9903109 v1 % Mar 1999

129 A. Gaeta, Collapsing light really shines, Science, 4 July 2003 vol 301 p54

130 H. Miesner, Bosonic stimulation in the formation of a Bose-Einstein condensate, *Science* 13 February 1998: Vol. 279 no. 5353 pp. 1005-1007

131 P. Engles *et al.* Observations of long-lived vortex aggregates in rapidly rotating Bose-Einstein condensates, arXiv:cond-mat/0301532 v1 28 Jan 2003

132 I. Coddington *et al*, Experimental studies of equilibrium vortex properties in a Bose-Einstein gas, arXiv:cond-mat/0405240 v4 5 Oct 2004

133 J. Abo-Shaeer, Observations of vortex lattices in Bose-Einstein condensates, *Science*, 20 April 2001 vol 292 p476

134 B. Anderson *et al*, Watching dark solitons decay into vortex rings in a Bose-Einstein condensate, arXiv:cond-mat/0012444 v1 22 Dec 2000

135 K. McDonald, Slow Light, arX:physics/0007097 v2 5 Aug 2000

136 I. Coddington *et al*, Experimental studies of equilibrium vortex properties in a Bose-Einstein gas, arXiv:cond-mat/0405240 v4 5 Oct 2004

137 R. Boyd *et al*, Limits on the time delay induced by slow-light propagation, Optical Society of America 2005

138 I. Fry, The emergence of life on Earth: a historical and scientific overview, Rutgers University Press 2000

139 Molecular Evolution Prebiological and Biological, ed. by D. Rohlfing and A. Oparin, Plenum Press, 1972 p 28

140 Beyond Neo-Darwinism, ed: M. Ho and P. Saunders, Academic Press 1984 p 72

141 H. J. Hamilton, A Thermodynamic Theory of the Origin and Hierarchical Evolution of Living Systems, Zygon, 12, no. 4(December 1977).

142 J. Wicken, Evolution, Thermodynamics, and Information, Oxford University Press 1987 p7

143 Communication in Development, ed: A. Lang, Academic Press 1969 p14

144 Towards a Theoretical Biology, C. Wadington, ed., Aldine Publishing Co.1970 p119

145 The Origins of Prebiological Systems, S. Fox ed. Academic Press 1965 p388

146 S. Fox and K. Dose, Molecular Evolution and the Origin of Life, Freeman and Company 1972 p238H. H.

147 H. Morowitz, Energy Flow in Biology, Academic Press, 1968 p45

148 H. Morowitz, Energy Flow in Biology, Academic Press, 1968 p55

149 Biology and the Physical Sciences, ed: S. Devons, Columbia University Press 1969 p332

150 H. Morowitz, Beginnings of Cellular Life, Yale University Press 1992 p105

151 J. Bronowski, New concepts in the evolution of complexity: stratified stability and unbounded plans, Zygon, vol.5 1970 p31

152 Beyond Neo-Darwinism, ed:M. Ho and P. Saunders, Academic Press 1984 p64

153 A. Szoke, W. Scott, J. Hajdu, Hypothesis Catalysis, evolution and life, FEBS Letters 553 (2003) 18-20

154 T. Hoang, et al, Geometry and symmetry presculpt the free-energy landscape of proteins, PNAS May 25 2004 vol. 101 no. 21

155 V. Avetisov, V. Goldanskii, Mirror symmetry breaking at the molecular level, Proc. Natl. Acad. Sci. vol. 93, p11435

156 D. Blackmond, Asymmetric autocatalysis and its implications for the origin of homochirality, PNAS, April 20, 2004 vol.101 no. 16

157 Molecular Evolution Prebiological and Biological, ed. by D. Rohlfing and A. Oparin, Plenum Press, 1972

158 M. Braham, A general theory of organization, General Systems, Volume XVIII, 1973 P13

159 K. Boulding, Toward a general theory of growth, General Systems, Reprinted from the Canadian Journal of Economics and Political Science, 19, 326 1953 p66

160 L. Orgel, The origin of life – A review of facts and speculations, TIBS 23 December 1998 p491

161 E. Imal, H. Honda, K.Hatori, K. Matsuno, Elongation of Oligopeptides in a simulated submarine hydrothermal system, Science, vol. 283 5 February 1999 p831

162 D. Braun and A. Libschaber, Trapping of DNA by Thermophoretic Depletion and convection, Physical Review Letters Vol 89, no. 18, 28 October 2002

163 D. Braun and A. Libschaber, Trapping of DNA by Thermophoretic Depletion and convection, Physical Review Letters Vol 89, no. 18, 28 October 2002

164 D. Braun, N. Goddard and A. Libchaber, Exponential DNA replication by laminar conviction , Physical Review Letters vol 91, no. 15 10 October 12 October 2003

165 D.Braum and A. Libchaber, Thermal approach to molecular evolution, Pnys. Biol. I (2004)

166 J. Amend and E. Shock, Energetics of amino acid synthesis in hydrothermal ecosystems,Science vol. 281, 11 September 1998

167 Beyond Neo-Darwinism, ed: M. Ho and P. Saunders, Academic Press 1984 p64

168 A. Szoke, W. Scott, J. Hajdu, Hypothesis Catalysis, evolution and life, FEBS Letters 553 (2003) 18-20

169 P. Wolynes, Symmetry and the energy landscapes of biomolecules, PNAS vol. 93, December 1996 p14249

170 D. Blackmond, Asymmetric autocatalysis and its implications for the origin of homochirality, PNAS, April 20, 2004 vol.101 no. 16

171 The Nature of Life, University Park Press ed: W.Heidcamp, 1977 p28-37

172 Molecular Evolution and Protobiology, ed: K. Matsuno, K. Dose, K. Harada and D. Rohlfing, Plenum Press 1984 p270

173 R. Fox, Energy and the Evolution of Life, Freeman and Company 1988 p 40

APPENDIX

A Review of the
CONCEPTUAL STEPS
TOWARD A SELF-CONSISTENT
RELATIVISTIC PARTICLE THEORY

STEPS PART I
GRAVITY
IN RELATIVISTIC THERMODYNAMICS

Chapters 1-5

1.1 A Thermodynamic System, taken as a particle gas, is said to be in an Equilibrium or Stationary State only as a Simplifying Approximation, as defined by the Calculus of Variation. That Calculus severely *constrains the formulation of Thermodynamic Theory*. Nonequilibrium Systems can be adequately comprehended *only* in *Evolutionary Terms defined by the Relativistic Flow of Energy*.

2.1 Where possible, Thermodynamic Systems approach a Stationary State via the Evolution of a Thermodynamic Minimal Gradient Structure (TMGS), as required by the 1st Law and in accordance with the proposed Principle of Thermodynamic Selection. In the linear regime this accords with Onsager's Principle of Minimum Energy Dissipation and Prigogine's Theorem of Minimum Entropy Production, though such evolution makes no dependence on the Calculus of Variation, requiring only that the Phenomenological-Functions (P-Functions) be taken as Constants.

3.1 Weakly Nonlinear Systems (well-behaved P-Functions) are characterized by Short Range Correlation or Dynamic Organization of Matter and Marginal Increase in Entropy Production but Marginal Decrease in Entropy Production per unit Energy-Flow. This reveals the more efficient evolution of a TMGS via the conversion of some small portion of the energy-flow to the Creation and Maintenance of Simple Molecular Structures and Dynamic Systems.

4.1 Strongly Nonlinear Systems are characterized by Long Range Correlation and Dynamic Organization of Matter-Energy that may lead to the Creation, Growth and Maintenance of Open Subsystems, marking a Transformation of the State of Matter-Energy in the Subsystem. Matter and Energy-Flow through a Real or Virtual Membrane Surrounding the Subsystem may lead to Minimal Density-Specific Entropy Production in the Subsystem (Maximal Energy-Density) and Minimal Flow-Specific Entropy Production in the System, in the Stationary State. All of the above has Profound Implications in Cosmology, Quantum Theory, Molecular Theory and the Life Sciences. Strongly Nonlinear Systems may result in the Creation and Maintenance of Multiple Subsystems of Similar Structure (daughter progeny) or Subsystems Within Subsystems, defining various hierarchical levels of energy-flow and stability (lifetimes) in Systems and Subsystems.

5.1 General Principles and Postulates of Nonequilibrium

Thermodynamics

- *General Principle of Thermodynamic Selection*
- *General Principle of Minimum Flow-Specific Energy Dissipation*
- *General Principle of Minimum Flow-Specific Entropy Production*
- *General Principle of Minimum Density-Specific Entropy Production*
- *Corollary Principle of Maximal Energy-Density*

5.2 Postulate:

In the evolution of a quasi-isolated thermodynamic system or of the Universe, taken to be an ongoing, adiabatic-expanding thermodynamic system in spacetime, entropy is positive and increases monotonically, on average, and at a minimal, monotonically-decreasing rate, i.e., $S>0$, $dS/dt>0$ and $d^2S/dt^2<0$. Thus, entropy increases without limit in all such systems and on average in the slowest possible way.

5.3 The foregoing extends the stature of the 1st Law of thermodynamics: in addition to its requirement on conservation of energy, it imposes a requirement on the 2nd Law that entropy in the Universe increase on average at a minimal pace, thus constituting an extended formulation of the 2nd Law. The above General Principles and (Extended) *2nd Law*, together with Heisenberg's (In)Determinacy Relation may be referred to as the Nonequilibrium Thermodynamic Imperative (NTI) governing the evolution of the system and its subsystems.

5.4 The foregoing also suggests that we may acquire a more complete understanding of Nature via the bundling of Nonequilibrium Thermodynamics, in Theory and Experiment, with a Relativistic Quantum (Particle) Mechanics. The farther the system is from equilibrium, the greater is the need to understand its origin and evolution and that of its environment. *In the large, self-consistency requires the principles and Laws of Thermodynamics to respect and serve to define the underlying suppositions and postulates in the various areas of science, as for example: No physical system can evolve faster than the ρ-defined speed of light* (see below).

STEPS PART II
GRAVITY
<u>IN</u> RELATIVISTIC COSMOLOGY

Chapters 6-9

6.1 Hot Big Bang Inflationary models of the origin and evolution of the Universe are fraught with numerous inconsistencies, including the requirement for some 20-orders of- magnitude adjustment in the speed of light from its quantum-state origin and ending in precise agreement with the present speed of light. Most troubling; "In order for this scenario [Inflation] to work, it is necessary for the universe to be essentially devoid of any strictly conserved quantities." Alan Guth

7.1 Our Most Fundamental Assumption requires the permittivity ε and permeability μ of space each to vary as ρ^{-3}, where ρ is the average energy-density of the Universe. This is equivalent to arguing that the distributed neutral mass-energy in the Universe (the average energy-density) acts as a dielectric to reduce the Coulomb coupling strength of charges q_1 and q_2, and also as a permeable medium to decrease the coupling strength between parallel paths of charge flow, as for example, in the early, high-density Universe.

7.2 Which in turn would require c, the velocity of light, to vary as ρ^{-3}, Maxwell's equations to be valid also in general relativity, the Universe to be causally-connected in all time and nearly flat ($\Omega \geq 1$) in all expansion, and Δc to dilate in special relativity, rather than Δt, where the unit of time advance Δt is taken as a universal constant.

7.3   And the above would further require $\sqrt{G}$ to vary as ρ^{-3}, to be self-consistent in the field equations; the electric, magnetic and gravitational forces $k\sqrt{\varepsilon}$, $k'\sqrt{\mu}$ and $m\sqrt{G}$, to be invariant over all time; and the Universe, to be nearly static in all time, with far reaching implications for fundamental physical theory.

7.4   We note also that the (In)Determinacy Relation $\Delta\rho\Delta t \geq \hbar$ and the NTI serve as external adjuncts to the field equations, with adiabatic quanta-number expansion and scalar-physical expansion and cosmic time closely connected with the first Law. Self-consistency with the NTI thus requires the Universe to be quasi-flat, homogeneous, isotropic and have nearly null angular momentum in all time.

8.1   The requirement on null angular momentum leads us to postulate a GSBP (Gravitational and Symmetry Bound Photon), whole-thirds-charge model of Primordial-quarks/Antiquarks, in an initial quantum Euclidean space, thereby directly and self-consistently linking the demands of quantum, gravity and relativity theory.

8.2   Quantum-state evolution proceeded via energy down-conversion (quanta number-doubling and energy halving), leading to a differential increase of c and $\rho...$, stress buildup and a critical phase transition culminating in the relief of stress, with energy-halving and c constant throughout the quantum era, if and only if Planck's parameter h halves with each doubling. That would require h_i to be $\sim 10^{80/3}h$ (present value of h) $\approx$ 10^{-7}Js in order for large fluctuations in the early extreme energy-density Universe to kick-start the quanta number expansion of the Universe in the nearly-infinite past.

8.3   Quanta doubling makes no demand for spatial expansion but defines the initial radius R_i of the quantum Universe $\sim 10^{-5}$m. Halving of h in the down-conversion process conserves E and ρ in the quantum state. (The present value of h holds as an absolute constant throughout scalar expansion). Termination of down-conversion led to the annihilation of all antimatter particles and a plasma quantum state of remnant matter particles.

Scalar Expansion, and Evolution of the Relativistic
Quantum/Cosmic Universe

9.1 Quanta-number expansion with each doubling (nearly dissipation-free) resulted in $10^{80 \times 3}$ composite matter quarks free in a 10^{28}K plasma, to create 10^{80} neutrons in the recombination epoch consistent with the extended 2nd Law and the NTI.

9.2 Opposite variations of ε and c with ρ^{-3} left the fine structure and Rydberg constants absolute and Universal.

9.3 Scalar parameters T, m, ρ, k, k' declined by ten as R, r_q, c, λ, $\sqrt{G}$, $\sqrt{\varepsilon}$ and $\sqrt{\mu}$ increased to their present values with each integer increase in N, as R, r_q, c, ρ and S changed little in the plasma epoch [Fig. 9-1 and Table 9-1], but T increased, by a factor of ten, 28 or more times to $\sim 10^{28}$K in about an epoch making this Grand Transition a Spacetime-Distribution of Little Bangs, all because c was $\sim 10^{-22}$m/s and free Matter quarks would have required 10^{17}s (an epoch) or more just to encounter antimatter quarks on the other side of the quantum Universe.

9.4 Further expansion and cooling after quark recombination led to n→p transmutation of the d-quarks in a neutron (by reversing their spin as explained later), resulting in the neutron transmuting to a proton in order to relieve stress in the neutron. The transformation made no requirement for a Weak Force or for the creation of W-, W+, Z⁰ Particles in that early time.

9.5 This enabled the NTI-Creation of atoms, as enduring open thermodynamic subsystems forever free of shear stress, as a 2nd major means of locking-up energy in stable subsystems and the coupling of neutrons and protons in nuclei. Thus, we conceptualize our Universe as a quasi-linear thermodynamic conservative system, from beginning to end, with minimum kinds of fundamental particles.

All of the above were driven by the NTI.

9.6 Further expansion and cooling led to nucleosynthesis of hydrogen, helium and lithium as the number of protons (electrons and antineutrinos) gradually increased, governed by the electromagnetic coupling forces in the nucleus. The neutrons and protons were in thermal equilibrium via the neutrinos, which interacted easily with the bosons in the existing high energy-density state.

9.7 Fluctuations following the plasma epoch led to early seeding and gravitational amplification of large scale structures in the cosmos (The time allowed for this process was at least six orders of magnitude greater than that allowed in current cosmology).

9.8 The foregoing argues: There is no need for a Higgs Boson to give mass to W or π mesons or a WEAK Force to bring about the transmutations. There is no need for DARK ENERGY as we show in Chapter 7. There is no need for DARK MATTER as we show in Chapter 9, since the *Force* of Gravity m$\sqrt{G}$ is an absolute Universal constant, accounting for the stability of gravitationally flat structures, e.g., Spiral Galaxies [refer to Begeman's Rotation Curve, Fig. 9-4 and there is no need for multiple Universes.]

9.9 All such spatially-flat structures evolve as subsystems progeny of the NTI, as quasi-rigid bodies, marginally stable in expansion and rotation from epoch to epoch.

9.10 Radiation decoupling occurred six epochs ($\sim 10^{60}$ years) ago, followed by the creation of all the cosmic structure we now see, though our present look-back capability is limited to a single epoch (about 10^{10} years).

9.11 Supernovae and Black Holes may have emerged only an epoch following radiation decoupling, as open subsystems of Galaxies which grew, maximizing their energy–density, while minimizing the flow-specific entropy production in the galaxies and the rate of entropy increase in the Universe.

Creation, and Evolution of a Relativistic Black Hole

9.12 Gravitational collapse of a star or stars to a Black Hole requires conservation of all mass-energy, momentum of stars, particles and bosons consumed to become embodied at quantum scales and temperature consistent with relativity and the NTI such that T rises inside the membrane (Horizon), lowering the entropy of the BH leading to a plasma state, freeing matter quarks with the high energy bosons giving rise to pair-production of matter/antimatter quarks in a local quiescent quantum state, epoch by epoch, as ρ_{BH} increases, so long as mass-energy flows through the BH membrane. The galaxy evolves toward a stationary state of minimum flow-specific entropy production.

9.13   The above and other extreme cosmic dynamics suggest that h may play an important role in the cosmos comparable to that in the quantum world, i.e. in characterizing the fine structure of the expanding Universe. Just as the electromagnetic fine structure constant $\alpha = e^2/\hbar c$ defines the relativistic velocity of the electron in the stable lowest energy (highest energy-density) state of the hydrogen atom, a gravitational fine structure constant $A \equiv (m\sqrt{G})^2/\hbar c$ (where $m\sqrt{G}$ is the gravitational coupling constant) might define the relativistic velocity of stars and gas in stable Spiral Galaxies and other cosmic bodies.

9.14   All of the above expansion-induced transitions are in our distant past driven by the NTI. We cannot create the high energy-density conditions in accelerators to reverse a transition, e.g., n←p without the creation of high energy-momentum extraneous particles, which greatly complicate the analysis and interpretation of events. Hyperons and other such high energy particles did not exist in the initial events, nor was there any need for them.

9.15   All such events and defining parameters are represented self-consistently in Fig. 9-1 and Table 9-1 [Fig 9-1 caption b], deriving from the relatively few fundamental assumptions presented above and in Chapters 7 and 8.

STEPS PART III
GRAVITY
<u>IN</u> RELATIVISTIC QUANTUM MECHANICS

Chapters 10-16

10.1   Chapter 10 traces the historical roots of quantum theory with particular focus on the extent to which the notions of "wave-particle duality," "point particles," "single particles" as thought to be emitted from quantum sources, "superposition" and associated terms which played such a powerful *external* (philosophical) role in setting current quantum theory apart from classical mechanics.

Internal Structure and Dynamics of the "Fundamental Massive Particles"

11.1   Since Primordial Quanta/AntiQuanta (GSBP physical particles) and their number expansion played an essential role in our scenario in the origin and evolution of the Universe, we see that gravity played a key role in quantum theory from the beginning, in contrast to current theory. Also we show that the muon and tau electron would be taken here simply as higher energy-density representations of the electron, existing naturally only in a time when the average or local energy-density of the Universe was much greater. The same arguments would apply to quarks (antiquarks) and to neutrinos (antineutrinos). We assume the latter particles to be distinguished principally by their very low energy-density (perhaps three orders of magnitude less than that of the electron) with their radius of photon gyration being proportionately larger. *Thus we make no reference to three "Generations" or "Families" of quarks and leptons as in present theory.*

11.2   The four fundamental forces in current theory are here reduced to two: gravity and the electromagnetic force, there being no need for the Weak and Strong Force. And the same form of gravity and the EM force $(m\sqrt{G}$ $k\sqrt{\epsilon}$ $k'\sqrt{\mu})$ reveals a common theoretical foundation, though the coupling strength of gravity was from the beginning and remains six orders of magnitude weaker than the EM force, since they govern the dynamics of subsystems of radically different energy-density (GSBP quarks in the one case and nucleons in the other). "Gravitons" or other "wave-particle" means of mediating the attractive force of gravity are likewise not needed, since here that force is mediated by the emission of real bosons from neutral or charged particles, resulting in an attractive force via impulse-recoil action on the emitting particle or body, whether at the cosmic, macro or quantum scale. (See chapter 11 accounting for the Casimir effect) The same holds for the EM force acting on fermions to attract or repulse them. In each case the emission of bosons is initiated in timely response to the buildup of differential energy-density gradients at the surface of the emitting particle. Similar action can occur between bosons under special conditions.

11.3   We assume that quarks in nucleons orbit axes of their local Euclidean space in lowest-energy ground states with higher energy metastable states defining spectral boson emission, though "Rest Mass" may have no comparable significance, as for the electron. Yet fluctuations in an early lower-energy-density time led to n→p transmutation and a spin reversal in the two d-quarks with each having a plus one-sixth charge, implying a breaking of symmetry in a quasi rest mass state sufficient for that transmutation to occur. Quarks in the light hadrons may thus appear to have quasi "Rest Mass" ~4.5 times that of electrons, in close agreement with recent lattice QCD studies [*Science* v322 p1224].

11.4   GSBP relativistic massive quarks and leptons and GSBP quasi-massless bosons (emitted from atoms and characterized by very small discreet mass) and positrinos having no measurable mass constitute the fundamental substance of the Universe for which the concepts of waves or fields may have at best only local philosophical relevance. (In a much earlier and greater energy-density time the positrinos also may have carried a very small discreet mass.) And the illusion of a "spreading wave" derives from the emission of a very large number of bosons from atoms in a localized space in a weakly controlled distribution of close trajectories. Spin-1 "massless" vector bosons like fermions also have GSBP internal dynamic structure though with the axis of photon gyration orthogonal to the boson's trajectory. Thus it is the internal symmetry that distinguishes bosons from fermions. The speed of the GSB photon gyrating in the boson and that of the boson in its linear travel is determined solely by the final state n_f of the electron transition (slowest for $n_f=1$ and progressively faster for $n_f=2$, etc. See Fig. 11-4). The gyrating photon serves as a frictionless flywheel, conserving the energy, momentum, spin angular momentum and polarization of the boson over the life of the boson, unless perturbed. In so characterizing the radiation of bosons, we highlight Maxwell's postulate of *displacement current* in deriving his field equations and we understand that the equations are wholly consistent with our strictly-particle theory, giving new insight into what constitutes mass and momentum.

11.5   **The quantum and cosmic worlds thus are shown to have a common origin and theoretical foundation.**

11.6   **The gravitational force acting on quarks and leptons, as on neutral massive particles or bodies, is defined in terms of the differential energy-density gradients $\rho\nabla$ at their surfaces and is mediated by massless spin-1 vector bosons that act through impulse-**

recoil to maximize such gradients. As such, quantum gravity may be seen as a gauge field and also as a quantum force that may be expressed in terms of a phase shift, as with the electromagnetic force.

11.7   The electric force acting on quarks and leptons is defined in terms of the differential electric energy-density gradients $\rho_e \nabla$ at their GSBP spin surfaces and similarly the force is mediated by bosons which act through impulse-recoil to minimize (same particle charges) or maximize (opposite charges) the differential gradient at their surfaces. All such forces act locally and timely on the boson-emitting particle and conserve some fundamental property such as charge and hence act as gauge forces.

11.8   The higher energy fermions, e.g., "2nd Generation" muons, found in accelerator detectors (or on mountain tops) are simply electrons that have been accelerated to a higher energy-density state and traveling at a correspondingly lower velocity. This follows since otherwise energy-momentum would not be conserved in experiments and in the expansion of the Universe. The same applies to quarks, neutrons and protons.

Hierarchical Order of GSBP Composite Particles

11.9   Three quarks are bound in a nucleon by the force of SU(3) symmetry, the electromagnetic force, the conservation of energy-momentum, angular momentum, requiring minimum flow-specific entropy production (maximum energy-density) in such transformations. General Relativity is inherent in that state and in the internal structure of the quarks, as first-order GSBP SU(2) composite-particles.

11.10   The charge-carrying electrons are composite-particles of GSB-photons in SU(2) symmetry, whose quantized radius of gyration varies as ρ^{-3} thereby conserving spin angular momentum. Like the quarks, the leptons are first-order GSBP SU(2) symmetry composite-particles, though characterized by lower energy (energy-density) and stability.

11.11   "Neutral" electron neutrinos are SU(2) symmetry composite-particles comprised of GSB photons, likely having very low average energy-momentum and spin angular momentum (positive charge density), but otherwise similar in structure to the electron. If so, the radius of photon gyration in the neutrino would be several orders of magnitude greater than that of the electron. The neutrino would perhaps orbit the electron for a time in its initial high energy-density state.

11.12   The Nucleons constitute second-order GSBP composite-particles characterized by higher levels of mass-energy, but lower energy-density and stability in the present low energy-density of the Universe. The nucleus of atoms may then be referred to as third-order GSBP composite massive particles, and the atoms themselves as fourth-order GSBP composite-particles of still greater mass, but lower energy-density. Higher orders of GSBP composite-particles and structures characterize the simple molecules and the much greater mass-energy (lower energy-density and stability) of organic macromolecules.

11.13   The GSB photon in fermions and bosons are thus seen in this view to be the most elemental of all particles, defined by energy, momentum and spin angular momentum and dynamically constrained for the life of the fermion by its own gravitational force. It is this quantized bounded-energy that sharply distinguishes the bound photon in fermions from free massless "Vector Spin-1 Bosons" or simply "Bosons," and it is in such bound photons in fermions that 99% of all the energy of the Universe resides, for all time.

11.14   The attraction between a neutron and a proton in the nucleus derives from opposite spin orientations in a d-quark in the neutron and the d-quark in the proton, resulting in a residual attractive force, while the like spins of the quarks in two protons in a nucleus would act to repel them via impulse-recoil action.

11.15   The above argues for a strictly particle representation of the organization and dynamics of the nucleus of atoms, in which the values of the parameters c, $\sqrt{G}$ vary as ρ^{-3} and m as ρ^{3} over four hierarchical energy-density ranges. The region of greatest energy-density and stability is occupied by quarks, which in current theory takes the light quark masses to be $\sim 10^9 m_e$, but are here seen simply as electrons whose mass increased by $\sim 10^3$ and volume decreased by $\sim 10^6$ resulting in an energy density $\sim 10^9 \text{kg/m}^3$. The lowest energy-density state, displaying "asymptotic freedom", is "confined" by a large negative energy-density gradient relative to that occupied by the electrons.

Particles Diffraction and Interference from First Principles

11.16   There is nothing that more uniquely defines and separates the quantum world from common sense reasoning than the phenomena of diffraction and interference. This ongoing concern derives from the century-old classical assumption that the intensity of energy-flow in a system (in particular light quanta) may be reduced to the point where we can be certain that only a single quantum is involved in the phenomena of interference. In chapter 10 we cite two recent experimental studies, designed to test the "single particle assumption. On the 200th anniversary of Thomas Young's demonstration to the Royal Society the interference of light, N. Garcia et al. showed, "within the 2ps accuracy of our streak camera resolution that the time delays measured correspond to straight line 'bullet' (impact of a wave packet) trajectories from the slit to the detectors…. We find it difficult to explain these experiments, if we cannot know the photon trajectory and if before wave collapse on the detector the diffraction picture is one wave. We believe that this is a manifestation of the particle character of the photons during their flight from source to detector describing straight lines." And they reason "with no overlap of the [femtosecond] pulses no interference can be observed no matter how good resolution we have, be the photons of the same energy or different. The fact that the overlap is needed may indicate that indistinguishability of the photons is necessary at the detector but certainly not at the slits. … the key requirement for interference is to have overlap of the pulses impinging the photo-cathode of the streak camera."

An internet paper, "Interference of a Single Photon" by Robert Austin and Lyman Page, illustrates further the extent to which we take for granted the "single particle" assumption. They illuminated a triple slit with a laser beam that was attenuated with a neutral density filter to one part in 10^{11}, such that the mean free path between photons (bosons) was about 2 kilometers, or about 7µs. They "show that despite this large distance, a diffraction pattern obtains – the photon has 'interfered' with itself." They assume each event to be the signature of a single photon. However the CCD pixel evidence strongly supports the interaction of multiple bosons, even in this extremely low intensity study.

11.17   The Coulomb force acts to join or separate two or more GSBP nearly-identical bosons (traveling on close extended trajectories and, according to their GSB photon relative phase) to account for their bunching and antibunching at detectors in Fraunhofer single-slit far-field interference. Similar action accounts for high-resolution, near-field *interference in double- or multiple-slit interference phenomena, making no dependence on classical or quantum wave or field theory, yet requiring the entanglement of at least two particles* in explicit denial of the "single-particle" presumption in quantum mechanics. Local energy-density slows the speed and bends the trajectory of bosons passing very near the edges of a slit in a light-blocking screen (gravitational lensing) and the Coulomb bundling and antibundling of bosons leads to Fraunhofer single-slit interference. Double-slit near-field interference occurs similarly and with higher resolution and intensity.

11.18   The above fails to account for interference in Thomas Young's initial experiment and demonstration to the Royal Society, two centuries ago involving light from the Sun. In particular we should have to account for the relatively high correlation of bosons (in energy, phase, polarization and momentum) emitted in the Sun's corona, perhaps resulting from some form of local cyclic undulation or fluctuation. Could there be clues from such a study leading to further understanding of the Sun? [Ch. 10]

A GSBP-Schröedinger's Relativistic Quantum (Particle) Mechanics

12.1   The Challenge in developing a relativistic particle mechanics is to meld the well-established formalism and power of Schröedinger mechanics with the inherently relativistic structure and dynamics of GSBP particles, so as to directly satisfy the demands of general and special relativity. This may be accomplished straightforwardly and simply in the one dimensional Schröedinger time-independent equation by reducing the electric charge e in the same proportion as we reduce the electron's mass. The result of this is to increase the radius of gyration of the GSB photon in the electron and a corresponding increase in the orbital radius of the electron about the proton in the ground state, as shown in Chapter 14. And the same reasoning should apply generally to relativistic corrections for the quarks in the neutrons and protons in the nucleus of atoms.

12.2   The above succinctly reveals the delay in the development of fundamental quantum theory resulting from the relatively little effort to endow the electron with dynamic internal structure and dimension.

12.3 We are challenged also to square GSBP-Schröedinger mechanics, self-consistently, with related fundamental theoretical issues including:

- The Lorentz Transformation

  In our scenario, the velocity of the electron v_e and that of the GSB photon v_{GSBP} are governed by the relation $v_e^2 + v_{GSBP}^2 = c^2$, as required in order to conserve spin angular momentum of the electron, and making it inherently relativistic. Thus the displacement dr of an electron can be described equally well in terms of $dr^2 = dc^2 t^2$ or $dr^2 = c^2 dt^2$. But in our schema it is the element of time advance, dt, that is taken as an absolute universal constant and the velocity of light c as a local or epochal parameter.

- The Physical Basis for the Indeterminacy Relations

  If we think of an electron as a cylinder of radius r_0 (radius of gyration of the GSB photon) and length r_0, we could expect to determine the electron's position in space only over the volume $\sim r_0^3$, when the electron's velocity is small compared with the speed of light. In this case the electron's energy, energy-density and momentum correspond closely with those parameters defined by its rest mass. However, when the electron's velocity approaches the speed of light, its energy and energy-density increase rapidly (volume decreases and mass increases) as does its momentum. The result of this is to increase the range of momentum Δp, and the range of position Δr, such that $\Delta p \Delta r \geq h$.

- The Uniformity and Constancy of the Velocity of Bosons Emitted in the Decay of Atomic States.

  In our scenario, the GSBP electron doesn't just jump from an excited state to a lower one, it travels down a potential energy spiral staircase at relativistic velocity, such that the period, corresponding to the wavelength of the emitted boson at velocity c exactly defines the distance traveled by the electron, at relativistic velocity, between the initial and final states of the electron.

- The Derivation of E=mc²

  Einstein's most famous equation can be derived straightforwardly from the relation $v_e^2 + v_{GSBP}^2 = c^2$, dependence of the speed of light and particle mass on the local energy-density and the requirement on conservation of energy-momentum and angular momentum (chapters 7 and 8)

- The Seamless Joining of Relativity and Quantum Theory with c and G Evolving as Parameters, leads to:

  - A rational cosmology with no requirement for dark energy or dark matter
  - Creation of all of the massive matter particles in the quantum Universe
  - Accounting for the present lack of antiparticles in nature
  - The inherent relativity of all particles
  - Rationalization of gravitational and electromagnetic forces
  - NTI creation of all organized subsystems ranging from galaxy clusters to quantum and living subsystems
  - Quantum theory freed from Copenhagen quandaries: complementarity, superposition theory, etc.
  - Gravitational explanation for the Casimir force

GSBP-Quark ElectroDynamics – The Atomic Nucleus

13.1   In chapter 13 we sketch the principal features of GSBP-Quark ElectroDynamics (GSBP-QED), a physical particle model of the structure and dynamics of the nucleus, and contrast that with the field-theory algorithm of Quantum ChromoDynamics (QCD). Toward that end, we seek first to underscore the importance of *energy-density states* in bound GSBP-QED nucleons, as contrasted with *energy states* in QCD theories. The latter refers to *excited* states of particles or *families* of particles found in high-energy accelerator experiments, while the former may refer to the higher *energy-density state* of a particle, e.g., an electron, in the local high *energy-density* time of its evolution in the expanding Universe. The transmutation of quarks, key to such evolution in GSBP-QED, will be seen not as the work of "weak *force*" particles (W^{+-}, Z^0), but simply as the underlying imperative of the *2nd Law*, as the average energy-density of the Universe declined to a critical enabling energy-density level. The action of force particles, bosons, which govern the dynamics of quarks in nucleons and bind nucleons together in the nucleus of atoms, will occupy our attention in much of the remaining part of the chapter.

GSBP-ElectroDynamics – The Hydrogen Atom

14.1 We build a GSBP-electron model of the hydrogen atom by first defining the parameters of the GSB Photon in an electron marginally-bound to a proton, in which the electron is nearly at rest (having near-zero KE and negative PE relative to the proton), with both its mass m_r and electric charge e_r reduced by the factor $m_p m_e / (m_p + m_e)$, where m_p is the mass of the proton and m_e the mass of the free electron). In that state the GSB photon must travel at constant velocity c in a circular orbit of radius r_0, such that the period of one complete orbit with exact phase closure is τ = $h / m_r c^2$, where h is Planck's Constant and $m_r c^2$ is the reduced energy of the gyrating photon. Energy conservation and stability of the ground state of the electron requires the orbital period of the electron, together with exact closure of the phase of the bound photon, to be exactly equal to τ times the constant α^{-1}, where α is the fine structure constant given by $e^2 / 4\pi\varepsilon_0 \hbar c$. That is, *the GSB photon must cycle exactly α^{-1} times, in exact phase closure with one complete orbit of the electron about the proton. Thus the relativistic radius of the electron orbit in the ground state $a_{rel} \equiv \alpha^{-1} r_0$. (The value of a_{rel} thus derived must also satisfy the relation $a_{rel} = h^2 / 4\pi^2 m_r e^2$ in order to account, together with the Rydberg constant, for the radiation spectra of the hydrogen atom.) The relativistic velocity of the electron in the ground state is given by $v_{rel} = (c^2 - v_{ph}^2)^{-1/2}$, where v_{ph} is the velocity of the GSB photon, which varies inversely with the third power of ρ_{local} in the immediate environment of the electron. In the ground state $v_{rel} \equiv \alpha c$. With these relativistic adjustments, the ratio $\mu_e / \mu_B = g_e$, where μ_e the angular momentum of the electron $= e v_{rel} a_{rel} / 2$, μ_B the Bohr magneton $= e\hbar / 2m_r$, and $g_e \equiv 1$ is the Lande g factor (orbital magnetic moment of the electron), giving $\mu_e / \mu_B \equiv v_{rel} a_{rel} m_r / \hbar \equiv 1$. This relation for the hydrogen atom is an absolute Universal constant, since v_{rel} and a_{rel} each vary as the inverse third-power of ρ_{local} and m_r as the two-thirds power of ε_{local}. In our scenario the gyrating GSB photon is the physical embodiment of a spin dipole moment, which must be defined similarly in relativistic relation to the Bohr magneton. Substituting μ_s for μ_e, v_{phrel} for v_{rel} and r_{orel} for a_{rel} in the above equation, we obtain $g_s \equiv$ 2, which is likewise an absolute Universal constant. Thus, the spin g_s factor has the* *identical value 2 as is the total angular momentum in the ground state of the hydrogen atom, since the orbital angular momentum in the ground state is 0. If this is so, the electron has no "anomalous magnetic moment," in accord with our reasoning and the Dirac equation, but contrary to QED theory.*

14.2 Missing from the above is any reference to the "Lamb shift," referring to a small energy difference between the $2S_{1/2}$ and $2P_{1/2}$ states of the hydrogen atom, which according to Dirac theory are degenerate. In a Classic mid-twentieth century experiment, Lamb and Rutherford demonstrated the existence of that energy difference. However we show in chapter 14 that the energy difference they measured is that associated with the energy well in which the metastable $2S_{1/2}$ state must briefly reside, the bottom of which is at the same energy level as the $2P_{1/2}$ stable state. Lamb sensed the core of the problem, as he said in his 1950 Nobel speech: "In order to obtain a finite mass of the electron on a purely electromagnetic basis, it was necessary to assign an extended structure to the electron (emphasis added). Attempts to form a satisfactory relativistic theory of an extended charged particle failed."

14.3 *All boson emission occurs as the electron spirals down from one allowed discrete energy level to another, such that the time required to make an exact integral number of GSBP orbits in the electron coincides <u>precisely</u> with the closure of one orbit of the electron at relativistic speed about the nucleus and is equal to the inverse frequency v^{-1} of the emitted boson.* That is, the distance traveled by the electron, taking into account its relativistic speed, is equal to the wavelength of the emitted boson. This ensures exact simultaneous closure of the orbits of the bound photon in the electron and of the bound electron in the atom, thereby minimizing line broadening of the emission of a boson, allowing only a ±h error in the energy budget, where h is the Planck constant. Thus the boson is not emitted as a wave packet.

14.4 In hydrogenic, hydrogen-like atoms and ionized high Z atoms, the coupling of the electron(s) to the proton(s) may be significantly increased, resulting in a decrease of the ground state radius *a (see Fig 8-2)*, depending on the value of Z. Such systems may involve a significant increase in the local energy-density of the system, thus requiring corrections to the parameters ρ, c, R_∞ and the GSBP radius r_o and electron orbital radius a and the inclusion of a GSBP charge radius r_p. What "precision physics of simple atoms" is all about, in QED terms, is patchwork attempts to achieve self-consistency among the various corrections, lacking any understanding of the internal structure and dynamics of massive fundamental particles.

14.5 Clearly the challenge for GSBP particle theory is to determine the manner in which the local energy-density of a system varies with the atom's immediate environment, enabling thereby the above corrections to the energy levels of bound electrons in a simple and straightforward way. And in the larger view the challenge is to determine how such corrections might vary consistently in such atoms in regions of high ρ_{local} (e.g., near the core of galaxies). This should greatly facilitate the precision physics of simple atoms.

An Inherently Rational Quantum (Particle) Mechanics

15.1 We show that quantum mechanics, contrary to its reputation, can be inherently intuitive, rational and consistent with the demands of relativity and thermodynamics, once we take into account the internal dynamic structure of the fundamental particles and shed our long-held commitment to the notions of wave-particle duality, wave packets, wave function reduction, the single particle assumption and superposition. Discarding that considerable baggage would in no way diminish the power of Schröedinger's equation to characterize the nonrelativistic dynamics of quantum particles, as those notions were philosophical appendages to the mathematical formalism. Indeed nothing is lost in the more fundamental strictly-particle GSBP-Schröedinger theory, while still greater precision is gained in its straightforward finite compliance (limited only by the Heisenberg indeterminacy relations) with both special and general relativity.

15.2 What evolved from the early days of QM was a powerful mathematical formalism and the solidification of attending non-formal assumptions and concepts which were conceived initially to support the new QM thinking and later the emerging field theory. Bohr and Einstein were both right in their fundamental disagreements, one on what was then possible to predict in quantum systems and the other on what should be possible eventually to predict. Neither could see further in the long-lasting debate.

15.3 As John Bell and others have said, what is needed to make the Schröedinger equation self-consistent with relativity theory is not some hidden variable but a more fundamental theory of the electron. We argue here it is the GSBP physical particle model of the electron (more generally of the leptons and antileptons) that provides a fundamental connection between quantum and relativity theory (general as well as special), as developed in chapters 7-13 and confirmed for the hydrogen atom in chapter 14.

15.4 Summing up, the irrationality of the assumptions underlying QM and QED/QFT-QM and the resultant conclusions has forced us to start from scratch – to reexamine the foundation of physical theory, with the requirement of self-consistency as our guide, in the development of a more fundamental physical relativistic particle theory of fermions and bosons. We can be satisfied in this effort, as applied in the theory of the hydrogen atom (chapter 14):

- to the extent that there is conclusive evidence that a bounded GSBP-electron makes no requirement for an anomalous magnetic moment; and

- in the larger view, to the extent that the theory applies to hydrogenic and hydrogen-like atoms and high-Z atoms taking into account the increased interaction of the nucleus with the electrons, and

- as the theory applies to weekly-bound electrons in molecules, the scattering of free electrons, and the dynamics of GSB bosons and their interaction.

This poses a considerable challenge to experimentalists to provide evidence that might invalidate the theory, reveal its limitations or support it in precision physics, and perhaps in the larger view as it applies generally to the dynamics of leptons and bosons in the nucleus of atoms.

15.5 But the above does not adequately deal with the two most counterintuitive pillars supporting QED/QFT theory: the "single particle" assumption and that, closely related, to the "superposition" of multiple "virtual-particles" in "Hilbert space." The latter derived, as if made to order by advances in pure mathematics, and served thereby to lend legitimacy to the strange new wave-mechanics theory. The former emerged as an external supposition which if anything added to the mystery associated with the new mechanics. Nevertheless, this assumption may play a critical role in judging the relative validity of QED/QFT versus GSBP-Schröedinger theory, since we are as yet unable to prove *conclusively* that only a "single particle" *need* be involved in experiments demonstrating interference, as claimed for the former, while two or more particles are required *beyond doubt* in the case of GSBP-Schröedinger theory.

15.6   The two publications cited in chapter 10 by N. Garcia *et al.* and by Robert Austin and Lyman Page provide fundamental evidence supporting the requirement that two or more quanta must necessarily be involved in experiments revealing interference phenomena. No less important Garcia shows that interference requires only a sufficient "overlap" of the two particle states. And the Austin experiments, designed to show Interference of a Single Particle, actually reveal evidence of multiple particle-pairs involvement in interference even when the laser beam was attenuated with a neutral density filter to one part in 10^{11}, such that the mean free path between photons (bosons) was about 2 kilometers, or about 7μs distant in time. Clearly such real photons are not in spacetime superposition, though they show, despite this large distance, that diffraction patterns obtain.

Quantum Molecular Theory

16.1   The problem faced by theorists in modeling molecular systems has turned on the question: How and in what form can we include the corrections, needed to square theory with experiment in the study of increasingly complex molecules, so as to pragmatically minimize computational costs, given "point particle" assumptions? Efforts toward this end have been focused on the development of Molecular Orbital Theory (MOT) in dealing with small molecules and on Density Functional Theory (DFT) to apply to more complex systems, including electron dynamics in solids. It soon became apparent that MOT was inadequate to model even relatively simple high-Z molecules. In an effort to circumvent these problems, density functional theory evolved in recent decades toward the development of a generalized Schröedinger-like equation to model the ground-state density of all the electrons in a molecule or solid. Considerable progress has been made in recent decades in this effort, though the method is limited to the "ground state" and hence excited states of such systems cannot be modeled in this way. The principal problem is the lack of a suitable exchange-correlation functional in any equation of the above form, applicable to a wide variety of molecules and solid-state systems. This is a many-body problem in which the motion of one electron affects, and is affected by, that of all the others and so we should expect that the fundamental problem traces to the lack of physical dimension and internal dynamic structure of the electron. We might expect the GSBP model of the electron to facilitate the development of MOT and a general exchange-correlation functional in DFT, but it is unlikely that a first-principle theory of molecular evolution may emerge from either of the above strategies, because neither theory takes account of the growing local energy-density of the molecule as other atoms are enabled to bind to it. Nevertheless there is much to be gained in the application of GSBP-Schröedinger theory to MOT-like studies in the short run and to retrodictive validation of nonequilibrium open thermodynamic molecular subsystems in the longer-term fundamental studies. The latter could benefit greatly from advances in four-dimensional ultra fast electron microscopy.

STEPS PART IV
GRAVITY
IN RELATIVISTIC MOLECULES
AND SYSTEMS OF LIFE
Chapter 17

17.1   A century of theoretical speculation and laboratory research on the origin of life has shown how relatively easy it is to create the amino acids we find in all cells today from available precursor molecules and how all-but-impossible it is to make nearly all such amino acids left-handed stereospecific, as in present cells. As a consequence laboratory progress has been largely stalled for several decades, there being no way to create the high-mass proteins and enzymes from the racemic amino acids. Fundamental theory has been stymied from the outset, there being no rigorous nonequilibrium thermodynamic theory to guide the development of research programs including the search for physical systems and processes in the early earth capable of selecting or creating enantiomers of amino acids, as Bretig counseled in 1924. This is not to say that theorists have been idle. Various biochemical strategies have been proposed whose primary focus is on either an early demonstration of molecular self-replication or the creation of high-mass proteins and enzymes, in each case commencing from abiotic amino acids. In regard to symmetry and symmetry-breaking in the origin of life, there are several which-came-first type of problems that have effectively blocked the advance of research. First and foremost of these is the need to demonstrate a straightforward path, consistent with geophysical and chemical assumptions, to the creation of stereospecific L-amino acids and their polymerization in the making of proteins and enzymes, on the one hand, and D ribose sugars, in the creation of nucleic acids and their polymerization into RNA and DNA, on the other hand, which we may refer to as the homochirality problem. Other needs vie for evolutionary urgency, which have to do with the requirement for a bounded, selectively open protein enclosure, in which there may then be demonstrated a straightforward path leading to the creation of template apparatus enabling self-replication and evolution of the cellular system via Darwinian natural selection and coevolving with a simple, limited metabolic pathway. A later requirement would be for the transformation of the protein enclosure into a lipid-protein membrane capable of more selectively constraining the flow of matter into and out of the proto-cell. Such seemingly-circular reasoning, suggesting the need to resolve multiple demands nearly simultaneously, as seen from our understanding of present-day cellular dynamics, can be resolved only via some external symmetry-breaking force or mechanism, as we shall argue later.

The Homochirality Problem

17.2   If we focus our attention on a hydrothermal vent, as a localized stable heat source and sink, we see that the vortex-flow of hot saline water appears in part as a mesoscopic, localized, single Bénard cell. It is a kinetic, open thermodynamic subsystem of the larger energy-flow, geophysical system, which serves to minimize entropy production and maximize the efficiency of heat transfer in the system. It does this through the very long-range organization of molecular motion. The open subsystem (vortex) stores some small portion of the energy flow (that would otherwise be dissipated) characterized by maximal average energy-density under given system constraints. As we speculated in chapter 5 about interpreting Bénard convection as a very primitive quasi-living system, so we might perceive the vent vortex subsystem in which billions upon billions of water molecules have been constrained to move in a strongly ordered or organized manner (quasi-laminar flow). But much more is going on in the "system" and the "subsystem;" the water cycles with it ions, other atoms, molecules and solutes, some small portions of which are newly added from within the earth in each cycle. The vortex is driven and constrained by mechanical (hydraulic) forces as well as thermal and gravitational gradients, and the existence and flow of ions results in electromagnetic forces and flows, giving rise to a very complex open subsystem involving the interplay of four forces and their respective flows.

17.3 In such a system chlorine and sodium ions would cycle in quasi-laminar flow and in vertical, bounded semi-planes, orderly displaced about the orifice axis, in which gravity plays an important role, as in the case of Bénard convection. (We neglect here the relatively small contributions of sulfate and magnesium ions). However, since the mass of Cl^- is about 1.5 times that of Na^+, we might expect that chlorine ions would be accelerated (by thermophoretic forces) to a slower average velocity and on smaller vortex orbits than for the sodium ions, in the stationary state. But the difference in masses and kinetics would be significantly reduced since each ion is enclosed in a "cage" of six water molecules, making the effective mass of the sodium ion ~6(2+16)+23= ~131amu and that of the chlorine ion ~143amu. The mass of the sodium ion complex being about 0.9 that of the chlorine ion complex should then result in the average velocity of the former being about 1.09 that of the latter, in the stationary state, still a substantial difference in computing their relative electromagnetic properties. But the net positive ion current and therefore the strength of any resultant magnetic field should be expected to increase as the square of their mass ratio, since the relative ion densities would vary also inversely with their mass ratio (due to the greater constraint on potential energy of the Cl^- complex). Such vortex kinetics would thus tend to break the local charge-null symmetry that would otherwise characterize a uniformly-heated reaction pot. The effect of this ion-flow asymmetry might be to create a moderate-to-strong magnetic field in the regions of relatively large net positive current flow. The asymmetry of reaction zones should play an important role in the simultaneous synthesis of reactants at one level and of reaction products at a higher level.

17.4 Such relatively simple, ubiquitous and asymmetric physico-chemical systems in the prebiotic ocean could account for a significant enantiomer excess (ee) of amino acid monomers if the magnetic field strength, produced by the cycling net charge, were sufficient, relative to thermal noise, to suitably align the spins in the C—N bond in the vast majority of amino acid reactions, whether via a single-pass through the reaction zone or multi-cycle amplification of stereoisomers. The same applies to the carboxyl and side-chain groups and the peptide bond.

17.5   The issue of homochirality would be so very important in an early period of transformation from inorganic to bio-organic chemistry because that option would enable the most efficient utilization of energy and precursor molecules, as compared with other later options in which the unwanted monomers would be selected out. No less important in this consideration is the realization that the greater the number and kind of L-amino acids available in the early phase of the polymerization process, the sooner would be the yield of stereospecific enzymes to catalyze other reactions leading to the creation of D-ribose sugars, bases, nucleosides, nucleotides and ATP-like energy carriers.

17.6   Estimates follow of the magnetic field intensity H (normal to the planes of ion cycling), the torque acting to align the spin of a carbon α-atom in the amino acids with the field, and the angular acceleration of spin rotation. And from the latter we can estimate the minimum time required to ensure the creation of each amino acid of a specified enantiomer excess (ee). We assume that the early ocean salinity was similar to the present value (about 30 each Cl- and Na+ for every thousand water molecules), which is relatively insensitive to that ratio, and we argue that the Na+ ions would travel, on average, ~1.1 times the velocity of the Cl- ions in response to electrophoretic forces. Also the charge density of Na+ would be about 1.1 that of Cl- (see above). And since the ion current would be the product of the ion velocity times the ion density, the net ion current density would be approximately $[(1.1)^2 \times 30-30] \times 10^{-3} \times 10^{24} \times 1.6 \times 10^{-19} \times 0.2 = $ ~200 Amperes/cm^2, assuming an average velocity for the Na+ complex of 0.2 meter per second. Finally, we find that the maximum magnetic field intensity H_{max} produced by this current density would be about $20 \times 60 \approx 1200$ Ampere-turns per meter, with the field oriented according to the right-hand rule, relative to the direction of travel of the Na+ to that of the magnetic field (see Fig. 17-1). Here we assume about 60 channels of 0.1cm^2 average cross section (0.1cm width×1cm height) displaced around the vent at a radius of 10cm. We note also that the right-hand rule in ion flow translates into the left-hand rule in electromagnetic interaction with the spin of charged particles and it is the latter that defines the orientation of atoms in molecular systems. The maximal torque T acting on the magnetic dipole moment m_c of a carbon atom would then be given by T_{max} = m_c × H_{max} and the maximum torque T_{max} = $m_c \times H_{max} = \mu_B \times H_{max}/(12 \times 1836)$, where μ_B is the Bohr magneton = 9.3×10^{-24} J/tesla × 1200AT/(12×1836) and the denominator is the factor representing the increase of the mass of the carbon atom over that of the electron. Thus, $T_{max} \cong 5 \times 10^{-25}$Nm. The maximum angular acceleration of the dipole moment would then be $a_{max} = T_{max}/C_{mass} \cong 5 \times 10^{-25}/(2.2 \times 10^{-26}) \cong$ 23 radian/s². In the first 0.1s (about one-fifth of one vortex cycle) $[\theta=1/at^2]$, a carbon atom could be maximally rotated about four radians in the hot thermal reaction region of the vortex, as a result of the differential velocities and densities of the sodium and chlorine ions. Thus, we might expect that many if not all amino acids would have left-handed asymmetry in one pass through a vortex cycle. Moreover, since the amino acids remain in the same high field environment before and during a polymerization process, there is every reason to expect the products to conserve or extend their enantiomer excess (ee), enabling the creation of natural proteins, enzymes, sugars and nucleic acids, etc.

Proposed "Fox" Research Under Simulated Hydrothermal Conditions

17.7   The renewal of Fox's origin-of-life research strategy would ask, with little presumption: What might result if we cooked various constituents of amino acids in an aqueous solution, under simulated hydrothermal-vent conditions and constraints in which the reactor (pot) is repeatedly cycled, between upper and lower temperature limits, in a kinetic open-thermodynamic-subsystem, as if it were driven by the Nonequilibrium Thermodynamic Imperative? Nature's laboratories at the bottom of the oceans would have had little need for the "hand of the chemist" to cycle the pots from one temperature limit to another. Indeed, these laboratories would have come equipped with a unique property, a magnetic field capable of biasing the spin orientation of some, many, or perhaps all, α-carbon atoms involved in the linking of amino acids into short- or long-chain life-like proteins or enzymes, depending on the strength of the magnetic field in the vortex. The same applies to D-sugars. Whereas the initial proteinoids were capable of combining to form microspheres, they were clearly limited in chain length and in their capacity to fold into highly specific and active catalysts. With those limitations decreased or removed, the protein vesicles might be free to continue to grow (mainly in density) in response to the thermodynamic imperative and the relatively rapid and extensive cycle of thermal energy in the vortex. Such growth in the form of long-chain proteins and their folding in the internal milieu to form enzymes could assist in the creation of phospholipid structures and signal a major structural advance, giving rise to a real membrane and the legitimacy of the enclosure as a real cell, as Morowitz emphasized in distinguishing living systems from Fox's microspheres. It is important to note that neither these or the previous "advances" would have occurred in anticipation of living cells; they would have emerged, driven by the Nonequilibrium Thermodynamic Imperative and the physical, chemical and electromagnetic forces associated with the hydrothermal vents. Had a sentient being observed Nature's hydrothermal experiments, it might have discovered cells and the beginning of living systems in hierarchical subsystems of earth's geological systems.

About the Author

Harold Hamilton was trained in radar theory and engineering at Stanford University in 1940 and subsequently served in the Army Signal Corps in the South Pacific in WWII, until his discharge in 1946, when he returned to Stanford to complete his undergraduate studies in 1949.

He joined the Librascope division of General Precision, Inc. in 1950 and briefly supervised the development of digital computers for the military.

In the early 1950s, he proposed and directed an advanced research program for the study of neural information processing systems. That effort led ultimately to sponsoring an international conference on Neural Theory and Modeling in 1962, supported in part by the Air Force Office of Scientific Research, and resulting in the publishing of contributed papers by Stanford Press, including Hamilton's report on gas-ion simulation of excitable membrane as a means for modeling neural systems.

In 1964 he left Librascope to pursue an interdisciplinary Ph.D. at the University of Southern California for the development of theoretical and pragmatic tools for coping with what Warren Weaver called "problems of organized complexity." In this effort he generalized Leontief's input/output analysis of economic systems, potentially applicable to a wide variety of systems of organized complexity, including those defined by Darwinian selection.

Along this path, Dr. Hamilton came to see a need to extend thermodynamic theory to apply to systems far from equilibrium. A first effort in this reasoning appeared in *Zygon* (1977): "A Thermo-dynamic Theory of the Origin and Hierarchical Evolution of Living Systems."

About the same time, he saw the need for computer systems to share mass information storage and to timely interact with multiple computers. This led to the creation of Censtor Corp., an early Silicon Valley startup in 1981, with Dr. Robert Noyce as Chairman of the Board. Harold's name is on many fundamental patents in perpendicular magnetic recording.

Hamilton retired from Censtor in 1994 to return to his original interests in thermodynamics and living systems and to pursue Alexandr Zotin's counsel. (see Preface) For the past nineteen years, his lifelong interest has from that date (when he was 70 years old) to the present, beckoned him to one rewarding insight after another in the reformulation of fundamental physical theory.

INDEX

Relevance of waves, fields, phase,
197
research proposal for investigation
of amino acids, 319

S

self-consistent theory, 16
Standard Model, the hierarchy and
other problems with, 220
strongly-nonlinear thermodynamic
system, 27
subsystems progeny of the NTI,
117

T

theoretical limitations in the
laboratory, 297
thermodynamic creation of more
complex organic components,
309

Tolman, 77
Transition from Quantum to
Classical Systems, 266

U

uniformity, constancy of boson
velocity, 195
Universe as an evolving, quantum,
nonequilibrium thermodynamic
system, 102

W

weakly-nonlinear system, 27
well-behaved systems, 27

Z

Zero-Point Energy, 202
Zotin, Alexandr, 15

www.ingramcontent.com/pod-product-compliance
Lightning Source LLC
Chambersburg PA
CBHW021548210326
41599CB00010B/364